# Understanding
# Media Convergence

# Understanding Media Convergence: The State of the Field

*Edited by*

August E. Grant
Jeffrey S. Wilkinson

New York Oxford
OXFORD UNIVERSITY PRESS
2009

Oxford University Press, Inc., publishes works that further Oxford University's objective of excellence in research, scholarship, and education.

Oxford New York
Auckland Cape Town Dar es Salaam Hong Kong Karachi
Kuala Lumpur Madrid Melbourne Mexico City Nairobi
New Delhi Shanghai Taipei Toronto

With offices in
Argentina Austria Brazil Chile Czech Republic France Greece
Guatemala Hungary Italy Japan Poland Portugal Singapore
South Korea Switzerland Thailand Turkey Ukraine Vietnam



Published by Oxford University Press, Inc.
198 Madison Avenue, New York, New York 10016
http://www.oup.com



**Library of Congress Cataloging-in-Publication Data**
Understanding media convergence : the state of the field / [edited by] August E. Grant, Jeffrey S. Wilkinson.
p. cm.
Includes bibliographical references and index.
ISBN 978-0-19-532777-9 (cloth)
1. Mass media. 2. Journalism. I. Grant, August E., 1956– II. Wilkinson, Jeffrey.
P90.U46 2008
302.23—dc22

2008012160

Printing number: 9 8 7 6 5 4 3 2 1

Printed in the United States of America
on acid-free paper

# CONTENTS

# ACKNOWLEDGMENTS

Finding and assembling the chapters in this book has been a labor of love for more than three years. First and foremost, we want to thank the authors, whose insight, patience, flexibility, and persistence were tested as the chapters were massaged to create the final product you hold before you. Their hard work set the stakes that are helping to define the convergence terrain, blazing trails for the research that is sure to follow.

The genesis of this volume is the research presented at the annual "Convergence and Society" conference that Augie has had the pleasure of chairing since 2002. We are grateful to Dean Charles Bierbauer and the University of South Carolina, as well as all of the presenters and attendees, for their support of this annual gathering of convergence scholars.

Peter Labella and the staff at Oxford University Press were a primary enabling force behind this volume. The professionalism of Chelsea Gilmore, Lisa Grzan, and Josh Hawkins has played a major role in helping us produce this volume and a companion textbook, *Principles of Convergent Journalism,* simultaneously. Others doubted the wisdom of editing a theoretically-based book such as this one at the same time we were writing a basic textbook covering the practical side of convergent journalism. We are grateful that they shared our vision of the need for both books to be marketed side-by-side.

On a personal level, Augie thanks Diane for giving up so much time over such a long period so that this book could be completed. Bobby and Jessica made sacrifices throughout the process, and he hopes that, as they start college themselves over the next few years, they will know how much time and effort goes into every book they read.

Jeff is indebted to his wife, Suet Fan, for her patience and support. Likewise, his parents, Betty and Bob, deserve thanks for steadfast encouragement throughout. This endeavor spanned (at times) 12 time zones, three states, and two countries. Communication technologies helped, but someone needs to invent a teleportation device.

The end product that you are holding in your hands has impacted us far beyond what we expected when we began this endeavor. Our lives have been enriched by the collegiality of all those involved in this project, and we can't wait to see the next generation of scholarly convergence that will follow this effort.

*Augie Grant*
*Columbia, South Carolina*

*Jeff Wilkinson*
*Zhuhai, China*

# Understanding Media Convergence

# Dimensions of Media Convergence

August E. Grant

Change is the coin of the journalists' realm. News is about change, and the bigger and more unexpected the change, the bigger the headline. It is no wonder that a profession fueled by change is itself changing. Since the advent of professional journalism, the tools, techniques, and scope of journalists have been evolving, but the pace of that evolution has been accelerated by a set of phenomena broadly referred to as "convergence."

The logical place to introduce this inquiry is by defining the term "convergence." Indeed, most academic research addressing convergence begins with one or more definitions (e.g., Castañeda, Murphy, & Heather, 2005; Huang et al., 2006; Lowrey, Daniels, & Becker, 2005; Smith, Tanner, & Duhé, 2007), and a number of works have been devoted to defining or explicating the term (Dailey, Demo, & Spillman, 2005; Gordon, 2003; Lawson-Borders, 2005).

As with other macro phenomena, these definitions have become more expansive over time, to the point that there is no single, agreed-upon definition of convergence. Some paraphrase Justice Potter Stewart: "I can't define it, but I know it when I see it." Others resort to analogy, including Jim Carroll's oft-quoted explanation of why convergence is like teenage sex:

- No one knows what it is but thinks that it must be great.
- Everyone thinks that everyone else is doing it.
- Those who say they are doing it are probably lying.
- The few who are doing it aren't doing it well.
- Once they start doing it, they realize that it's going to take them a long time to do it right.
- They'll also soon start realizing that there is no "right" way to do it

(*Poynterextra, 2002*)

Two of the most oft-cited "definitions" of convergence have been provided by Dailey et al. (2005) and Gordon (2003). Dailey et al. (2005) proposed a "convergence continuum" for news organizations that starts with cross-promotion, and advances to "cloning," "coopetition," "content sharing" and, finally, to true "convergence."

Gordon (2003), on the other hand, identified five dimensions of convergence: ownership, tactics, structure, information gathering, and presentation (storytelling). Although the Dailey et al. (2005) continuum may be more illustrative of the progression that media organizations can go through to achieve "convergence," Gordon's (2003) concept of distinct dimensions of convergence better facilitates analysis of the wide range of convergence activities, especially because these activities occur in so many different combinations.

The impetus for this book was to assemble a broad range of research, inquiry, theory, and speculation regarding the nature and practice of convergent journalism. This introduction will explicate a number of the dimensions that are explored throughout this book, followed by an overview of the individual chapters and how each adds to our understanding of "convergence."

As broad as these dimensions are, they cannot fully cover all meanings because the term "convergence" has become a buzzword across industries as well. The processes that journalists refer to in using the term "convergence" (explored throughout this book) are only tangentially related to those referred to by those in the telecommunications industry, where "convergence" subsumes technological integration and marketing of a host of technologies ranging from wired to wireless and from telephone to television.

Lacking a single definition, "convergence" has become a construct that subsumes a host of meanings. The purpose of this chapter is to introduce the field by exploring the most relevant dimensions of convergence, including technological convergence, convergent journalism, coordination of media content, collaboration, and consumption of media content. These dimensions should not be considered exhaustive or mutually exclusive; indeed, every new medium or application offers opportunity for the creation of a new dimension of convergence. But these dimensions provide an illustrative set of frameworks from which media practices may be analyzed. Later chapters in this book address specific dimensions in greater depth, with the last section of this introduction devoted to presenting the overall structure of the book.

## Technological Convergence

Many definitions of "convergence" focus on technology. For example, Burnett and Marshall's (2003) exploration of impact of the Web defines convergence as the "blending of the media, telecommunications and computer industries, and the coming together of all forms of mediated communication in digital form" (p. 1). Although technology itself is rarely as important as either organizational factors, social factors or user factors in the analysis of media (Grant, 2006), two specific technological developments are at the core of media convergence: digital technology and computer networks. As discussed in Grant (2006), these technological innovations are not the motivating force behind convergence, but they do serve to enable most forms of media convergence.

### Analog to Digital

The first innovation is the transition of virtually all media from analog to digital technology. The physical world is an "analog" world, in which our senses perceive the environment by receiving and interpreting waves of sound and light. Accordingly, the first electronic media were designed to capture, record, transmit, receive, and amplify these waves. The different wavelengths for sound and light, along with the different bandwidth needed for light versus sound led to the creation of technologies that dealt with the two in different ways. A microphone could no more pick up images than a radio could display them.

Once sound and light waves are transcoded into electronic waves for transmission and reception, the differences start to blur, but the vast differential in the amount of information needed to represent an image versus the much smaller amount needed to represent sound largely kept the two separate. Even when sound was added to pictures in the movies and television, the methods of coding the pictures and sound were completely different. (For example, analog television uses FM transmission for audio and AM transmission for video.) Transmission of text was completely separate, using its own, unique coding.

The advent of the computer and digital technology yielded a common format for transmission and manipulation of media content. All computers are based upon binary technology, where all messages and instructions are encoded as a series of "0"s and "1"s (or "on" and "off"). Text was the first message form to easily be transcoded to digital form, with early computer scientists developing a common code called ASCII that allowed a series of eight "bits" of data—simply eight on/off switches—to represent 256 different characters in the Western alphabet, including letters (upper and lower case), numbers, symbols, spaces, and so forth (For a detailed discussion on Claude Shannon's role at Bell Labs in laying the foundation for digital signals, see Fitchard, 2008.)

Images and sound needed a more complex coding, but it was not long before technology was developed to "sample" an analog audio or video signal, representing the wave for that fraction of a second as a number. With enough samples—and a medium that could store and transmit all of the numbers for each sample, video and audio signals now shared a common, digital coding scheme. (If this explanation sounds simple, consider that the "sampling frequency" for an audio CD is 44,100 samples per second [Carlin, 2006]. Each sample has 16 bits of data, so one second of audio requires transmission of more than 700,000 numbers!)

The major revolution resulted from the fact that text, audio, video, and images were now in a common form, a stream of "0"s and "1"s. Once problems of transmission capacity were solved, a single connection could transmit any type of information that could be coded in digital form. It was a small step to then program computers and storage technology that could store and manipulate each of these forms.

## Computer Networks

The second innovation is the diffusion of the Internet and related computer networks, ranging from corporate LANs (local area networks) to WiFi and home networks. In our networked society, almost any message available from any medium is available any time, anywhere. The ubiquitous availability of media messages has served to erase distinctions among media that traditionally were limited to time or place availability (e.g., newspaper delivery at home in the morning, radio listening in automobiles, television news at home in the evening, etc.). Any medium can now transmit almost any type of message, any time, to almost any audience. Newspapers can easily add video when that content is distributed digitally, just as television stations can convey any amount of text on their Web sites.

The final technological dimension of convergence is that virtually any digital signal can be stored, manipulated, and edited on a computer. Over time, digital encoding algorithms have evolved with similarities in encoding that allow a single computer (with the proper programming) to control virtually any type of content, including text, audio, video, and graphics.

One of the challenges with the term "convergence" is that it has been applied to characterize a variety of similar phenomena related to communication technology. For example, the telecommunications industry refers to the transmission of voice, data, and video over a single connection (such as DSL and fiber optics) as "convergence" (McElligott, 2006). Similarly, the term is used in a variety of studies of new media to represent the manner in which boundaries between these media are disappearing. For example, the academic journal *Convergence* lists the topics covered by the journal as "Video games; Cable and telecomms; Mobile media/content; Internet studies; Digital/new media art; Digital photography; VR;

Control and censorship of the media; Copyright/intellectual property; New media policy; New media industries/institutions; New media history; New media in cross-cultural/international contexts; New media products; Digital TV; DVD; Digital music—recording, production, distribution, file formats/file sharing; Cinema; Gender and technology" (Sage, 2008, n.p.). For this journal, "convergence" subsumes the study of virtually all new media.

This wide-ranging list of applications for the term "convergence" includes a spectrum of distinctly different phenomena. The common thread is the replacement of analog transmissions that were unique to each application with digital transmission systems that can convey, store, and manipulate virtually any type of message so long as the message has been converted to a binary bit stream.

The spread of digital technology is therefore a major enabling force for the boundary-crossing phenomenon that is represented by the term "convergence," but the convergence trend represents a much broader set of phenomena. The next sections of this chapter will explore the dimensions of media convergence that are the subject of this book, including the advent of multimedia content and a variety of organizational behaviors ranging from collaboration among media organizations to co-ownership and cross-media production of content.

## Multiple-Media Content

The simplest form of media convergence takes advantage of technology to add a Web site to the output of the newsroom. The practice is so common that it is easy to ignore, but it must be noted that the practice of producing content for multiple media in a traditional newsroom is revolutionary. What is more remarkable is that almost all traditional media outlets—from daily newspapers and television stations to magazines and radio stations—have embraced the Web as a means of expanding the reach of the publication or broadcaster.

Another simple form of convergence is taking the content from the traditional outlet and repeating it on the Web site. This repetition of content is the most convenient, least time consuming and least costly, and it reinforces the branding of the media outlet (Wilkinson, Grant, & Fisher, 2008). Once an organization has such a Web site, the site has the potential to extend the ability of the organization to publish content. Broadcasters can provide text, tables, and background information that could never be part of a newscast. Newspapers can provide video, photo galleries, and longer versions of the stories that appear in print. Furthermore, they both can provide new types of content, including interactive narratives, blogs, and hyperlinks.

The barrier to the introduction of these new types of content is typically financial—Web sites are expected to pay for themselves. The early forays of news organizations onto the Web in the 1990s were notorious for losing money because of the high cost of creating the sites and the low initial revenue from Internet advertising. Since then, design tools, templates, and commercial services have dramatically reduced the cost of building a Web site, while revenue streams from classified advertising, banner ads, and text ads have increased the revenue opportunities for these sites.

The end result is that almost every media organization maintains a Web presence. Therefore, if "distribution of content through multiple media" is a dimension of "convergence," then every news organization that has a Web site can be defined as engaging in convergent journalism.

## Ownership

A different dimension of convergence is the co-ownership of two or more media outlets serving the same market by a single entity. Notable examples include Media General's "News Center" in Tampa, Florida (discussed in detail in Chapter 11), which includes *The Tampa Tribune,* WFLA-TV, and Tampa Bay Online (www.tbo.com); Belo's *Dallas Morning News* and WFAA-TV; and Gannett's *Arizona Republic,* KPNX-TV, and azcentrl.com in Phoenix.

Note that this dimension of convergence does not require anything such as coordination or sharing of content. Rather, this dimension simply captures common ownership. It may definitely be easier for organizations with common ownership to share content or coordinate coverage.

The question is whether cross-ownership involving only a traditional media outlet and a Web site would fall under this dimension. Using a strict interpretation of the dimensions, it would, but the fact that Web sites are most commonly an ancillary distribution source suggests that the field may want to focus discussion of this dimension upon cross-ownership of two or more traditional media outlets.

In the U.S., cross-ownership of newspapers and broadcast stations has been limited by the Federal Communications Commission (FCC) since 1975 (Gomery, 2002). A number of cross-owned media properties were "grandfathered" by the FCC because they predated these regulations, and a few have been allowed by special exemption. Still, the general rule since 1975 has been that newspapers and television stations in the same market cannot be co-owned.

In 2003, however, the FCC proposed relaxing these cross-ownership rules, allowing cross-ownership in the largest markets in the United States (Ahrens, 2003). That proposal failed, but a new proposal was introduced in late 2007 by FCC Chair Kevin Martin (Eggerton, 2008). As of this writing, no decision has been made, but the history of cross-ownership regulations indicates that a political battle may swirl around these regulations for years.

The United States is an anomaly in global communication regulations in prohibiting newspaper and television cross-ownership. Indeed, nascent media organizations, especially those in developing nations, thrive on the economies of scale, sharing of content, and promotional advantages afforded by distributing content simultaneously in multiple media.

## Collaboration

Even when media outlets are not co-owned, many publishers, editors, and news directors have grasped the opportunity for collaboration with media outlets that were formerly seen as competitors. Rather than seeing other media as threats that must be excluded from a newsroom, a number of media organizations enjoy the benefits of working with these former competitors.

Examples can be found on late evening television newscasts where newspaper staffers tease the next morning's top stories. This segment provides television stations (which typically have smaller reporting staffs) with additional news content and also helps to brand television news as timely and immediate. Similarly, many newspapers include forecasts from (and photographs of) television meteorologists on their weather pages in order to add personality and credibility to their forecasts.

As with other dimensions, there are multiple levels of collaboration. The most basic form of collaboration involves one-time cooperative relationships, articulated either formally or informally. As cooperative relationships develop, the frequency and scope of collaboration may increase, involving multiple stories and types of content.

Collaborative relationships must be mutually beneficial to persist. External forces also impact the beginning and ending of these collaborative efforts; these include the goals of the corporation and overtures from the competition. In cases where collaboration provides too great a competitive advantage, the relationship can be limited by regulation.

## Coordination

Media outlets that are not co-owned with other media outlets serving the same market can engage in the same or similar convergent journalism practices as co-owned media outlets. Most commonly they simply agree to share news, personnel or other resources. The motivation for these cooperative endeavors is usually not to achieve economies of scale but rather to attain greater visibility in the market through cross-promotion or to access resources that would otherwise not be available.

Because there are so many more television stations than newspapers in most markets, the local newspapers have an edge in negotiating these cooperative arrangements and generally wield greater power to choose a partner and set the terms.

But because television news departments usually have a fraction of the number of reporters as the daily newspaper(s) in the same city, the advantages from such coordination are limited.

One motivation for these cooperative endeavors is the desire of individual media outlets to leverage the strengths of their competitors. In most cases, the newspaper has a much larger reporting staff, while the television station staff includes more "personalities" that are known and respected by the public.

Enterprise reporting is a case in point. Rather than hold a hard-hitting story until the morning edition can be published, a newspaper can release part of an enterprise series or story to a television station, using the coverage of that story to draw readers to the rest of the story in the next morning's newspaper. Similarly, during breaking news, a newspaper might choose to work with a specific local television station to get a story on the air before other television stations break the story in advance of publication of the morning edition of the newspaper. (It is also common for newspapers to break such stories on their Web sites, but the reach of these stories is one to two degrees of magnitude greater when released on a top-rated television newscast compared with being released on the newspaper's Web site.)

## Exploring the Dimensions of Convergence

The purpose of this book is to provide insight into convergent journalism by exploring the dimensions discussed in the first half of this chapter. Each chapter provides a snapshot or exploration of a dimension of convergence. Some chapters provide a broad overview, and others analyze a narrow slice of the phenomenon. These chapters were chosen because, taken together, they provide an in-depth understanding of the dimensions of convergence and the opportunities for research and analysis of media organizations engaged in the practice of convergent journalism. The remainder of this chapter briefly introduces each of these dimensions.

### The State of the Field

A fitting place to start this exploration is by providing an overview of the practice of convergence in U.S. media. In the next chapter, Camille Kraeplin and Carrie Anna Criado provide this overview with a snapshot of convergence efforts among newspapers and television stations. Their research provides numerous insights into the effectiveness of these efforts.

As the authors indicate, this research is an update from an earlier study. By the time of publication, further updates are certain to have been published. But it is critical to start the exploration of convergence with one of these snapshots, and the strength of this chapter is the manner in which it addresses the interplay of newspapers, television stations, and the Internet.

## The Journalist's Perspective

Numerous textbooks provide instruction for journalists on how to report in a converged newsroom. Those practices are summarized in Chapter 2 by Janet Kolodzy, who authored one such textbook, *Convergence Journalism*. This chapter provides the practicing journalist's perspective on convergence, contributing the practitioner's dimension. In the process, she addresses the challenge faced by changing audience behaviors, exploring the evolving interplay between journalists, news organizations, and the public.

## Convergence in Consumption

Most studies of media convergence focus on the technological, organizational, and structural issues discussed earlier. It is equally important, however, to consider convergence from the perspective of consumers, who may see little or no difference among content offerings that media practitioners consider to be widely disparate.

Among the issues that must be explored in the convergence of media consumption are the types of content consumed, the place where the content is accessed, and the trend toward simultaneous consumption of multiple media. Two separate chapters address these issues.

First, in Chapter 3, Robert A. Papper, Michael E. Holmes, and Mark N. Popovich report the results of their second multimethod inquiry into media use, "Middletown Media Studies II." This chapter provides one of the most detailed pictures of individual media use that has been conducted, providing remarkable insight into simultaneous media use and issues with traditional measures of media use.

The evolving role of consumers is then explored by Jeffrey S. Wilkinson, Steven R. McClung, and Varsha A. Sherring in Chapter 4. This chapter begins with a detailed conceptualization of the changing role of consumers, many of whom are taking on the role of content creators. Wilkinson et al. characterize the process by which individuals may evolve as content creators, using three case studies to illustrate this evolutionary process.

Implicit in these two chapters is the need to reconceptualize consumers. Rather than being passive consumers of media content, they actively layer consumption of multiple media with the opportunity to contribute as well as consume media messages. The lesson from these chapters is that greater attention must be given to the complexity of consumer behaviors in a converged media marketplace.

## The "Dark Side" of Convergence

As Rogers (1986) indicates, the anticipated effects of new technologies tend to be those that are positive and direct, but most innovations also have negative implications that are unanticipated and indirect. Convergence is no different. The positive implications of convergence presented in many of these chapters are balanced somewhat by a look at possible negative implications in Chapter 5.

In this chapter, Van Kornegay applies historical precedent to discern the possibility that the proliferation of new media could result in an increasingly fragmented society rather than the "global village" envisioned by McLuhan (1962). Kornegay also discusses the potential loss of archived content related to the relatively rapid turnover in storage technologies.

## The Dissolution of Media Boundaries

One characteristic of convergent media that has received little attention is the increasing ease of use of the underlying technologies, which has led to a proliferation in media production in fields far removed from journalism. This proliferation portends a challenge to the traditional role of journalism and mass communication educators to provide skills training in media technologies.

This trend is explored in detail by Jeffrey S. Wilkinson in Chapter 6. His discussion explores the implications of the widespread application of skills traditionally confined to mass communication to fields as disparate as law and medicine. One implication of this discussion is that journalism and mass communication educators must consider what our unique role is in educating the next generation of media practitioners.

## The Organizational Dimension: Managing Convergence

Just as convergence demands new journalistic practices, it also demands a new approach to managing newsrooms. Synergy is not possible without addressing newsroom management along with daily news routines. Two chapters explore issues in managing converged newsrooms, taking a theoretical approach and a practical approach, respectively.

The theoretical perspective is offered by Vincent F. Filak in Chapter 7, where he applies theories of organizational structure to identify processes and barriers that must be considered in the management of converged newsrooms. In addition to providing significant explanatory power regarding the integration of multiple media routines in a single newsroom, this chapter provides an exemplar of how organizational theory can be applied to help explain journalistic processes and practices.

This theoretical approach is complemented by a practical perspective on media management from Holly A. Fisher in Chapter 8. In this chapter, Fisher synthesizes a wide range of research into converged newsrooms to create a set of recommendations for management processes that would facilitate the management of converged newsrooms.

## Theoretical Approaches to the Study of Convergence

As with the study of most emerging technologies, most of the research on media convergence is descriptive, with little attention to theoretical perspectives. But, as with the study of any media phenomenon, the application of theory allows us to draw lessons from other times and other media technologies to inform the diffusion, processes and effects of media convergence.

Rogers's diffusion of innovations theory is the most frequently used theoretical perspective for the study of technologies that have just been introduced. In preparing this text, the editors looked for researchers who have applied a wide range of theoretical perspectives to aid our understanding of convergence. The ones included in this book are not comprehensive, but instead provide an indication of the range of theories that can be applied to the phenomenon, along with an indication of what we can learn from each of these applications.

As discussed earlier, Vincent F. Filak applies theories of interpersonal communication to explore intra- and intergroup processes in the adoption of convergence in newsrooms in Chapter 7. Feminist theory is applied by August E. Grant, Jennifer H. Meadows, and Elizabeth Jordan Storm in Chapter 9 to illustrate issues in power, structure, and control that are implicit in the organization and information flows in a newsroom.

A range of theoretical perspectives are then applied by George L. Daniels in Chapter 10 in a study of the impact of converged practices upon news products. This research adds an important dimension to our understanding of the impacts of convergence by addressing differences in content that can be directly attributable to the structure of the news organization. The utility of his findings is further enhanced by the theoretical underpinnings that enable application of these results to other news organizations and content.

## Lessons from the Field

The exploration of media convergence in this book would not be complete without including a set of case studies that illustrate the practice of convergent journalism. Chapters 11 and 12 each analyze a specific set of operations to illustrate the day-to-day issues implicated in the practice of media convergence.

The number of cases that could be studied is much too large to allow coverage of even a fraction of the number of cases in this book. These chapters therefore provide a set of snapshots that illustrate a number of dimensions of convergence rather than exploring all possible converged journalistic environments.

In Chapter 11, Michel Dupagne and Bruce Garrison provide their take on what is arguably the most-studied converged news operation in the United States—Media General's Tampa News Center. In Chapter 12, Tony DeMars then provides a similar glimpse of convergence practices in small markets.

There are a number of lessons that can be drawn from these two studies and other similar studies. Susan Keith and B. William Silcock pull together many of these lessons in Chapter 13, drawing conclusions from the wide range of case studies of converged newsrooms. Their synthesis of these field studies results in a set of observations that is equally relevant to academics and practitioners seeking to understand and practice convergent journalism.

## Converging with New Forms

The evolution of media forms that began with print has continued through the introduction of electronic media, including radio, television, and, most recently, the Internet. This evolution should be expected to continue, as new forms are developed that will merge many dimensions of media, content, and creator.

One of the most interesting new forms to emerge is the Web log, or blog. In Chapter 14, Bryan Murley provides insight into the history, functions, and impacts of blogs, illustrating this tool as a new type of convergence.

## Global Insights

American media play a unique role in world media. On one hand, the United States is seen as a leader in technology, distribution, and control of global media flows. On the other hand, U.S. media are uniquely prevented from experimenting with or enjoying many prospective dimensions of convergence because of cross-ownership limitations that prevent most companies from publishing a newspaper and operating a television or radio station in the same market.

There is much to be learned from the study of convergence in other countries. In Chapter 15, Kenneth C. Killebrew attempts this daunting task, providing a global perspective that gives context to many of the dimensions of convergence explored in other chapters of this book.

## The Past and Future of Convergence

Context is critical in analysis of any phenomenon as complex as media convergence. The final two chapters serve to provide two, complementary, contexts for the information presented in the preceding 15 chapters.

The process of teaching convergence is contextualized in Chapter 16 by Timothy E. Bajkiewicz. This discussion provides a detailed history of journalism education, illustrating the manner in which issues related to "new media" and "convergence" have been a factor in journalism education since the founding of the first journalism schools more than a century ago.

No tome such as this one would be complete without peering over the horizon to speculate on the future of convergent journalism. Long-time journalist-turned-academic Charles Bierbauer knits together history, trends, and speculation in Chapter 17 to provide a set of observations that conclude the book by steering our attention toward the implications of recent history and the impact of this history upon the future of journalism in the United States.

## Summary

The premise of this book is that "convergence" is a multidimensional construct. From a research perspective, the term is probably too nebulous to be used to identify specific variables, processes, and media-related phenomena. On the other hand, the construct is useful to capture multiple dimensions of the revolution that is sweeping the media industries.

This discussion suggests that use of the term "convergence" should be limited to broad discussions that do not require precision in terms. But for research and analysis, scholars must identify the specific, operational dimension(s) of convergence being considered. Specificity brings clarity and facilitates comparisons and analysis across studies.

This book explores these individual dimensions of convergence. The goal is to provide a broad understanding of the impact (and potential impact) of these trends upon the media. In the process, these chapters seek to clarify specific practices and operationalizations that characterize this trend.

It is generally dangerous to analyze a trend that is still in process. Trends accelerate, decelerate or sometimes disappear altogether. The attention of the contributors to this book is therefore focused as much on the underlying processes and pattern of impacts as upon the convergence trend itself. However long the convergence trend lasts and whatever trend comes next, these processes and patterns should serve to facilitate understanding of the role of media in society.

## References

Ahrens, F. (2003, June 3). FCC eases media ownership rules *The Washington Post,* p. A1.

Burnett, R., & Marshall, P.D. (2003). *Web theory: An introduction*. London: Routledge.

Carlin, T. (2006). Digital audio. In A. E. Grant & J. H. Meadows (Eds.), *Communication technology update* (10th ed., pp. 235–257. Boston: Focal Press.

Castañeda, L., Murphy, S., & Heather, H. J. (2005). Teaching print, broadcast, and online journalism concurrently: A case study assessing a convergence curriculum. *Journalism and Mass Communication Educator, 60*(1), 57–70.

Dailey, L., Demo, L., & Spillman, M. (2005). The convergence continuum: A model for studying collaboration between media newsrooms. *Atlantic Journal of Communication, 13*(3), 150–168.

Eggerton, J. (2008, February 4). FCC publishes final orders on newspaper-broadcast cross-ownership rules. *Broadcasting and Cable*. Retrieved March 6, 2008, from http://www.broadcastingcable.com/article/CA6528516.html?q=cross%2Downership.

Fitchard, K. (2008, February 11). Bell Labs: Reviving an icon. *Telephony Online*. Retrieved February 14, 2008, from http://telephonyonline.com/belllabs/telecom_reviving_icon/

Gomery, D. (2002). *The FCC's newspaper-broadcast cross-ownership rules: An analysis*. Washington, D.C.: Economic Policy Institute.

Gordon, R. (2003, November 13). Convergence defined. *Online journalism review*. Retrieved October 30, 2008, from http://www.ojr.org/ojr/business/1068686368.php.

Grant, A. E. (2006). The umbrella perspective on communication technology. In A. E. Grant & J. H. Meadows (Eds.), *Communication technology update* (10th ed., pp. 1–6. Boston: Focal Press.

Hollander, B. (2005). Late-night learning: Do entertainment programs increase political campaign knowledge for young viewers? *Journal of Broadcasting and Electronic Media, 49*(4), 402–415.

Huang, E., Davison, K., Shreve, S., Davis, T., Bettendorf, E., & Nair, A. (2006). Bridging newsrooms and classrooms: Preparing the next generation of journalists for converged media. *Journalism and Mass Communication Monographs, 8*(3), 221–262.

Lawson-Borders, G. L. (2005). *Media organizations and convergence: Case studies of media convergence pioneers*. Hillsdale, NJ: Lawrence Erlbaum.

Lowrey, W., Daniels, G. L., & Becker, L. B. (2005). Predictors of convergence curricula in journalism and mass communication programs. *Journalism and Mass Communication Educator, 60*(1), 32–46.

McElligott, T. (2006, November 29). When universes collide. *Telephony online*. Retrieved March 6, 2008, from http://telephonyonline.com/software/technology/telecom_media_convergence_112906/index.html

McLuhan, M. (1962). *The Gutenberg galaxy: The making of typographic man*. Toronto: University of Toronto Press.

Poynterextra (2002, May 2). Thursday E-media tidbits. Retrieved March 27, 2008, from http://poynterextra.org/extra/tidbits/2002_04_28_tidbitsarchive.htm

Rogers, E. M. (1986). *Communication technology: The new media in society*. New York: Free Press.

Sage Publications (2008). *Convergence: The international journal of research into new media technologies* [Advertising copy]. Retrieved March 6, 2008, from http://www.sagepub.com/journalsProdDesc.nav?prodId=Journal201774.

Smith, L., Tanner, A. H., & Duhé, S. F. (2008). Convergence concerns in local television: Conflicting views from the newsroom. *Journal of Broadcasting and Electronic Media, 51*(4), 555–574.

Wilkinson, J. S., Grant, A. E., & Fisher, D. J. (2008). *Principles of media convergence*. New York: Oxford University Press.

CHAPTER 1

# The State of Convergence Journalism Revisited

## Newspapers Take the Lead

Camille Kraeplin
Carrie Anna Criado

### Introduction and Literature Review

When asked in November 2004 what they believed would be the most important issue within their industry in the coming years, a number of managing editors at the nation's largest newspapers mentioned convergence. "An essential theme of convergence: Counting eyeballs rather than hard copies," wrote one in the online survey. Another editor saw convergence journalism as a way to attract younger readers, a common concern: "Using convergence techniques to branch and appeal to more demographics."

As these comments suggest, convergence has become part of the media landscape. Attitudes toward the role convergence journalism will play in the coming years run the gamut. At a recent international conference, executives heard that newspapers have no future without online and digital services, one significant segment of

This article originally appeared in *Newspaper Research Journal*, Vol. 27, No 4, Fall 2006, pp. 52–65.

convergence journalism ("Newspapers have no future," 2005). On the other hand, convergence may play a more limited role in future media organizations than we can imagine at this point, and in forms we cannot anticipate. This longitudinal study examined where convergence journalism has been and where it is going. It includes two surveys—the first conducted in 2002–3, the second in 2004–5. The latter, which reexamined the growth of convergence journalism, demonstrated that newspapers in this country's top 210 markets appear to be maintaining convergence partnerships with both Web sites and TV stations. This article describes those partnerships and uses the economic theory of product differentiation to explain their success.

In 2002, Criado and Kraeplin (2003) conducted a baseline survey that determined the extent to which convergence journalism had taken hold in U.S. news media industries. It concluded that most of the news organizations sampled viewed convergence as important to the future of the profession. It also noted that the majority of both newspapers and TV stations surveyed had forged convergence partnerships with another media platform—around nine in 10 newspapers and eight in 10 TV stations.

There has been much discussion over the precise meaning of convergence and a fully converged newsroom. According to Killebrew (2002), convergence means that all platforms available for delivery to a Web-based operating system contribute to the final information product. It also suggests that information sharing and enhancement take place in the process. Singer (2004) writes that convergence journalism combines news staff, technologies, products, and geography from previously distinct print, broadcast, and online media. Dailey, Demo, and Spillman (2004) examined the role of convergence in both TV and newspapers. They defined convergence simply as newspapers and television news staffs working together. For this study, convergence journalism was defined as print, broadcast, and online news staffs forging partnerships in which journalists often work and distribute content across several news platforms. The only real difference between this definition and the one used by Dailey et al. is that the latter does not explicitly highlight content sharing, although it does not preclude it. Clearly, "working together" could encompass content sharing. In both cases, the focus is on collaboration across media.

In recent years, convergence has been driven by a combination of economic, regulatory, cultural, and technological forces (Pavlik, 2001). The trade press closely covered the first industry experiments in convergence journalism. Much of this coverage presented the philosophical questions surrounding any news industry trend: Will convergence work, and how (Barnhart, 1999)? Will it help the industry or hurt it? And if it does help, who will benefit most, company executives and shareholders, or front-line journalists and the public they serve (Tompkins, 2001)? Other reports told the stories of news organizations that had embraced the convergence model, the steps they had taken, and the lessons they had learned (Harvey, 2000).

A number of recent convergence studies by media scholars have taken a more systematic approach to surveying the new landscape. At least one examined convergence strictly from the perspective of television practitioners. This survey explored how television news directors defined convergence and how it was being practiced in the nation's television newsrooms. Researchers found that nearly nine of 10 of these affiliates were practicing a type of convergence (Duhe', Mortimer, & Chow, 2002). Respondents listed sharing content, sharing staff or promoting another partner with whom they shared content as practices that defined convergence.

A more common research method has been to interview both broadcast and print journalists. For instance, an online national survey of daily newspapers and commercial television stations noted the need to update news staff, production quality, compensation for multiplatform production, and the legitimacy of media convergence among the chief concerns of media professionals when it came to working across media platforms (Huang et al., 2003).

Likewise, as was noted earlier, Dailey et al. (2004) used similar survey instruments and approaches to interview both newspaper editors and broadcasters. Their nationwide study of 372 newspaper editors found that about 30 percent were involved in news-gathering relationships with TV stations. In addition, newspaper partners were frequently found to participate in functions associated with convergence, such as sharing news budgets, promoting the partner's content and requiring staff to appear on TV broadcasts. In a separate study, Dailey, Demo, and Spillman (2005) found that half of the television news organizations in the United States had partnered with a newspaper. As reflected in the 2004 study, these partnerships were characterized by cross-promotion of the partner's content and some sharing of daily news lineups.

Research questions for Phase II of this study included: (1) What changes in convergence journalism have taken place since Phase I? (2) Have television stations or newspapers generally done a better job of sustaining convergence partnerships? (3) Which convergence journalism partnerships, or which specific convergence models, appear to be most successful?

## Product Differentiation

In perfect competition, firms all supply an identical, standardized product. In monopoly, one firm sells a unique product, although that product may have indirect substitutes. Monopolistic competition, as the term suggests, represents a mixture of these two situations. The primary feature of monopolistic competition is product differentiation, where firms compete by selling products that differ slightly from one another (Chyi & Sylvie, 1998; Powers, 2001). Product differentiation occurs in most consumer markets. However, media products have often been very standardized. For instance, Powers (2001) noted that it is difficult to distinguish among different local newscasts if one is not familiar with the stations' news personalities. Thus, she reported, even when viewership of local news was falling nationwide, many stations were reluctant to experiment with product differentiation. For instance, in smaller markets, only the most financially secure, number-one-rated stations would risk adding more time for news to their programming. But in top-10 markets, where the financial stakes are higher, trailing stations compete by increasing their news presence throughout the day.

Convergence partnerships also appear to offer an attractive approach to product differentiation for local broadcast organizations, or at least the stations in the top-200 market surveyed for this study. For these broadcasters, the appeal of partnerships with newspapers is the papers' reporting depth. The definition of "partnership" can extend from sharing reporting tips to working together on investigative projects, which most likely would be promoted extensively on the broadcast medium as well as by the newspaper partner. In addition, newspaper reporters may appear on their partner station's broadcasts as "expert" sources, while popular newspaper sportswriters or columnists may be featured on-air in special segments or shows. This differentiates the station's newscast and associates the station with favorite local media personalities. For newspapers in these partnerships, broadcasters provide a high-profile platform for both individual staff and the organization as a whole as well as an element of immediacy. This clearly differentiates the product in the public eye as well.

The Internet, however, may offer even more attractive ways of differentiating a newspaper brand. It provides the same competitive edge as broadcast, in addition to new storytelling possibilities. Chyi and Sylvie (1998) examined the role of the electronic, or online, newspaper in the media marketplace. They suggested that from the company's perspective such competition makes sense in terms of profit making only if the two media can attract mutually exclusive audiences. To accomplish this, electronic newspapers should try to reduce substitutability with their print counterparts, or to differentiate themselves.

Chyi and Sylvie (1998) write that online papers should avoid relying on shovelware—moving whatever is published in the newspaper onto the Web without further developing the information. Shovelware provides inexpensive news but also may cause two problems. First, when online newspapers fail to distinguish themselves from their print editions, people may not bother to read online for the same information available in the paper's print edition. The authors argued that electronic newspapers distinguish themselves from traditional media by developing Internet-related features:

> As different media satisfy medium-specific needs, any new communication technology trying to survive must successfully define itself by fulfilling pre-existing needs or creating new needs.... Thus services such as interactive forums, searchable news archives, online surveys, online transactions, up-to-the-minute information, live chat rooms, and audio/video presentations would make the electronic newspaper a distinct medium and give Internet news—as opposed to TV news, cable news, radio news, or print news—a technological definition. (p. 13)

Another important factor, according to Chyi and Sylvie (1998), involves how print newspapers think about their online experiments. Due to the inevitable interrelationship with their print counterparts, many Web newspapers have not become independent entities, not even conceptually. However, implementing intermedia product differentiation requires new ways of conceptualizing news. Chyi and Sylvie cite the example of how, in covering a spectacular murder case, Denver-area newspapers used their Web sections to provide "a mix of articles, sound files, discussion groups, and other material" (p. 13). So in the context of online newspapers, product differentiation refers to the development of Internet-specific features. As for what features will suit online newspapers best, experimentation must continue, the authors suggest. But the general principle remains—online papers must distinguish their product from print versions.

## Methods

Researchers followed the same sampling methodology for both phases of this study. In Phase I (2002–2003) the 210 largest U.S. media markets were identified based on rankings provided by Nielsen Media Research (2003). Then one TV station from each market was selected randomly. The largest daily newspaper, based on circulation, from the same 210 markets (there was only one newspaper in the majority of markets) was also selected. In Phase I, surveys were e-mailed in October 2002 with a reminder following several weeks later. Follow-up calls to the majority of selected television news directors and newspaper managing editors who had not responded were completed through early 2003. For Phase II, the surveys were e-mailed in mid-November 2004; a reminder e-mail was sent in early December. However, due to a focused telephone campaign that lasted through the following spring, the response rate for Phase II jumped from 40 percent to 53 percent for newspapers and from 38 percent to 47 percent for television respondents. In Phase II, all television news directors and newspaper managing editors who had not responded to the e-mail survey were contacted by telephone.

In Phase I, the survey that was e-mailed to selected news managers consisted of 18 items; only one of those items, the question that asked the name of the respondent's news organization, was open-ended. Most of the 18 closed-ended items were designed to measure level of involvement in what might be labeled as "convergence practices" or attitudes toward those practices (the latter was not addressed in this study). And most responses could be measured on the nominal or interval level.

In Phase II, five open-ended questions were added. Respondents did not answer these questions as consistently as they did the closed-ended questions. This was particularly true among television news directors. To answer the study's research questions, researchers compared the results of Phase I with the results of Phase II.

The television respondents in the two samples were demographically similar. In Phase I, just over a third (37.25 percent; $N = 51$) were in large markets (1–50 based on Nielsen ratings). Just under a third (29.41 percent; $N = 51$) represented mid-sized markets (51–100). And a third (33.34 percent; *N* = 51) were in small markets (101 or higher). In Phase II, 39 percent of respondents (*N* = 92) were in large markets, 30 percent in mid-sized markets and 32 percent in small markets. The two newspaper samples differed somewhat. In both cases, mid-sized papers represented the bulk of the sample. However, newspapers with a circulation of 150,000+ were better represented in Phase II. In Phase I, just under half of respondents (45.59 percent; *N* = 69) were from papers with circulations between 25,000 and 74,999. Nearly that many (44.12; *N* = 69 percent) claimed circulations between 75,000 and 149,999. Around 6 percent (5.88 percent; *N* = 69) of the sample comprised newspapers in the 10,000 to 24,999 circulation category, and another 4.41 percent were in the 150,000+ category. In Phase II, the largest group of respondents were from mid-sized papers: 43 percent (*N* = 105) represented papers with circulations of 25,000 to 74,999, and 20 percent represented papers with circulations of 75,000 to 149,999. However, papers with circulations of 150,000+ were also well-represented (26 percent; *N* = 105). Just 9 percent of respondents represented papers with circulations between 10,000 and 24,999.

## Results: Changes in Convergence Journalism

The first research question addressed in this study asked, "What changes in convergence journalism have taken place since Phase I?" By Phase II, 68 percent (*N* = 105) of managing editors said their newspaper had a Web partner, down from 95 percent (*N* = 69) in Phase I.[1] In contrast, although the number of newspapers that said they had a partnership with a TV station remained fairly stable, at 70 percent in Phase I (*N* = 69) and 67 percent in Phase II (*N* = 105), more TV stations claimed newspaper partners in Phase II, up to 56 percent (*N* = 92) from 41 percent (*N* = 51) in Phase I.

In most of the newspaper-Web relationships identified in Phase II (78 percent; *N* = 105), the Web sites were directly associated with the newspaper. In most instances, the site was owned by the newspaper, but had a separate staff. These partnerships were often characterized by reporters providing versions of stories they had written for the newspaper to the Web site. Around eight in 10 respondents in Phase II (*N* = 105) said this took place at their organizations "frequently," versus 61 percent in Phase I (*N* = 69). Around seven in 10 newspaper respondents in Phase II said reporters would frequently provide their Web partner with briefs to update breaking stories they might be covering. This was also higher than in Phase I, when 56 percent of editors said this practice occurred frequently in their newsrooms. Only 29 percent of respondents in Phase II said their reporters frequently wrote stories exclusive to the Web. This number did not change much from

1. More complex statistical analysis was not completed on this data because of small sub-samples.

**Table 1.1** ▰ Charcterization of Web partners by newspaper respondents

| Charcterization | Phase I (%) | Phase II (%) |
|---|---|---|
| Reporters provide version of their newspaper stories to the Web site. | 61 | 80 |
| Reporters provide Web partner with briefs to update reporters' breaking stories | 56 | 70 |
| Reporters frequently write Web-exclusive stories. | 26 | 29 |

Phase I, when 26 percent of editors reported frequently providing exclusive content to the Web. (See Table 1.1.)

The open-ended questions also detected a range of additional activities undertaken by the newspapers online in Phase II that had not been asked about in Phase I. These included Connective elements (online chats, invitations to opinion polls, question-and-answer forums and Web logs) and Supplemental efforts (a range of activities that involved posting material to the Web that space requirements kept out of the newspaper). Around 60 percent of the newspaper editors had said staff had made supplemental contributions to their Web partners. For instance, just over six in 10 respondents ($N = 105$) said they featured additional photos, photo galleries, slide shows, or video and audio on the Web. As one respondent noted: "Photographers do slide shows, sometimes with accompanying digital sound recordings, using photographs that did not get printed in the newspaper because of space considerations." A number of respondents mentioned sports: "College town: extra football photos during the season." In other cases, the audio and video elements focused on a range of content: "Increasingly we are asking our reporting teams to think of ways they can use on-line strengths to add a little extra to long-term packages. A recent package on faith healing included slide shows and recordings our reporter did with subjects who were featured in his newspaper stories." Some respondents described more technically sophisticated efforts, including multimedia presentations, content-driven Flash programs and podcasting. A number of editors also said they used the Web site to accommodate longer textual elements such as databases, listings, crime reports, copies of source documents, movie reviews, and longer versions of newspaper stories.

More than a third of newspaper respondents in Phase II ($N = 105$) said their online efforts focused on interactive Connective elements. As one editor noted: "We use the site to find people and sources for stories that we're working on. So while we supply our site with most of its content, we also use it as a reporting tool to find subjects and get more interaction going." Another noted that the paper's reporters "participate in on-line chats," while a third said staff members "develop questions for readers/reader participation."

## Television-Newspaper Model

Unlike the newspaper-Web model, most of the television-newspaper partnerships include two independent news organizations. For instance, 89 percent of the TV respondents in Phase II had partnered with a local paper that was not owned by their station's parent company. In contrast, only 5 percent were working with a co-owned paper. Among Phase II newspaper respondents with a television partner, 74 percent were working with an independent station, while 19 percent had partnered with a co-owned station.

Among newspapers, some convergence practices remained fairly consistent from Phase I to Phase II of the study (see Table 1.2). For instance, 36 percent ($N$ = 105) of newspaper editors said staff were frequently interviewed on-air in Phase II, versus 40 percent ($N$ = 69) in Phase I. The number of newspaper staff frequently writing for TV broadcasts remained at around 10 percent in both phases of the study. The number of newspaper journalists frequently hosting TV shows or segments dropped slightly, from 13 percent in Phase I to 8 percent in Phase II.

Many of the convergence practices that television newsrooms were trying on for size during Phase I, however, seem to have been abandoned (see Table 1.3). Not one television news director said that members of his or her staff frequently wrote for their newspaper partner in Phase II ($N$ = 92), compared with 17 percent ($N$ = 51) in Phase I. Also in Phase I, nearly 30 percent of news directors said that reporters, anchors, and other staff were frequently quoted in the newspaper. In contrast, during Phase II, 5 percent responded the same way. And 11 percent of respondents in Phase II said staff frequently had a column in the paper, versus 25 percent in Phase I.

The open-ended questions in Phase II also identified some practices characteristic of the TV-newspaper model that were not evident in Phase I—Collaboration (information sharing and cooperation on stories or projects) and Cross-Promotion

**Table 1.2** ■ Description of Convergence Interactions as Reported by Newspaper Editors with a Television Partner

| Type of Interaction Reported | Phase I (% editors reporting interaction) | Phase II (% editors reporting interaction) |
|---|---|---|
| Staff journalists frequently interviewed on air | 40 | 36 |
| Staff journalists frequently writing for television | 10 | 10 |
| Staff journalists frequently hosting television shows or segments | 13 | 8 |

**Table 1.3** ▰ Description of Convergence Interactions as Reported by Television News Directors with a Newspaper Partner

| Type of Interaction Reported | Phase I (% directors reporting interaction) | Phase II (% directors reporting interaction) |
|---|---|---|
| Staff journalists frequently write for newspaper partner | 17 | 0 |
| Reporters, anchors, and other staff frequently quoted in the newspaper | 30 | 5 |
| Reporters, anchors, or other staff had their own column in the newspaper | 25 | 11 |

(promoting the partner or the partner's content in the other medium). Efforts at Collaboration were reported by 23 percent of convergent television respondents ($N = 92$) and 80 percent of convergent newspaper respondents ($N = 105$) in Phase II: "We provide briefs and tips," "Exchange news budget," "We supply information for them" and "Share news, projects together" were typical comments from newspaper editors about Collaborative practices. Here is one common response: "We let them know the stories we're working on and give them copies. We give them info on how to get info. Generally cooperate with them." Television news directors offered many of these same answers: "Share leads on breaking stories," wrote one. Explained another, "We often share information, especially when covering local politics." In some cases, Collaboration involved more elaborate forms of cooperation. This could include anything from working together on special projects or enterprise stories to partnering on town hall events to tapping newspaper staff to host shows or appear on air. One newspaper respondent wrote: "Our movie critic regularly contributes.... Television station also comes to photograph us when we have special projects." A television news director wrote: "We use NP sports reporters in a weekly part of our sports block to discuss local sports and make predictions."

Likewise, Cross-Promotional activity was reported by 26 percent of convergent newspaper respondents ($N = 105$) and 6 percent of convergent television respondents ($N = 92$). One typical description came from a newspaper respondent: "We provide a daily overview of the following day's headlines that appear in graphic form during the broadcast with a short explanation by the anchor. The TV station tells viewers they can read the story in the next day's edition." TV respondents provided a forum to publicize the next day's newspaper stories: "We air one of their headlines 'Tomorrow' on our 10 p.m. news," wrote one. Broadcast staff "write teases for paper, preview articles," wrote a newspaper editor.

A second research question asked, "Have television stations or newspapers generally done a better job of sustaining convergence partnerships?" Newspapers seem to have been more successful at maintaining these relationships than TV stations. For example, seven in 10 newspaper editors said their papers had maintained a partnership with a Web site and/or a TV station from Phase I to Phase II. However, among TV news directors, the number saying they had a Web partner dropped from 83 percent (N = 51) in Phase I to 45 percent (N = 92) in Phase II. Both groups seek out the TV-newspaper partnerships, but around seven in 10 papers had maintained them; the number of TV stations with partners, while growing, was still lower, at six in 10.

The third research question asked, "Which convergence journalism partnerships, or which specific convergence models, appear to be most successful?" The concept of "success" was operationalized simply as continued existence/continuation as a convergence model. As suggested earlier, the two models that meet this criterion are the newspaper-Web model and the TV-newspaper model. Both models continued to thrive through two phases of the study, providing benefits for each partner.

## Discussion

During the 1990s, many media analysts recognized the value of convergence journalism partnerships. The rate at which they moved online shows how much emphasis newspapers placed on the importance of a converged future (Singer & Thiel, 2002). In 1993, only about 20 papers had ventured online; a decade later, virtually every major U.S. newspaper had a Web presence. This study suggests a strong move toward convergence as well. During Phase I, nearly all sampled newspapers had Web partners, as did eight in 10 TV stations. In addition, a majority of papers had forged convergence partnerships with TV stations. However, interest in convergence seems to have slowed.

One explanation for the apparent decrease in Web partnerships could be that respondents in both groups have become more careful about what they define as "convergent." For instance, news directors from stations with Web sites that simply repurpose material may no longer define that site as a convergence partner, whereas they may have done so just a few years ago, using a looser definition of convergence. This change suggests a more sophisticated understanding of convergence among journalists, particularly in the larger markets. At a 2005 meeting of the Newspaper Association of America, the organization's director of electronic media communications, Rob Runnett, said that newspapers should give as much thought to their online product as they do to their print product: "Online is not just a second home for your print edition" (Christopher, 2005).

As this comment suggests, and this study confirms, media organizations increasingly recognize the value of emphasizing the unique contributions of each partner in a convergence alliance, or of using these partnerships to differentiate their product. As consultant John Morton (2000) wrote in *American Journalism Review:*

> The chief goal of the [convergence] alliances for newspapers is to expand their presence and brand names online at a time when the Internet poses a potential, though as yet unrealized, threat to core newspaper businesses. For the non-newspaper partners, the attraction is access to the papers' large and superior newsgathering forces. (p. 88)

Broadcast organizations clearly look to newspapers for depth of reporting; one way a station differentiates itself from other stations is by taking on serious in-depth investigative projects, the kind of projects that a newspaper partner can facilitate. In addition, featuring well-known local newspaper columnists and reporters on your newscasts can only boost credibility with the audience. For the newspaper, both the staff and product receive increased visibility, often among a younger, more attractive audience, from the partnership. So these papers are engaging in product diversification as well, especially in terms of market placement. But newspapers seem to benefit even more from the product diversification available through online partnerships. The Internet makes available whole new modes of storytelling that are not at all restricted by the spatial constraints that limit print. As prior research suggests, such intermedia competition makes sense from a business perspective only if the two media—online and print newspaper partners—can attract mutually exclusive audiences. To accomplish this, real diversification must exist between the online and print versions of the paper. Around 70 percent of the sampled papers were involved in each of the successful convergence alliances identified in the study—newspaper-Web and TV-newspaper. The newspaper-Web model has stabilized some since Phase I, but it is still a highly successful model, with much to offer both partners. Robinson (2005) defines Connectivity, which characterizes this partnership, as using the characteristics of the Web to "create and maintain social, political, and other kinds of cultural and democratic ties." Under Deuze's (2003) Connectivity model of journalism, the monitorial, dialogic, orienting, and other functions of a news medium function side by side, so that a strict division no longer remains between producers and consumers of news content. All become "prosumers." This occurs on the Internet in a way not possible in other media. Thus, these partnerships utilize the Web's connective strength, allowing newspapers to connect with readers–and contributing to the distinctiveness of the newspaper's online version.

When partnerships between newspapers and TV stations take place, they are most often characterized by Collaboration and Cross-Promotion. Study results suggest that newspaper journalists participate more in Collaboration than their television counterparts. They were more likely to be the ones sharing tips and information with their television partners. And four in 10 of them were frequently interviewed on air. Clearly, this contributed in a positive way to product diversification for their television partners. However, because of the low response rate among television news managers to open-ended survey questions, the real contributions of television staff to Collaboration efforts was difficult to gauge. Likewise, although TV stations might seem able to Cross-Promote because they are a high-profile news platform, the low response rate among television news managers made it difficult to determine how often stations promoted their partners.

## Limitations and Future Research

This study offers an overview of where convergence journalism has been and where it is headed. It also provides a theoretical explanation for why some partnerships seem to be working better than others: Within those partnerships, each partner provides something to the other that it cannot provide for itself easily or at all. In this way it diversifies the product or the brand. However, the study did not closely examine individual relationships. Nor did it gather extensive information about how successful partnerships operate on a day-to-day basis. In addition, a lower than ideal response rate overall, especially among television executives, limited the representativeness of the findings. Future research might seek to take a closer, more detailed look at how some of these successful convergence journalism models function. That would help us understand not only what is happening in convergence journalism and why, but also precisely how.

## References

Barnhart, A. (1999, October). News revolution: Join or you'll be left behind. *RTNDA Communicator, 53,* 40.

Christopher, L. C. (2005). *Association presents strategies for struggling newspaper industry*. Vienna, VA: Newspaper Association of America.

Chyi, H. I., & Sylvie, G. (1998). Competing with whom? Where? And how? A structural analysis of the electronic newspaper market. *Journal of Media Economics, 11,* 1–18.

Criado, C. A., & Kraeplin, C. (2003, August). *The state of convergence journalism: United States media and university study*. Paper presented at the annual meeting of the Association for Education in Journalism and Mass Communication, Kansas City, MO.

Dailey, L., Demo, L., & Spillman, M. (2004). *Newsroom partnership executive summary*. Muncie, IN: Center for Media Design, Ball State University.

Dailey, L., Demo, L., & Spillman, M. (2005). *Newsroom partnership executive summary*. Muncie, IN: Center for Media Design, Ball State University.

Deuze, M. (2003). The Web and its journalisms: Considering the consequences of different types of news media online. *New Media and Society, 5(2),* 203–230.

Duhé, S. F., Mortimer, M. M., & Chow, S. S. (2002, October). *Convergence in television newsrooms: A nationwide look*. Paper presented at The Dynamics of Convergent Media Conference, University of South Carolina, Columbia, SC.

Harvey, C. (2000, December). New courses for a new media. *American Journalism Review, 22,* 26.

Huang, E., Davison, K., Shreve, S., Davis, T., Bettendorf, E., & Nair, A. (2003, August). *Facing the challenges of convergence: Media professionals' concerns of working across media platforms*. Paper presented at the annual meeting of the Association for Education in Journalism and Mass Communication, Kansas City, MO.

Killebrew, K. (2002, October). *Distributive and content model issues in convergence: Defining aspects of "new media" in journalism's newest venture*. Paper presented at The Dynamics of Convergent Media Conference, University of South Carolina, Columbia, SC.

Morton, J. (2000). The emergence of convergence: The Web drives alliances like the one between the Washington Post Co. and NBC. *American Journalism Review, 22,* 88.

Newspapers have no future without online and digital services, media executives heard at a World Association of Newspapers meeting in Madrid. (2005, November 15). *The Australian.*

Nielsen Media Research (2003). *2002–2003 DMA ranks*. Retrieved April 15, 2003, from http://nielsenmedia.com/DMAs.html.

Pavlik, J. V. (2001). *Journalism and new media*. New York: Columbia University Press.

Powers, A. (2001). Toward monopolistic competition in U.S. local television news. *Journal of Media Economics, 14,* 77–86.

Robinson, S. (2005, August). *Experiencing journalism: A new model in online newspapers*. Paper presented at the annual meeting of the Association for Education in Journalism and Mass Communication, San Antonio, TX.

Singer, J. (2004). More than ink-stained wretches: The resocialization of print journalists in converged newsrooms. *Journalism and Mass Communication Quarterly, 81,* 838–856.

Singer, J., & Thiel, S. (2002, August). *In search of the forest amid a growing number of trees: Online journalism scholarship at the 10-year mark*. Presented at the annual meeting of the Association for Education in Journalism and Mass Communication, Miami, FL.

Tompkins, A. (2001, February 28). *Convergence needs a leg to stand on*. Retrieved April 15, 2003, from http://www.poynter.org/centerpiece/022801tompkins.html.

CHAPTER 2

# Convergence Explained

## Playing Catch-up with News Consumers

Janet Kolodzy

In opening his epic, *A Tale of Two Cities,* Charles Dickens wrote of the era that it was the best of times and the worst of times. Those words aptly describe the state of journalism at the beginning of the 21st century. It is the best of times because of the wide range of news outlets, news operations, and venues available for getting information about the rapidly changing world. A visit to a bookstore or newsstand finds a display of dozens of magazines and newspaper titles. Basic cable television opens up a hundred or more channels for news, entertainment, and information, and the Internet provides thousands of sites. The choices seem endless.

Yet it is also the worst of times for journalism. Audiences are becoming more fragmented, while news media ownership becomes more concentrated. Daily newspapers see a decline in readers, as well as a decline in advertising. The nightly network news sees its viewership decline, as the age of its audience rises. Journalism itself is being redefined. Anyone with a Web site and information can have access on a Web log (blog) to an audience greater than many daily newspapers or monthly

Adapted from "Why Convergence?" in Janet Kolodzy's *Convergence Journalism* (2006), published by Rowman & Littlefield. Reprinted with permission.

magazines. Anyone with a cell phone or camera can take photos or video of a news event or newsmaker and post the pictures online. Bloggers are challenging the traditional media's role as gatekeepers of news and information.

As a result, the news industry is in a state of flux and, some might say, a state of disarray. Technological, social, and economic changes are challenging traditional news organizations to develop innovative ways to attract new readers and viewers and to hold on to current ones. Convergence is one strategy being tried in several newsrooms across the United States. Yet convergence in journalism is ill-defined, misunderstood, and misrepresented. It has been used to explain everything from computer use to corporate consolidation.

When it comes to journalism, convergence means a new way of thinking about the news, producing the news, and delivering the news, using all media to their fullest potential to reach a diverse and increasingly distracted public. Convergence refocuses journalism to its core mission—to inform the public about its world in the best way possible. But nowadays the best way is not just one way: newspaper or television or the Internet. The best way is a multiple media way, doing journalism for a public that sometimes gets news from newspapers, at other times gets news from television and radio, and at still other times seeks news online. To be successful at convergence, journalists need to understand the strengths of each news medium or outlet and work to develop and provide news stories that dovetail with those strengths. Convergence requires journalists to put the reading, viewing, and browsing public at the center of their work.

However, convergence in journalism has many interpretations and definitions. Most journalists "know it when they see it" but really cannot describe convergence or explain its application in the newsroom. Although a 2002 survey of journalists indicates that nearly 90 percent of the newsrooms in the United States claim they are practicing some form of convergence, those same survey respondents were unable to define just exactly what they are doing that is convergent (Criado & Kraeplin, 2003).

More often than not, journalists distrust convergence. They view it as a marketing ploy, a way to promote the news as a "product," emphasizing the business rather than the journalism in the news industry. They also view it as a management ploy, a way to get fewer journalists to do more work with fewer resources.

Thus it is necessary to examine the different definitions of convergence: technological, economic, and journalistic. It is important to look at what has happened technologically, socially, and economically that led to this buzzword for all that is new in the news media. Finally, it is useful to examine why convergence is being tried in journalism, in response both to changes in news audiences and changes within the news industry.

## Defining Convergence

Dictionaries provide a simple definition of convergence: Convergence means the coming together of two or more things. In discussing convergence in the news media, however, the definition gets tricky because of disagreement over what exactly is coming together.

Massachusetts Institute of Technology professor Henry Jenkins (2001) provides a simple framework for defining convergence. "Media convergence is an ongoing process, occurring at various intersections of media technologies, industries, content and audiences; it's not an end state." Applying Jenkins to the news business, convergence of technologies involves the coming together of different equipment and tools for producing and distributing news. Think about computers and software. Convergence of industries involves consolidation of businesses and companies producing and distributing news. Think about Disney and News Corporation. Convergence of journalistic content involves journalists working in different media coming together to provide different content for different audiences. Think about newspaper Web sites and news organizations text-messaging the latest sports scores or stock quotes.

## Technological Convergence

The original discussions of media convergence focused on the technological: computers and digitization. Anyone who has sent an e-mail on a computer or used a cell phone that takes pictures and sends text messages is taking for granted technological convergence. But less than 30 years ago, the digitization of words, pictures, and sound for access by a variety of electronic devices seemed like science fiction.

The founder of MIT's Media Lab, Nicholas Negroponte, argued in the late 1970s that three industries that were separate at the time—computers, broadcast/film, and printing/publishing—would overlap and merge by the start of the 21st century (Brand, 1987). In the late 1970s, technology was challenging the status quo in each of those areas, so the idea of those industries coming together seemed radical. Thirty years ago, the personal computer, which would revolutionize an industry, was being developed. IBM, the powerhouse of the time, would be upended by Microsoft. Broadcast and film were being challenged by the technology of videotape. Video would not only change the way television news, entertainment shows, and movies were seen and distributed but also how they were made. The printing/publishing industry was moving from "hot," or lead, type to "cold" type, computer-produced layouts and printing without heavy lead plates. Cold type brought color and revolutionized the look of printed news.

The development of digitization would set off a new debate about technological convergence. Ithiel de Sola Pool (1983), a communications scholar, pointed out that this next wave of convergence would involve a merging of electronic devices, "the convergence of modes." He noted that "electronic technology is bringing all modes of communications into one system" (p. 28). Everything would come down to one device—computer + TV + telephone + stereo + movie player + organizer. This all-in-one "mega-device" may soon become a commonplace reality, if the iPhone is any indication.

This new wave of technological convergence can be found by digging into purses and backpacks and looking in our living rooms and dorm rooms at our information gadgets. The cell phone, personal digital assistant (PDA), digital music player, computer and camera have become a single device. Additionally, telephone companies have worked out deals with news organizations for sending news updates and headlines to cell phone users.

Anyone can download a story, article, even a book off the Internet and read it on an electronic device. Songs and video are downloadable and portable. In 1990, when Roger Fidler spoke about a portable electronic newspaper when he was new media director for the Knight-Ridder newspapers, that idea was considered crazy. He told groups of very skeptical journalists that, rather than buying a newspaper, subscribers would be able to plug their tablets into cable or phone lines, download newspaper content, and read it on portable electronic devices. Those devices now exist, as technology has made screens easier to read and easier to carry around without wires and heavy batteries.

Technological convergence of the stereo, CD player, and the computer, thanks to digitization, has brought about the iPod, MP3 and on-demand music. The DVD is revolutionizing access to entertainment, first with players and now with recording devices, such as TiVo, that put the consumer in charge of determining content. Broadband cable service has merged the computer and the television into a viewing and interactive device. And the Internet, the World Wide Web, is the technological convergence of computers, satellites, and digital technology. It has opened up new ways of getting and exchanging information, destroying geographical and political boundaries in the process.

On one level, technological convergence means the coming together of formerly distinct electronic devices or media delivery systems, changing the equipment used to get information and to access it. But technological convergence has also opened up new ways of presenting that information. Technological convergence has led to multimedia information presentation. The Internet allows formerly separate and distinct storytelling media or platforms—the text of print; the audio of radio, pictures and graphics of visual design; and the moving pictures of animation, film, and television—to be combined into a new way of providing information.

Trying to pin down a name for this new, evolving type of journalism that comes together via the Internet has added confusion to the definition of convergence. Journalism distributed on the Internet has been called new media, online news, multimedia journalism, digital news. But it also has become known as convergence journalism, since it marks the coming together of different elements of storytelling. The merger of AOL and Time Warner in 2000 helped solidify the definition of convergence to mean electronic content delivery, because that merger was the coming together of a content company, Time Warner, with an online delivery company, AOL. Yet that merger also created confusion over the definition of convergence, because AOL Time Warner (in 2003 the company name changed back to Time Warner Inc.) became the largest media conglomerate in the world. Thanks to the merger of AOL and Time Warner, and the mergers of other media companies, convergence came to mean media consolidation.

## Economic Convergence

Using the definitional framework of MIT's Henry Jenkins, AOL Time Warner represents economic convergence, or the coming together of media industries. When it was announced in 2000, the merger was the flashiest and most ambitious example to date of economic convergence. And it brought another buzzword to the fore, "synergies." AOL Time Warner executives talked about ways to work across the different media, taking advantage of all the company's different properties—online, television, magazines, films, and books.

## Cross-Promotion

A precursor of that synergy involved the handling in 1999 of the last movie directed by Stanley Kubrick before his death, *Eyes Wide Shut.* The Warner Bros. movie earned a big write-up in *Time* magazine; its stars, Tom Cruise and Nicole Kidman, appeared on the magazine's cover and conducted promotional interviews on shows such as *Larry King Live* on CNN, a Time Warner company.

The *Eyes Wide Shut* example demonstrates the most common and most visible aspect of economic convergence: cross-promotion of properties or brands. In the 1980s, at Turner Broadcasting, advertising that promoted CNN programs would appear on *Headline News,* TBS, and TNT. Ads for TNT and TBS programs would air on CNN commercial breaks. Additionally, an hour news special slated for the 8:00 p.m. CNN newscast would be cross-promoted in earlier newscasts throughout the day on the 24-hour news channel.

These days, cross-promotion means cross-media promotion. At MSNBC, it not only means commercial advertising and promotions within MSNBC programs, but it includes promotion of NBC news programming as well as special slots on the MSNBC and MSN Web site for NBC news shows. Such extensive cross-promotion raises issues of independence, diversity, and control. These issues have bubbled over when news organizations of a media company end up reporting and cross-promoting properties of its entertainment divisions.

In the case of *Eyes Wide Shut,* the movie received considerable attention by a variety of news media, not just those associated with Time Warner. But media critics raised questions about the prominence of the Time Warner coverage—a cover story, promotional interviews—with regard to journalistic independence and control.

In the summer of 2000, CBS and CBS News's *The Early Show* drew criticism for reporting and cross-promoting the hit reality show *Survivor.* The show, which aims at winnowing down a group of people facing extreme circumstances and challenges, ends each installment with a contestant being voted out of the group and off the show. On the morning after each episode, *The Early Show* would interview the person voted off the show. Since the show was so popular, newspapers, magazines, and online news sites were carrying articles about the show as well as weekly polls and analyses. On the day of the *Survivor* season finale, all the network newscasts ran stories about the show or the reality-TV phenomenon it had sparked. But the consistent use of *The Early Show* to promote the *Survivor* series came under the most frequent attack because it lowered the acceptable standard for the use of news programs to promote entertainment. "There seems to be no limit to what CBS News will do to shill for the network's prime time 'reality' entertainment shows in the vain hope that their popularity will rub off on its offerings," complained former NBC News president Lawrence K. Grossman (2000, p. 70).

Four years later, that standard of acceptable cross-promotion in news was under attack again, as NBC News devoted two episodes of its prime-time newsmagazine, *Dateline,* to the finales of two long-running popular NBC comedies, *Frasier* and *Friends*. NBC's *Today Show* anchors Katie Couric and Matt Lauer hosted these two *Datelines,* further blurring the separation of news and entertainment. Associated Press television writer David Bauder (2004) questioned whether NBC News had "besmirched its reputation" with such blatant promotion on its primetime newsmagazine. *Washington Post* television critic Tom Shales (2004) referred to *Dateline* as "NBC's so-called 'news' magazine," and NBC executives countered by saying the newsmagazine's audience understands that it offers a range of stories and that very little of it is promotional. They pointed to other hard-hitting journalistic reporting on the show, such as a look at racial profiling. Both the *Friends* and the *Frasier* shows on *Dateline* scored larger audiences than typical *Dateline* episodes, reinforcing the financial benefit of cross-promotion available through economic convergence, while undermining the journalistic value of untainted, independent news judgment.

## Consolidation

This weakening of journalistic in favor of marketing values has added fuel to the fire of protest against the other prominent aspect of economic convergence in the news media: media consolidation through corporate mergers. The 2000 merger of AOL and Time Warner created the world's largest media company and resurrected new fears about the lack of diversity of opinion and independent access to those opinions. But AOL Time Warner was created after a long line of mergers that mixed different media industries: film, music, television, cable, and news. The Time Warner half of the 2000 merger, in fact, was created in the wake of two previous mergers—the Time, Inc., publishing empire and the Warner Bros. entertainment empire in 1990, and the merger in 1995 with Turner Broadcasting, the cable news and entertainment company created by Ted Turner.

A few months before the January 2000 AOL and Time Warner merger announcement, Viacom, an entertainment and cable production and distribution company, bought CBS, creating yet another mega-media company. Before that, the

Tribune Company, which owned several newspaper and television stations throughout the United States, merged with the Times-Mirror Company, a prominent newspaper chain. Don Hewitt, executive producer of the CBS *60 Minutes* program, joked to the *Wall Street Journal,* "I'm convinced that before I die, one person will own everything" (Pope & Peers, 1999, p. A1). Hewitt was referring to CBS president Mel Karmazin and the Viacom deal. Yet his sentiment echoed criticism from media and political commentators Robert McChesney and Noam Chomsky that too much of the media was being controlled by too few companies. "We should deplore this concentration of media power," McChesney wrote in 2000. "It is dangerous when so few people control what we see and hear. And these giants have enormous power not only over the economy but the political system as well" (p. B7).

Media companies, however, argued that the economic realities of a multimedia world required mergers. Consolidation would allow them to respond to the demands of audiences using different media, seeking different information. As the competition became bigger and more diverse, media companies argued that they had to get bigger and more diverse to keep up. "Owning television, radio, and newspapers in a single market is a way to lower costs, increase efficiencies, and provide higher-quality news in times of economic duress," said Jack Fuller, *Chicago Tribune* publisher (Gordon, 2003, p. 64).

## Journalistic Convergence

News organizations that are experimenting with the notion of convergence aim to achieve Fuller's goal of "higher-quality news" in all the formats available: print, online, and on radio and television. The problem comes when convergence is seen as a benefit for media company stockholders and not as a benefit for journalists or for readers, viewers or browsers.

Convergence in journalism requires changes in how news organizations think about the news and its coverage, how they produce the news and how they deliver the news. Most convergence in journalism today focuses on the last of those areas, delivering the news. It involves a newspaper's daily edition or a newscast's scripts being placed online, a newspaper reporter appearing on television for a "talkback" or interview on his or her story, the television weather caster developing the weather page for the newspaper.

However, dozens of news organizations are trying to also think about and produce news differently. They are trying to ensure that the news they are providing is best suited for the audiences of each medium or format being used to distribute the news. These organizations realize that newspaper readers want more context and detail to their stories; online browsers are looking for quick hits of information, interactivity, and the ability to seek out other information; and broadcast listeners and viewers are looking for the latest information that puts them at the scene. Convergence in journalism means the coming together of journalists and certain types of journalism that have been operating in separate spheres—newspapers, magazines, radio, television, and online—to provide quality news in all those different formats. That coming together can involve shared resources and information. It can involve joint reporting and production on projects. It can involve "one-man bands" or "backpack" journalists—one person doing the reporting and producing of news for all the different formats. It can involve multimedia storytelling online or what could be called "converged presentation." It can involve some or all of these variations.

Convergence journalism is happening in a variety of newsrooms, in a variety of manners. No one form of convergence journalism has risen to be the best template for doing convergence. What has emerged among news organizations aggressively pursuing convergence is a mind-set. In operations as disparate as the *Lawrence (KS) Journal-World,* the *Columbus (OH) Dispatch,* the Ohio News Network, MSNBC, and ESPN, convergence is as much a way of thinking as it is a way of working. The *Journal-World*'s former director of New Media/Convergence, Rob Curley, says simply that convergence is just good journalism. Executives at ESPN say that doing journalism for a magazine, television, and Web site just makes sense and is the way they approach sports news.

While economic convergence has often pushed journalistic convergence, like at the Tribune Company at Media General's *Tampa Tribune,* tbo.com, and WFLA; and Time Warner's NY1, journalists are defining what convergence means for them. Journalist Chindu Sreedharan (2004) calls it "layering." Journalists "understand the possibilities of other mediums (sic), contribute across platform when called upon, and begin to layer their stories" (Sreedharan, 2004). Convergence is "new journalism" that is evolving to keep up with the times.

Convergence is one answer to the question of where journalism should be headed in the 21st century. It is a response to the convergence of lifestyle, business, and technological trends that are forcing a change in the relationship between the people who make the news—journalists—and the people who use it—the public. Convergence is a response to two seemingly dichotomous trends—the fragmentation of the news audience and the consolidation of news ownership.

## Why Convergence? Because of News Audience Fragmentation

Changes in lifestyle and advances in technology have been working in tandem to affect the kind of news people want and the way they want to get it. American news consumers are individualistic, and their news interests are scattered. They use different types of media at different times of the day to get news and information. The single mass audience is giving way to multiple niche audiences.

## Lifestyle

The audiences for news are becoming fragmented because their lifestyles are fragmented. People today talk about multitasking, doing more than one thing at one time. They talk about being time-starved, not having enough time to do everything they want to in a day. People discuss compartmentalizing their lives, trying to separate work, leisure, family, and other aspects of their world. They complain about information overload, having to digest too much information at one time. All these issues impact the way people choose news and the way news should be presented. Convergence, putting news out to audiences in different formats, at different times of the day, attempts to respond to the lifestyle changes of the news audience.

Americans are also working longer and taking more time to get to and from their jobs. The U.S. Census Bureau (2004) reports that more than a quarter (28 percent) of all workers over the age of 16 put in more than 40 hours a week. And 8 percent work 60 or more hours. The Organization for Economic Cooperation and Development (OECD) has noted that while workweek hours in many European countries have decreased in the past 25 years, the average hours worked per person in the United States has increased 20 percent (Organization for Economic Cooperation and Development, 2004). The OECD study of 19 countries found that more Americans are working, and they are working longer.

Not only are Americans spending more time working, they are spending more time getting to work, driving alone in their cars. The Census Bureau found that 77 percent of all workers drive to work by themselves these days, up from nearly 65 percent in 1980. And it is taking them longer to get to work. While the national average for the work commute is about 25 minutes, about 7 percent of Americans take more than an hour to get to work. And about a fifth of all workers are heading out to work between midnight and 6:30 a.m. (U.S. Bureau of the Census, 2004).

Journalists understand that this increased time at work and getting to work affects how much time people spend catching up on the news and how they go about it. People do not have time in the morning before they head out the door to read a 96-page morning paper. Many get their local news, particularly weather and traffic reports, from the car radio or by checking a morning newscast on television. Some are leaving for work before the morning newscasts begin or before the newspaper is delivered to their doors. At work, they are not reading the newspaper or watching television, but they are checking the latest news online. By the time their workday is over and they get home, they may have missed the evening news on network television, and they may fall asleep before the late-night local wrap-up at 10:00 or 11:00 p.m. The news cycle has become 24/7—nonstop, 24 hours a day, seven days a week—because American lives have become 24/7.

The biannual survey of news habits by the Pew Research Center for the People and the Press finds that people go after news morning, noon, and night. Thanks to the Internet, more and more of them are following it during the day. The 2004 survey found that nearly three-quarters of Americans keep track of news during the day. However, the average amount of time Americans spend with the news media each day has dropped about seven minutes in the past 10 years, the result of time-starved, busy schedules (Pew Research Center, 2004).

Increasingly, people under the age of 25 maintain that they are the most time-starved in terms of keeping up with the news (Pew Research Center, 2004). As a result, they spend the least amount of time getting news. They do not watch the network news, read a daily newspaper or spend as much time getting the news compared to every other age group. Older Americans spend more time with the news (85 minutes for the over-65 crowd) than younger Americans (35 minutes for the under-25 crowd). Furthermore, younger news audiences are not regular news consumers. More than two-thirds of the Pew respondents under the age of 25 said they check the news from time to time. When the Pew survey asked why they do not follow the news regularly, half of the under-25 respondents said they are too busy. When people do not have enough time for news, or only have time for it at irregular hours and intervals, it affects when and how the news media interact with their audiences. Convergence—providing news in more than one medium, when it is convenient for people to get it—responds to the need for a different paradigm of interaction.

That interaction is also being transformed by the lifestyle trend of multitasking. Although its official definition involves a computer running more than one program at a time, Americans have taken over the word to describe doing several tasks at once. From driving while talking on a cell phone, to listening to music while making dinner, to checking e-mail while grabbing a bite to eat or even talking to the office while watching a child's soccer game, Americans are doing more things at one time.

More and more, news audiences are multitasking when they are getting information. A study of media use conducted in 2003 by researchers at Ball State University found that while people think they multitask with media about 12 percent of the time, when those people were watched throughout a day, they were using more than one medium close to a quarter of the time. The study found that "the most common media multitasked with reading were TV (well ahead of all others) then radio and music" (Papper, Holmes, & Popovich, Chapter 3, this volume). People who were using more than one medium were most often were reading while watching television.

Additionally, Americans are multimedia, getting news and information from not just broadcast television or daily newspapers or a news Web site, but from various sources. This is especially true for young audiences, who are shying away from traditional news media (like newspapers and network newscasts) and are more comfortable with new media. The Online Publishers Association found that 18-to-34-year-olds consistently own more new media gadgets, from video games to Blackberry wireless PDA/e-mail devices (Carey, 2004). The study also found that young news consumers use television and the Internet in tandem when there is breaking news. These consumers also prefer news in a more visual format. Nearly two-thirds of those in the 2004 Pew survey say their best understanding of the news comes from pictures or video.

Surveys of the next generation of news audiences show they are even more likely to be multimedia multitaskers. For example, the Kaiser Family Foundation examinations of media use in both 1999 and 2003 found that computer and television use is common in the under-age-six set, and many youngsters are using more than one medium at a time sometime during their typical day (Rideout, Foehr, U., Roberts, D., & Brodie, 1999; Rideout, Vandewater, & Wartella, 2003).

Multitasking and communicating with visuals as well as with words has a profound impact on how news is presented, especially if news organizations want to maintain an audience for their news or expand their audience. Convergence, trying to take advantage of various formats to reach various audiences, allows traditional media to adapt to these new audience preferences.

Convergence is also a strategy that could allow traditional news media to adapt to how news audiences are reworking the notion of community. Common interests are often topical rather than geographical. Currently the traditional news media respond to local audiences and local interests, but, increasingly, the news audience is fragmenting into special areas of interest that are not limited by geographical boundaries.

For example, an audience interested in all things Rhode Island can come together on projo.com, the Web site for the *Providence Journal* newspaper. CNET.com, a Web site dedicated to new technology news, attracts an audience from around the world with an interest in everything digital.

This phenomenon is not just a result of the Internet. Audience interests have been fragmenting for years, as is evident by the plethora of special-interest magazines, from *MacWorld* for Macintosh computer aficionados to *Popular Mechanics* for do-it-yourself mechanics. In broadcast, audiences are dividing along special interest lines by tuning into niche cable channels like OLN, the Outdoor Life Network, or the SciFi Channel, which is devoted to science fiction entertainment. Satellite radio allows audiences to choose music formats from hip-hop to country, regardless of geographical location.

## Technology

The rise of the Internet, satellite global connections, and wireless communications has hastened audience fragmentation and has raised expectations about how and when people can get news and information.

Today, no one thinks twice about being able to get information and news at any time of the day or night. But a generation ago, that was not possible. If you wanted to get the up-to-the-minute score of your favorite major league baseball team 30 years ago, you had few options: You could listen to the game on the one local AM radio station allowed to carry it (if it carried it); you could wait until the news wrap-up at the top of the hour at another station; or you could hope for an update of game score during the one televised game of the week, which often did not feature your team. Today, you can check sports scores online, have them text-messaged to you on your cell phone or PDA, check the graphic crawl of scores on numerous cable and broadcast channels on television and listen for scores on the radio. Sports fans today rarely have to wait more than a few minutes to find out a score, and they have numerous places to go to find it. The technological advances in communications equipment have allowed news and information to be delivered instantly. As a result, news audiences expect instant news.

The use of satellites for transmitting information and images brought about the first wave of instant, specialized news in the late 1970s and early 1980s. Networks such as ESPN for sports and CNN for news could not have developed without satellite technology. That technology allowed those networks to get video (also a new technology in the 1970s) and send it out to audiences. The true measure of the impact of that technology came during the Gulf War in 1991. Audiences around the world heard and watched bombs falling on Baghdad in real time. They did not have to wait for the evening news; they could watch it, unfiltered, as it was happening.

Within the decade, the technology of the Internet provided another outlet for satisfying a demand for news at any time of the day or night. The only limitations on up-to-the-minute news online are the speed with which news organizations can update their information and user overload, when too many people try to go to one site. Wireless technology and cell phones have taken instant news to yet another level by allowing it to be accessible anywhere. Cell phones were used by a group of college-age journalists to report on the 2004 Democratic and Republican political conventions and to send those reports—both text and images—to cell phone users. Technology has not only broken down time limitations to getting news and information, it has broken down geographical and cultural ones.

During the military offensive in Iraq in the spring of 2003, American news audiences who wanted perspectives other than those offered by U.S. newspapers and television networks went to non-U.S. news media Web sites such as english.aljazeera.net, the Web site for the Arab news network Al-Jazeera, and guardian.co.uk, the Web site for Great Britain's *Guardian* newspaper. A poll by the Pew Internet and American Life Project in the spring of 2003 found that 10 percent of Internet users checking out war news went to sites by "foreign news organizations." Another 8 percent checked on nontraditional news or alternative commentary, such as Salon.com (Rainie, Fox, & Fallows, 2005).

The Internet also broke down barriers to the display of graphic images of war and terrorist violence in Iraq. While newspapers and television stations declined, for example, to show the beheading of an American contractor or the charred remains of ambushed U.S. contractors in the spring of 2004, audiences who wanted to see those images could find them on Internet sites. Technology once again created a niche interest in particular information not available from mainstream media.

The technological transformation of the ways to get news and the rapid pace of the work world and family life are bringing about the transformation of the audience for news. People are not looking for news at the same time, from the same place, with the same outlook, in the same format, or on the same communications device. The audience for news is shifting away from one-size-fits-all. News organizations trying convergence see it as a strategy to respond to this shift by news consumers.

## Why Convergence? Because of News Media Consolidation

The economic landscape for news organizations is shifting as well. That shift has profound implications for how journalism is produced and distributed. Consolidation has become commonplace within the news industry. So has vertical integration, which in business means having control of different companies along the production line between supplier and consumer. In the news media, vertical integration can mean a company like Viacom has control of the supply of raw materials (reports, stories, scripts), products (newscasts, movies), and distribution (outlets or platforms—television, cable, movie theaters).

When media critic Ben Bagdikian (1997) published his book *Media Monopoly* in the 1980s, he warned of ownership consolidated among about 50 companies. At that time, the fear focused on the demise of independent voices in newspapers. By the 1980s, hundreds of afternoon newspapers had folded or merged with their morning counterparts in joint operating agreements allowed by the 1970 Newspaper Preservation Act. Newspaper chains such as Gannett bought out smaller, individual- or family-owned newspapers, moving control of those papers out of town and into a corporate boardroom. Radio and television stations were gobbled up just like newspapers, and in 1975 the Federal Communications Commission (FCC) initiated the Newspaper/Broadcast Cross-Ownership Rule which prohibited one owner from having both media in any one market.

Despite such limitations, media consolidation kept rolling along. At the beginning of the 21st century, the majority of the media in the United States are owned by a half dozen major international media corporations. Yet media companies have been arguing for more consolidation to remain profitable in an era of scattered audiences, outlets, and interests. Their arguments have focused on rewriting federal legislation and regulation to allow them greater flexibility in handling the economic challenges of a world of instant messaging, wireless and satellite technology, and the Internet.

As the nation headed into the booming dot-com era in the mid-1990s, the Telecommunications Act of 1996 opened up all sorts of possibilities of mergers and consolidation. The act required the FCC to review its regulations every two years, to ensure that the rules were keeping pace with the technological advances in telephones, broadcast, and telecommunications. It was a much- needed update to a 1934 law, a law drawn up at a time when telephones, radio, and film were the new media technologies. In the meantime, federal courts ruled that the FCC needed to develop a modern-day justification for its ownership limits, which were anywhere from 25 to 50 years old. Media businesses pushed for the FCC to loosen if not eliminate those rules. FCC chairman Michael K. Powell told senators in 2003 that keeping the rules as they were was not an option, but that the FCC would make rule changes that would fulfill the FCC mandate to protect diversity, promote competition, and foster localism.

Consumer, antitrust, and civil liberties proponents, however, felt that the Telecommunications Act and further relaxation of FCC ownership rules would give corporations too much license to swallow up small competitors at the expense of diversity and local control. As evidence of "merger mania," anti-consolidation forces point to what has happened within the radio industry. Hundreds of independent radio stations and small radio station ownership groups have disappeared following the February 1996 enactment of the new telecommunications law. Within months, newspapers in cities like Pittsburgh; Minneapolis; Columbus, Ohio; Boston; and Washington reported on local stations being bought up and big radio chains like Westinghouse merging with other big radio chains like Infinity. An FCC study (Williams & Roberts, 2002, p. 3) of the radio industry from 1996 to 2002 noted a 34 percent decline in the number of commercial radio owners, creating more group-owned stations and bigger radio chains. That consolidation has led to smaller radio news staffs and the elimination of radio news at many commercial stations. Music formats are limited and are no longer programmed locally or with local on-air talent.

The largest owner of U.S. radio stations, Clear Channel Communications, has been at the forefront of criticism over consolidation. One oft-cited example of central control of music came in the wake of the September 11, 2001, terrorist attacks. The nearly 1,200 Clear Channel stations received a memo suggesting a list of songs that should not be played. Songs on the list included R.E.M's "It's the End of the World as We Know It" and "Crash into Me" by Dave Matthews. While the list was offered as a guide and not a corporate mandate, journalists on several newspapers suggested that the guide was an example of corporate censorship. Two years later, Clear Channel helped organize and promote Rallies for America, in support of the Bush administration's war in Iraq, in more than a dozen Clear Channel markets. With more than 110 million listeners, Clear Channel was attacked by media critics for supporting the rallies, and antiwar activists decried Clear Channel's use of the airwaves to promote a political position.

The demise of much independently owned radio through consolidation became the leading argument against further efforts to allow more consolidation in ownership throughout all media. A second wave of media consolidation had been anticipated if and when federal regulations are loosened. In June 2003, the FCC approved new rules allowing more concentrated ownership of the news media nationally and more cross-media ownership locally. But those rule changes were preempted by legislative and legal challenges.

In 2004, five companies dominated most broadcast media nationally, and some of those companies exceeded the FCC limit as to the amount of the national television market they could control. Before 2004, FCC rules said that no one broadcaster could reach more 35 percent of the national audience. A new law passed by Congress and implemented in 2004 changed that to 39 percent, bringing News Corp. and Viacom into compliance. The media conglomerates and the FCC were pushing for a 45 percent cap. A federal appeals court rejected that higher limit, and the 39 percent limit is now the law of the land.

The appeals court also rejected a planned FCC rules change eliminating the 1975 ban on cross-ownership of newspapers and television stations in a local market. The ban on cross-ownership has been eroding for years, following several national media mergers. Media companies such as the Tribune, Gannett, and Media General own television and radio stations and newspapers in major cities such as Los Angeles, Chicago, Dallas, Tampa, Hartford, and Phoenix. Many of these companies anticipated the end of the newspaper-broadcast ban after the Telecommunications Act became law in 1996. The FCC has been told by the courts to review this cross-ownership ban.

Journalists, civil liberties activists, and consumer groups have been fighting these ownership rule changes, fearing they will lead to fewer jobs, fewer voices, less local news and programming, and thus less public service. A review of recent history following media mergers and corporate consolidation lends credence to some of those fears:

- Within months after the AOL Time Warner merger, CNN eliminated 10 percent of its employees from news operations, the largest staff cut in the cable news network's history. Part of the job cuts came amid consolidation of CNN Interactive, the award-winning, trend-setting CNN Web site, with the AOL interactive operations.
- One study found that children's television programming in Los Angeles, the second largest TV market in the country, became less diverse in the wake of consolidated ownership of local stations (Children Now, 2003). The study by Children Now (2003), a children's research and advocacy group, reported that local, original children's programming was cut almost in half between 1998 and 2003 and that "the number of the same shows repeated between cable and broadcast channels increased almost fourfold between 1998 and 2003."
- Sinclair Communications, which owns more than 60 television stations around the country, has implemented News Central, a newscast production format in which weather, national and international news, and sports segments are put together in a central production facility in Hunt Valley, Maryland, near Baltimore. In several Sinclair markets, local newsroom jobs or unprofitable news operations were eliminated. *Television Week* reported that two-thirds of Sinclair's stations were not doing local news when this "central casting" format began in 2002 (Barnhart, 2004). Sinclair officials maintain that News Central allows for more local news at its stations that cannot afford

full-blown local news operations. And they point out that the format allows for longer reports on local news stories during the one hour-long News Central newscast of the day. News Central critics argue that the format takes away local control and local voices from the news, noting that the News Central newscast includes a nationally aired Sinclair political commentary.

Cross-ownership and cooperative relationships within local markets have raised concerns about the lack of diversity of opinion. If the newspaper's movie critic also serves as a TV station's movie critic, the local audience is getting only one opinion on whether the latest offering at the box office is worth seeing. Or the television critic for the local newspaper might think twice about criticizing the news team of the television station owned by the same company he or she works for.

Proponents of cross-ownership, particularly former FCC chairman Michael Powell, dismiss the diversity argument by noting the wide variety of outlets now available to anyone online, on television, and at newsstands and bookstores.

> Today, news and public affairs programming—the fuel of our democratic society—is overflowing. There used to be three broadcast networks, each with thirty minutes of news daily. Today, there are three twenty-four-hour all-news networks, seven broadcast networks, and over three hundred cable networks. Local networks are bringing the American public more local news than at any point in history, and new tools such as the Internet are becoming an increasing and diverse source of news and information for our citizens. There has been a 200 percent increase in outlets. But, more importantly for diversity, there has been a 139 percent increase in independent owners. In sum, citizens have more choice and more control over what they see, hear, or read than at any other time in history. (Powell, 2003, p. 3)

In the city of Boston alone, news audiences can choose from six local news stations, including a regional cable news network, New England Cable News, for newscasts in the mornings and evenings. They can pick up the *Boston Herald* and *Boston Globe* newspapers, the free *Metro* newspaper on Mondays through Fridays, alternative papers such as the *Boston Phoenix* and the *Improper Bostonian,* suburban and neighborhood newspapers such as the *Daily News Tribune* in Waltham and the *Beacon Hill Times,* ethnic papers such as the *Boston Irish Reporter,* specialty papers such as the *Boston Business Journal* and *Boston Law Journal,* and dozens of other publications. Web logs of local interest, such as the H2Otown-info.bryght.net about the city of Watertown, Massachusetts, have also expanded the marketplace for news and information.

The ethnic press, such as magazines, newspapers, radio, and television outlets in Spanish, Chinese, and other languages are "growing rapidly," according to the Project for Excellence in Journalism's State of the News Media 2004 report. The report noted that the number of Hispanic daily newspapers grew from fourteen in 1990 to 35 in 2002 (Project for Excellence in Journalism, 2005). The Latino Press Network lists nearly 150 newspaper publications nationwide. The Magazine Publishers Association annual handbook reports that more than 17,000 periodicals were available for public consumption in 2003, or an addition of 3,000 magazines in a 10-year period. The big leap came in 2003 with 440 new titles.

News Web sites are so plentiful that Google, AOL Yahoo! and others are providing news site indexes and compilations. More than 90 percent of newspapers and television stations in the United States have Web sites, and hundreds of publications and broadcast outlets around the world; media outlets from regional Arabic broadcasters like Al-Jazeera to national newspapers like the *Manila Times,* put their news online. Additionally, millions of Americans are using the Internet to publish their own content online. A March 2003 survey by the Pew Internet and American Life Project determined that 44 percent of adult American Internet users put some type of material online. That material ranged from e-mail to personal Web pages to Web diaries. A November 2004 Pew survey found that 7 percent of Internet users say they have a Web diary or Web log online, and more than a quarter of Internet users were reading blogs (Rainie, 2005).

Audiences have more sources than ever for news and information, yet, at the same time, financial control of those sources is being concentrated in fewer entities. The news industry is evolving in two seemingly contradictory directions—the mass media are getting more massive, while niche media are developing even smaller niches. Both have profound ramifications for the practice of journalism that news organizations cannot ignore or avoid. Being too small, being too limited a niche, causes the journalism to get lost in a sea of information. Being too big causes the journalism to get diluted. Journalism convergence, doing news in more than one medium, attempts to ride these waves of change in the news business.

## Convergence: Taking Journalism Back to the Future

Convergence journalism is providing news and information in more than one format, using the strengths of each format to best serve news audiences. It aims to respond to the fragmentation of the news audience while acknowledging the economic reality of consolidation of media ownership. The fear in newsrooms about convergence journalism centers on the notion that convergence may make sense from a corporate standpoint but not from a journalistic one. The concern is that convergence is responding more to the profit-making, cost-saving reality of consolidated ownership of news and less to the public-interest aspect of addressing audience needs and demands for news and information.

News organizations that are trying convergence journalism, however, argue that for convergence to work well, audience needs must supersede corporate strategies. They see the public as the ultimate winner of their efforts. A 2003 survey of journalists in four convergence-oriented newsrooms found that those reporters, editors, producers, and managers believed that the people reading, listening, watching, and using their news were better served by their convergence efforts (Singer, 2003).

Newsroom managers, that is, editors and producers, of many convergence-oriented operations such as MSNBC, ESPN, Ohio News Network, and the *Lawrence (KS) Journal-World* acknowledge the corporate aspects of convergence, but they are unwilling to concede to being overtaken by them. Instead, they are seeking opportunities to redirect attention to public service. As a result, they have had to redefine how they think about their work and how they do it. They have had to develop a different mind-set about journalism.

"We have to get away from this notion that we put stuff out there and news consumers can take it or leave it," said Dean Wright, editor-in-chief of MSNBC.com until mid-2005. "We have to serve the public but we can't give them just the candy they want...we have to tell them things they don't know and things they already know but in different ways. And we have to make it accessible" (personal communication, April 24, 2004).

The two current news media audience leaders—television and newspapers—have developed mind-sets about audiences that need updating. Over the years, newspapers have developed a reputation of being detached and arrogant. Many reporters and midlevel editors at newspapers had little exposure to, and perhaps even little use for, readership surveys and other audience information. That is changing. For example, many papers now list e-mail addresses along with reporter bylines. Less than a decade ago, many journalists opposed such an opportunity for direct reader feedback. Still, while many newspapers' marketing and advertising departments collect readership information, that information about the people reporters and editors are serving—or not serving—often does not make it to their desks. Newspapers can seem a bit patronizing or condescending toward their public. Newspapers can come across as out of touch with regular people.

Television, on the other hand, could be considered on the other end of that spectrum, fretting too much about giving the audience what it wants so the viewers will keep watching and keep stations' ratings high. Television journalists know all too well their audience's likes and dislikes, as the Nielsen ratings' estimates of viewers' interest is broken down minute to minute. Television news uses focus groups and other marketing research to determine everything from the likeability of an anchor to the types of stories to conduct during key ratings weeks (known as sweeps weeks). Television can seem to pander to news audiences, seeking ways to manipulate them to be interested.

With all the fragmentation of news audiences today, journalists need to strike a better balance. They cannot be so arrogant about their work that they forget who they are doing journalism for, or so driven to be popular that they forget their responsibility to provide useful journalism. In that sense, convergence, in which journalists are working in different media to address the changing information needs of the public, is an opportunity for traditional media who have strayed from their basic focus and purpose. By adopting a new approach, convergence, they can return to their traditional values of serving the public.

"The reasons for convergence have to be based on true journalistic collaboration," says Jon Schwantes, corporate director for news convergence for the Dispatch Group in Columbus, Ohio. "The reasons for convergence have to be rooted in good journalism and in serving the news consumers" (personal communication, July 5, 2003). Schwantes argues that if convergence is aimed at responding just to corporate concerns, such as cross-promotion of stories and efficiencies in staffing, then it will fall short. Cross-promotion has its limits with audiences, as Lawrence, Kansas, 6 News television learned in its convergence efforts (Associated Press Managing Editors, 2004). A credibility roundtable in the city found disagreement over the use of cross-promotion references in the news, so its news director decided to cut back on them. Managers at other convergence-oriented operations consistently point out that convergence requires a commitment to more staff, better training, and management flexibility (personal communication, July 5, 2003).

Wright, formerly of MSNBC.com, has argued that the online news medium is in the best position to take advantage of emerging technologies to develop a new way of doing journalism. He sees convergence with television and print partners, such as the NBC and MSNBC cable networks and *Newsweek* magazine, as a way to use new technologies to develop "new ways of looking at things," new ways of providing information and telling stories to better inform the public. "It *is* going back to the future," contends Wright (personal communication, April 24, 2004).

Newsroom leaders at ESPN, which has proven to be successful in broadcast (ESPN cable channels, ESPN radio), online (ESPN.com), and in print (*ESPN The Magazine*), profess a goal of serving and satisfying sports fans in whatever format those fans want sports news. John Walsh, ESPN executive vice president and executive editor, explained that "convergence is just another 'how can we do it better?' " He added that he believes in the "everybody is in" theory, which maintains that journalism thrives by trying to reach as many people in as many different ways as possible. "There are people who prefer the little screen to the big screen. There are people who still read the daily newspaper, still read magazines, still listen to the radio." Add them together, he said, and journalism gets an audience that he calls "astonishing" (personal communication, May 6, 2004).

ESPN, which is owned by Disney, one of the Big Five media conglomerates, has used the notion of journalistic convergence to respond to the fragmentation of media audiences and the consolidation of media ownership. It has come closest to a convergence goal in which all media partners view each other as equals, share their information and ideas, and work together to find the best way to present stories to their audiences using the strengths of any and all outlets available to them. Other organizations are achieving a lesser level of convergence, one that could be called content sharing, in which stories and ideas are freely exchanged among the different news outlets (Dailey et al., 2003). By using convergence, news organizations are innovating, taking risks in trying to scope out a new role for journalism in the digital age.

Nicholas Negroponte (2003), who as a founder of MIT's Media Lab is an expert on the digital age, has argued that "without innovation we are doomed—by boredom and monotony—to decline." (p. 34). New ideas, he has written, need diversity, risk, openness, and idea sharing to thrive. Convergence is a new idea in journalism that relies on the diversity, openness, and idea sharing among different media and that takes a risk on that mix to better serve news audiences.

## References


Associated Press Managing Editors. (2004, April 19). APME national credibility roundtables project. Retrieved November 29, 2005, from www.apme-credibility.org/members/2003/6NewsReport.html

Bagdikian, B. (1997). *The media monopoly* (5th ed.). Boston: Beacon Press.

Barnhart, A. (2004, January 5). Sinclair fortifies station newscast: Local-national hybrid "News Central" to counter "biased" reports. *Television Week*, p. *4*.

Bauder, D. (2004, June 1). *Dateline NBC: How is "news" defined?* Retrieved June 19, 2004, from www.cnn.com.

Brand, S. (1987). *The media lab*. New York: Viking Penguin.

Carey, J. (2004, April 20). *Media lifestyles of the 18- to 34 year-old consumer. Ethnographic study for the Online Publishers Association*. Retrieved December 5, 2005, from www.onlinePublishers.org/?pg=press&dt=042004.

Children Now (2003, May 21). *Big media, little kids: Media consolidation and children's television programming*. Retrieved April 2004 at www.childrennow.org/newsroom/news-03/pr-05–21–03.cfm.

Criado, C.A., & Kraeplin, C. (2003, April 4). *Convergence journalism: Landmark U.S. media and university study*. Retrieved November 29, 2005, from www.convergencejournalism.com

Dailey, L., Demo, L., & Spillman, M. (2003, August). *The convergence continuum: A model for studying collaborations between media newsrooms*. Paper presented at the Association for Education in Journalism and Mass Communication, Kansas City, MO.

de Sola Pool, I. (1983). *Technologies of freedom*. Cambridge, MA: Belknap Press of Harvard University Press.

Gordon, R. (2003). Convergence defined. In K. Kawamoto (Ed.), Digital journalism: Emerging media and the changing horizons of journalism (pp. 57–74). Lanham, MD: Rowman & Littlefield.

Grossman, L. K. (2000, September/October). Shilling for prime time: Can CBS News survive Survivor? *Columbia Journalism Review*, 70–71.

Jenkins, H. (2001, June). Converge? I diverge. *Technology Review*, p. 93.

McChesney, R. (2000, January 19). A media deal with plenty of bad news. *San Diego Tribune*, pp. B-7; B-9.

Negroponte, N. (2003, February). Creating a culture of ideas. *Technology Review*, p. 34.
Organization of Economic Cooperation and Development (2004). *Striking facts. OECD Employment Outlook 2004*. Retrieved December 5, 2005, from www.oecd.org/document/62/0,2340,en_2649_201185_31935102_1_1_1_1,00.html
Pew Research Center for the People and the Press. (2004, June 8). *News audiences increasingly politicized; Where Americans go for news*. Retrieved November 29, 2005, from http://people-press.org/reports/.display.php3?PageID=834
Pope, K., & Peers, M. (1999, September 8). Merging moguls: Redstone, Karmazin. Both like to be boss; now, they must share. *Wall Street Journal,* p. A1
Powell, M.K. (2003, June 4). Oral statement on broadcast ownership biennial review before the Committee on Commerce, Science, and Transportation, United States Senate, 3.
Project for Excellence in Journalism (2005). *State of the news media 2004*. Retrieved November 30, 2005, from www.stateofthenewsmedia.org/2004/narrative_ethnicalternative_ethnic.asp.
Rainie, L. (2005, April 1). *The state of blogging. Survey by Pew Internet and American Life Project*. Retrieved November 29, 2005, from www.pewinternet.org/pdfs/PIP_blogging_data.pdf
Rainie, L., Fox. S., & Fallows, D. (2005, June 3). *The Internet and the Iraq war. Survey by Pew Internet and American Life Project*. Retrieved November 29, 2005, from www.pewInternet.org/pdfs/PIP_Iraq_War_Report.pdf
Rideout V., Foehr, U., Roberts, D., & Brodie, M. (1999, November). *Kids and media @ the new millennium*. Retrieved December 1, 2005, from http://www.kff.org/entmedia/upload/Zero-to-Six-Electronic-Media-in-the-Lives-of-Infants-Toddlers-and-Preschoolers-PDF.pdf
Rideout, V., Vandewater E.A., & Wartella, E. (2003, Fall). *Zero to six: Electronic media in the lives of infants, toddlers and preschoolers*. Retrieved December 1, 2005, from http://www.kff.org/entmedia/upload/Zero-to-Six-Electronic-Media-in-the-Lives-of-Infants-Toddlers-and-Preschoolers-PDF.pdf
Shales, T. (2004, May 7). A big hug goodbye to "Friends" and maybe to the sitcom. *Washington Post,* p. C1.
Singer, J. (2003, August). *The sociology of convergence*. Paper presented at the meeting of the Association for Education in Journalism and Mass Communication, Kansas City, MO.
Sreedharan, C. (2004, May 11). *The "C" word*. Retrieved November 29, 2005, from www.poynter.org/content/content_view.asp?id=65471
U.S. Bureau of the Census (2004, March). *Journey to work: 2000*. Retrieved November 29, 2005, from www.census.gov/Press-Release/www/releases/archives/facts.
Williams, G., & Roberts, S. (2002, September). *Radio industry review 2002: trends on ownership, format and finance, 3*. Retrieved July 16, 2007, from www.fcc.gov/ownership/materials/already-released/radioreview090002.pdf.

CHAPTER 3

# Middletown Media Studies II

## Observing Consumer Interactions with Media

Robert A. Papper
Michael E. Holmes
Mark N. Popovich

The Middletown Media Studies are a comprehensive attempt to better inform our understanding of how consumers interact with all major media and the relative role that media play in their daily lives. In several research studies spanning almost 80 years, "Middletown" has been used to describe the community of Muncie, Indiana, and surrounding Delaware County.

Media have been in a state of change. The variables driving media change continue to hinge on its form, speed, extent, and—perhaps most importantly—the impact of change on the underlying business models that sustain the media landscape. Much of our understanding of the impact of changes in media lies in how well we understand corresponding changes (or the lack of them) in consumer behavior. The rate of media change has significantly accelerated over the last decade or so, driven by factors like continuing media proliferation, the advent of interactivity, and mobile media. However, while the media landscape may have changed, what we know about how consumers respond is limited and based upon research methods developed during a far simpler media age.

For example, much of what we think we know about how people use the media comes from asking them. Unfortunately, as demonstrated in the first of the Middletown Media Studies (Papper, Holmes, & Popovich, 2004), people are poor witnesses of their own behavior. In many cases–especially regarding television, radio, and computer use–people's estimates of how much time they spend with media bore little relationship to what was observed. Many people could not even recall whether they used a medium, much less how much time they spent with it.

The problem is amplified when researching use of more than one medium. Asking people to recall using one medium is hard enough, but when investigating multiple media use and multiple locations (home, work, car, and other), self-reporting becomes untenable. The problem becomes even more convoluted if we want to understand the context in which people use media, when they use more than one medium and which other activities coincide with media use.

## Background and Method

Since the 1960s, the search for more accurate means of measuring media activity has led researchers to test a variety of methods in diverse environments across a range of media. Unfortunately, the approaches have been piecemeal as investigators have typically focused on one medium at a time.

Media researchers have long known the weaknesses of audience measuring instruments. For example, Bechtel, Achelpohl, and Akers (1972) compared results of 52 subjects from media self-report questionnaires and observed family behavior during television program and commercial viewing. In comparison with filmed observation data, subjects tended to over-report commercial viewing by about 25 percent in diaries and 40–50 percent in questionnaires. While most studies involving audience measurement examine use of a single medium, theirs is one of the few to compare different methodologies.

Investigators have argued that media usage occurs in a complex environment, but little, if any research adequately tests that proposition. Stempel, Hargrove, and Bernt (2000) measured 10 different types of media use, but paid little attention to interactions between media. Advertising researchers reported that complexity appeared in the form of supplementary activities, like sewing, reading or writing, while their subjects attended to either television programming or commercials. However, Coffey and Stipp (1997) suggest another kind of complexity—as interactions that take place between computer use and other kinds of media use.

The study reported here measures these types of interactions. In particular, our study examines how often concurrent media exposure happens in the everyday life of the media user, and whether there are patterns to how the media user employs multiple media sources.

Non-home locales have often been overlooked when media researchers measured media use. Our study identifies how media users utilize media in their cars, in their workplace, at other locations, and also in their homes.

The first Middletown Media Studies (MMS I) involved three companion research projects: a telephone survey of 401 people in Middletown, a diary study involving 359 people, and the direct observation of 101 people. For the second Middletown Media Studies (MMS II), we wanted to observe 400 people, half from Muncie and Delaware County, Indiana, and half from Indianapolis. For MMS I, data were collected using paper forms; for MMS II, software was developed to run on "smart" keyboards (small laptoplike computers running the Palm OS™). The touch-screen software allowed observers to log 15 media and 17 life activities from four locations. The program wrote the current state of media exposure, life activities, and location to a data file every 15 seconds. Almost 150 observers (mostly graduate and undergraduate students) were recruited in Indianapolis and Muncie. Each observer was trained in naturalistic research techniques, institutional research board rules and guidelines, and the operation of the data-gathering equipment. Oversight operation centers were established in both Muncie and Indianapolis.

A targeted, demographically balanced population was recruited by Ball State University staff and Strategic Marketing and Research, Inc., in Indianapolis. Observer training was conducted in February, March, April, and May 2005. Observations were conducted in March, April, May, and early June 2005.

A total of 412 full-day observations were scheduled, starting as soon as the person awoke in the morning and would allow an observer in. A shift change occurred at approximately 3 p.m., and observation continued until as close to bedtime as the person would allow. Of the 412 arranged observations, 394 were successfully completed. For data analysis purposes, a number of observations were discarded because the observer either arrived too late or left too early, an observer became uncomfortable with the situation, the participant refused to allow entry until too late or asked the observer to leave too early or because of equipment failure. All told, Middletown Media Studies include over 5,000 hours of observation and nearly 1.2 million data records.

- 18–24: 13.9 percent
- 25–34: 19.2 percent
- 35–44: 25.4 percent
- 45–54: 19.5 percent
- 55–64: 12.7 percent
- 65+: 9.2 percent

Somewhat less than half of the participants (45.9 percent) were in the 18–39 age group; 54.1 percent were 40 or older (all age-category percentages and age-related analyses were based on the 338 cases with known ages). Education levels varied, but 84.9 percent had some college or more education. The ethnic breakdown matched regional profiles: 81.4 percent Caucasian, 10.5 percent African-American, 2.3 percent Hispanic/Latino, and 5.9 percent representing one or more other group. Income levels were slightly below national profiles but typical of the area: 26.9 percent earned less than $30,000, 38.8 percent earned $30,000–$60,000, 22.9 percent earned $60,000–$100,000, and 11.3 percent earned more than $100,000.

## Results

The average length of the observational day was 12.9 hours. The average length of the observed day (resulting in complete, valid data) was 12.6 hours. In terms of media use, more than two-thirds (68.9 percent) of that time—nearly 9 hours—was spent using one or more media. Of that nearly 9 hours, 2.75 hours (30.7 percent) involved concurrent exposure to two or more media. For 3.8 hours (42.0 percent of media usage), people consumed media only without engaging in any other life activity.

At 225.6 minutes, media-only activity (not involving any other life activity) was the number one activity during the day. That time amounted to just 43.3 percent of the 521.1 total minutes the average person spent with all media during the observed day (225.6 minutes of media alone plus 295.5 minutes of media activity occurring simultaneously with another life activity). On the other hand, some of the former minutes (225.6) are spent with two or more media at the same time, so while it may be exclusively media time, it is not necessarily time spent exclusively with one medium.

Similar to the first Middletown Media Studies, both the incidence of media use (the percentage of people out of the total observed) and the average amount of media use across all the people observed are staggering (Table 3.1).

### Media and Location

One of the strengths of an observational study is that it is people-based rather than media- or location-based. While this consumercentric approach to research may be complex and difficult to administer, the level of insight gained may be the best way to address the parallel complexity of people's media lives. Having adopted this method, we were able to observe all media in all contexts.

The data indicate just over half the media-use day (56.9 percent) is spent at home. Work locations account for 21.1 percent, cars for 8.3 percent, and other locations (friends' homes, libraries, churches, retail shops, etc.) for 13.7 percent. Television dominates the home, radio dominates the car, and the computer dominates the workplace (Table 3.2). Almost 30 percent of the people watch television outside the

**Table 3.1** ▰ Overall Incidence and Amount of Daily Media Use and Concurrent Media Exposure

| | Percentage who used the medium | Average minutes of use | Average minutes in CME | Average percent of time in CME |
|---|---|---|---|---|
| Television | 90.6% | 240.9 | 68.7 | 28.5% |
| Radio | 73.4 | 80.0 | 23.8 | 29.7 |
| Any computer[a] | 72.9 | 135.8 | | |
| All Internet | 68.0 | 93.4 | | |
| Web | 62.3 | 67.3 | 53.8 | 80.0 |
| Email | 60.6 | 46.4 | 39.2 | 84.6 |
| Instant Messaging | 13.7 | 9.3 | 9.1 | 98.5 |
| Computer software | 51.4 | 73.6 | 51.6 | 70.1 |
| All phone[b] | 94.6 | 42.2 | | |
| Land line | 82.6 | 29.9 | 22.2 | 74.2 |
| Mobile phone | 58.6 | 11.8 | 8.1 | 68.7 |
| Music[c] | 60.0 | 65.1 | 24.4 | 37.5 |
| Print media[d] | 62.6 | 32.8 | | |
| Newspaper | 39.1 | 12.2 | 8.5 | 70.1 |
| Magazine | 30.6 | 7.3 | 5.2 | 71.4 |
| Book | 28.3 | 12.7 | 7.3 | 57.5 |
| All video[e] | 32.3 | 32.6 | | |
| VCR | 13.4 | 9.3 | 4.3 | 46.2 |
| DVD | 22.0 | 23.3 | 3.6 | 15.4 |
| Game console | 15.4 | 11.6 | 5.1 | 43.9 |
| Other Media | 40.9 | 21.3 | 12.7 | 59.6 |

[a]Any computer includes all Internet (uses) and software. All Internet includes Web browsing, e-mail, and instant messaging.
[b]All phone includes standard land line and mobile phone.
[c]Music includes cassette, CD, MP3, and other sources but does not include music on the radio.
[d]Print media combines newspaper, magazines, and books.
[e]All video includes both VCR and DVD.
*Note*: The enormous overlap of time within "any computer" use reflects extensive concurrent use for software, Web browsing, e-mail, and instant messaging.

home, and out-of-home viewing accounts for 9.4 percent of all television use. Time spent with television is the highest—three times that of the second most heavily used medium (any computer). Overall, computer use is highest at work—primarily because of software, but e-mail is also at its highest there. Reading occurs mostly at home, but a quarter of magazine reading and a third of book reading takes place outside the home.

**Table 3.2** ▰ Incidence and Average Minutes of Media Use by Location

| Medium | Home | | Car[a] | | Work | | Other | |
|---|---|---|---|---|---|---|---|---|
| | Incidence (%) | Min. | Incidence (%) | Min. | Incidence (%) | Min. | Incidence (%) | Min. |
| TV | 89.8 | 217.4 | 4.6 | 1.2 | 21.1 | 9.8 | 26.7 | 11.5 |
| Radio | 41.3 | 27.0 | 74.8 | 28.7 | 43.4 | 17.1 | 46.9 | 7.2 |
| Any computer | 56.3 | 72.8 | 6.8 | 0.8 | 69.7 | 85.3 | 13.2 | 8.0 |
| All Internet | 51.8 | 47.4 | 4.3 | 0.5 | 63.2 | 40.5 | 8.9 | 4.9 |
| Web | 46.1 | 36.4 | 2.9 | 0.3 | 54.6 | 26.7 | 7.8 | 3.9 |
| E-mail | 43.7 | 19.7 | 2.2 | 0.3 | 57.2 | 24.5 | 7.4 | 1.9 |
| IM | 9.9 | 4.9 | 0.4 | 0 | 12.5 | 3.5 | 1.2 | 0.8 |
| Software | 29.9 | 25.4 | 4.0 | 0.3 | 62.5 | 44.8 | 7.4 | 3.1 |
| All phone | 85.9 | 23.6 | 33.8 | 2.4 | 81.6 | 13.9 | 31.0 | 1.9 |
| Landline | 71.9 | 17.5 | 3.2 | 0.1 | 73.0 | 12.0 | 7.0 | 0.3 |
| Mobile phone | 39.8 | 5.9 | 31.7 | 2.3 | 46.1 | 2.0 | 27.1 | 1.6 |
| Music | 30.2 | 23.3 | 32.7 | 8.4 | 34.2 | 18.9 | 41.1 | 14.6 |
| Print media | 50.6 | 24.9 | 4.0 | 0.2 | 30.9 | 4.1 | 19.4 | 3.6 |
| Newspaper | 34.7 | 10.7 | 2.5 | 0.2 | 14.5 | 0.8 | 4.7 | 0.5 |
| Magazine | 20.1 | 5.3 | 1.4 | 0.1 | 15.8 | 0.7 | 9.3 | 1.1 |
| Book | 20.1 | 8.2 | 0 | 0 | 13.2 | 2.6 | 7.4 | 2.0 |
| All video | 29.0 | 30.4 | 2.5 | 0.1 | 5.3 | 0.9 | 4.3 | 1.2 |
| VCR | 10.5 | 8.3 | 1.1 | 0 | 3.9 | 0.5 | 2.3 | 0.5 |
| DVD | 21.0 | 22.2 | 1.4 | 0.1 | 2.0 | 0.4 | 2.3 | 0.6 |
| Game console | 14.7 | 11.3 | 1.8 | 0 | 2.0 | 0.1 | 1.9 | 0.2 |
| Other | 26.6 | 9.7 | 4.7 | 0.4 | 42.8 | 8.3 | 9.7 | 2.9 |

[a]Person observed was not necessarily the driver.

## Media and Gender

Overall time watching television is comparable between women and men, with women watching more on average (254.7 minutes compared to 227.1). Conversely, men spend more time listening to the radio (87.5 minutes compared to 72.6).

Computer use among men and women is also different. Both men and women spend about the same amount of time in front of a computer, but they do different things. Men used software (79.4 vs. 67.8 minutes) and instant messaging (12.6 vs. 6.0 minutes) more often than women. Women showed higher use of the Web (70.8 vs. 63.9 minutes) and e-mail (53.6 vs. 39.2 minutes).

Overall, women were observed spending almost twice as much time reading as men (41.6 minutes to 24.1 minutes). Women generally spend more time reading newspapers (13.2 minutes to 11.1 minutes), magazines (10.4 minutes to 4.2 minutes), and books (16.6 minutes to 8.8 minutes).

Women spend more time on the phone, overall (46.4 minutes to 37.9 for men), but men use mobile phones slightly more than women (12.1 to 11.5 minutes). Men dominate the use of game consoles, spending 18.9 minutes per day compared to 4.3 minutes for women.

## Media and Age

There are even more significant differences in media use by age (Table 3.3). People under the age of 35 tend to be more involved with electronic media, while those over 55 tend to be more involved with print media. The exception is television. Media usage increases with age, ranging from a low of 182.6 minutes (25–34 age group) to a peak of 364.9 minutes per day for those 65 and older.

The 18–24 age group spends more minutes per day than any other group on instant messaging (18.3), mobile phones (20.4), music (100.0), VCR use (47.4), and game consoles (36.4). The 18–24 year-olds scored lowest in radio use and print media. Although the group was the highest in mobile phone use, it was so low in using landlines that overall phone use was the second lowest of all groups. The 18–24 year-olds spend a fair amount of time online, but they are still below all other age groups except 65+. The youngest group, 18–24, watches less television than most groups, but even for them, television is still clearly the dominant medium.

The 25–34 age group spends more minutes per day than any other group in total computer use and DVD use. The group also scores high in specific computer applications along with mobile phone use, music, and video. The 25–34-year-olds were the lowest in television usage and among the lowest in all phone and all print media.

The 35–44 group scores the highest in e-mail, software, and phone use, but it is also high in all computer-related areas. In the middle of the age groupings, 35–44 tends to come in near the middle of most media activities.

The 45–54-year-old group comes out on top for Web browsing (barely ahead of ages 25–34) and magazine use (14.7 minutes). The group places in the bottom for book reading and game console use.

The 55–64-year-olds are the dominant radio listeners and the biggest readers of newspapers (34.7 minutes) and users of VCRs. But the group also scores high in most computer applications, phone, and books. The group listens to far less music than any other age group.

The oldest group (65+) was the highest in both television viewing (364.9 minutes) and overall print media use (67.2 minutes). Conversely, it is the lowest in virtually every electronic media category.

Total media use varies by an hour a day from the lowest (25–34-year-olds at 753.1 minutes) to the highest group (55–64-year-olds at 808.1 minutes).

**Table 3.3** Incidence and Average Minutes of Media Use by Age

| | Incidence (%) | | | | | |
|---|---|---|---|---|---|---|
| | 18–24 | 25–34 | 35–44 | 45–54 | 55–64 | 65+ |
| TV | 83.0 | 89.2 | 90.7 | 92.4 | 97.7 | 93.5 |
| Radio | 72.3 | 72.3 | 76.7 | 81.8 | 72.1 | 58.1 |
| Any computer | 59.6 | 80.0 | 79.1 | 75.8 | 79.1 | 51.6 |
| All Internet | 57.4 | 72.3 | 72.1 | 74.2 | 76.7 | 48.4 |
| Web | 48.9 | 70.8 | 68.6 | 68.2 | 67.4 | 38.7 |
| E-mail | 48.9 | 64.6 | 67.4 | 65.2 | 67.4 | 38.7 |
| IM | 23.4 | 16.9 | 16.3 | 4.5 | 14.0 | 3.2 |
| Software | 40.4 | 61.5 | 59.3 | 48.5 | 53.5 | 29.0 |
| All phone | 95.7 | 93.8 | 97.7 | 93.9 | 95.3 | 90.3 |
| Landline | 66.0 | 72.3 | 90.7 | 89.4 | 93.0 | 90.3 |
| Mobile phone | 78.7 | 67.7 | 61.6 | 54.5 | 48.8 | 19.4 |
| Music | 72.3 | 67.7 | 60.5 | 63.6 | 37.2 | 48.4 |
| Print media | 42.6 | 60.0 | 62.8 | 62.1 | 74.4 | 87.1 |
| Newspaper | 14.9 | 27.7 | 33.7 | 48.5 | 58.1 | 71.0 |
| Magazine | 8.5 | 24.6 | 36.0 | 31.8 | 39.5 | 45.2 |
| Book | 23.4 | 33.8 | 33.7 | 22.7 | 25.6 | 35.5 |
| All video | 44.7 | 41.5 | 40.7 | 27.3 | 16.3 | 9.7 |
| VCR | 17.0 | 10.8 | 16.3 | 16.7 | 7.0 | 6.5 |
| DVD | 34.0 | 33.8 | 26.7 | 15.2 | 11.6 | 3.2 |
| Game console | 23.4 | 24.6 | 14.0 | 4.5 | 18.6 | 6.5 |
| Other media | 27.7 | 32.3 | 47.7 | 45.5 | 55.8 | 32.3 |
| | Average Minutes | | | | | |
| TV | 203.1 | 182.6 | 212.5 | 248.5 | 306.4 | 364.9 |
| Radio | 53.9 | 72.9 | 81.0 | 76.4 | 128.5 | 85.9 |
| Any computer | 131.5 | 154.6 | 150.8 | 128.8 | 140.5 | 84.1 |
| All Internet | 85.1 | 105.3 | 97.8 | 105.4 | 98.8 | 46.1 |
| Web | 64.1 | 78.9 | 65.2 | 83.6 | 67.5 | 24.9 |
| Email | 33.4 | 47.6 | 59.8 | 50.0 | 47.0 | 24.6 |
| IM | 18.3 | 17.7 | 10.1 | 1.2 | 3.7 | 0.2 |
| Software | 75.5 | 89.5 | 97.3 | 51.5 | 62.4 | 50.5 |
| All phone | 36.9 | 37.4 | 50.0 | 44.8 | 48.4 | 26.4 |
| Land line | 14.6 | 21.4 | 39.9 | 35.0 | 39.6 | 24.1 |
| Mobile phone | 20.4 | 16.0 | 9.4 | 9.8 | 8.8 | 1.3 |
| Music | 100.0 | 75.6 | 67.6 | 64.1 | 24.5 | 45.1 |
| Print media | 14.6 | 23.3 | 27.3 | 32.3 | 59.8 | 67.2 |
| Newspaper | 2.3 | 3.6 | 8.2 | 11.6 | 34.7 | 27.6 |
| Magazine | 2.4 | 3.7 | 6.0 | 14.7 | 5.1 | 13.7 |
| Book | 9.9 | 16.1 | 13.1 | 6.0 | 18.5 | 20.1 |
| All video | 47.4 | 46.7 | 35.3 | 22.5 | 25.9 | 7.1 |
| VCR | 12.2 | 4.8 | 10.2 | 7.3 | 13.8 | 3.7 |
| DVD | 35.1 | 41.8 | 25.1 | 15.2 | 12.1 | 3.4 |
| Game console | 36.4 | 14.3 | 5.6 | 2.6 | 8.9 | 3.9 |
| Other media | 12.5 | 14.2 | 37.1 | 18.9 | 24.8 | 6.7 |
| Total media use[a] | 755.7 | 753.1 | 790.5 | 776.3 | 808.1 | 762.7 |

[a]Includes concurrent media exposure.

## Concurrent Media Exposure

Researchers cannot help but see concurrent exposure to multiple media as an emerging complication in media research. A starting point for understanding concurrent media exposure is to establish its empirical profile, including features such as how much of it occurs in a media user's typical day and what patterns of media groups are commonly found in concurrent exposure in different circumstances.

We use the phrase "concurrent media exposure" as a carefully considered alternative to the commonly used label of "media multitasking" or the more recent "simultaneous multiple media usage" (Pilotta, Schultz, Drenik, & Rist, 2004). The former is potentially ambiguous because it is often applied to the combination of media use with other life activities (such as watching television while eating); the latter begs the question of what "use" means. Both erroneously imply active engagement. Media use may range from fully engaged attention (intently watching a new episode of a favorite television show or playing a video game) to incidental exposure to a largely unnoticed medium (the background presence of recorded music while shopping in a retail store). "Use" is uncomfortably problematic in the case of simultaneous multiple media use, as the phrase implies task-driven attention to multiple media. Such simultaneous use can range from full engagement in multiple media (an office worker on the telephone, describing to the "help desk" the error messages displayed on a computer screen), to a clear foreground-background relationship (playing a challenging computer game while listening to the stereo), to restless attention-shifting among multiple candidates for attention (a person reading a newspaper with the radio providing background music or talk, and occasionally glancing at a muted television to check the progress of a sports event).

We adopt the phrase "concurrent media exposure" (CME) to avoid such ambiguous or potentially misleading characterizations, and define it as *exposure to content from multiple media simultaneously available through shared or shifting attention*. For example, a book on a nearby shelf does not constitute media exposure. Holding an open book in a reading position constitutes an exposure because the content is immediately available. An audible radio broadcast is a media exposure for someone in the room, even though we may not know if the radio is subject to conscious attention. Reading the book while the radio plays is an instance of CME.

This study demonstrates both the tremendous amount of media exposure and the prevalence of CME. As presented in Table 3.1, almost a third or more of the time spent with any one medium, with the exception of DVD use, was time also spent with another medium. It is important to note the especially high CME incidence among computer-based media; this is not surprising given that modern computer operating systems are designed to support multitasking. The support includes the ability to display multiple media sources or genres simultaneously (by having two or more windows open on-screen) and the ability to switch rapidly back and forth between overlapping windows.

The data presented here are also a reminder that media vary considerably in their patterns of use. Mobile phones are similar to the Web in that both are used by the majority of the study participants (59 percent vs. 62 percent, respectively) and have high percentages of their total time spent in CME (69 percent vs. 89 percent). Their total minutes of use, however, differ markedly. Mobile phones accrue only about one-sixth as many total minutes as the Web (11.8 minutes vs. 67.3 minutes). Some media, such as both forms of telephony and instant messaging, are characterized largely by intermittent use in relatively short episodes; other media such as television, radio, and the Web are characterized by sustained use in longer but perhaps less frequent episodes.

The contrast in time spent in CME for DVD use (the medium with the least percentage of CME, at 15.5 percent of total time) and its equivalent analog medium, VCR (at 46.2 percent) suggests that medium content alone cannot account for differences in CME.

## Discussion/Implications

The Middletown Media Studies give us an unprecedented look at how people actually spend their time with the media: not what they say they do, but what trained observers saw them doing. This provides an important validity check on self-report research.

For example, telephone surveys and focus-group research regarding computer use have supplied a number of popular notions such as the following:

- Young people are getting most of their information from the Web.
- Young people spend more time online than with television or other media.
- People, generally, are spending more and more time online and less and less time with television.

Using this observational methodology, the findings of this study raise serious questions and cautions about each of these assertions:

- The youngest group we observed, ages 18–24, were not the biggest online users. They were actually among the lowest users, and little of their Internet use involved informational Web sites.
- Television remains the dominant medium—with all age groups—including men ages 18–24. The only issue is how dominant.
- People spend over two hours a day at a computer, but that only tells part of the story. More than an hour of that involves computer software; almost an hour involves e-mail and instant messaging. Thus, people's perceptions of spending more time online may well be fed by spending more time at the computer. But most of that time is actually engaged in software tasks and

communication, and almost all of it involves concurrent media exposure within the computer arena and typically with other media as well.

It is evident that concurrent media exposure cannot be disregarded as an insignificant behavioral aberration on the part of a subset of consumers. It is also clear that CME is not the sole preserve of the young and media-savvy. Rather, it occurs across the entire sample to a surprising degree. This is perhaps an indication that CME is merely an extension of other forms of multitasking; many of us who were teens in the 1960s and 1970s can remember our parents being incredulous that we could do our homework while listening to The Doors or The Sex Pistols. Today, our incredulity extends to our own offspring interweaving their homework with television, the Web, e-mail, and instant messaging. But today's multitasking is just a 21st-century echo of our own. While the specific combinations and overall amounts of CME may vary, its significance does not.

As this report is not the result of a tracking study, it is impossible to say with any certainty if the overall amount of CME is increasing or if specific examples are coming to prominence while others fall away. Only consistent research over time can reveal such things reliably, but as new media-based devices come to the marketplace, as existing devices offer new capabilities and as compelling content is repurposed to pursue the opportunities these create, it would seem that consumers have both the will and capacity to respond.

Any number of studies, including work by the Pew Media Research Center, have reported on the Internet equaling if not supplanting so-called traditional media, including television. But none has reconciled those self-reported (telephone) studies with the home-only (metered) results from A.C. Nielsen, which reported 2005–2006 as the biggest year ever for television viewing. After two rounds of the Middletown Media Studies, it seems that people are simply unable or hesitant to report their media consumption behavior from memory. Instead, we have found through direct observation that it is possible to get a reasonable handle on the increasingly complex world of concurrent media usage that allows computer use to rise even as television use reaches new pinnacles.

## References

Bechtel, R.B., Achelpohl, C., & Akers, R. (1972). Correlates between observed behavior and questionnaire responses on television viewing. In Rubinstein, E.A., Comstock, G.A., & Lloyd-Jones, S. (Eds.), *Television and social behavior, reports and papers: Vol. IV. Television in day-to-day life: Patterns of use, a technical report to the Surgeon General's Scientific Advisory Committee on Television and Social Behavior*. U.S. Department of Health, Education, and Welfare, Health Services and Mental Health Administration,

National Institute of Mental Health, Rockville, MD. Washington, DC: U.S. Government Printing Office.

Coffey, S., & Stipp, H. (1997). The interactions between computer and television usage. *Journal of Advertising Research, 37*(2), 61–67.

Papper, R.A., Holmes, M.E., & Popovich, M.N. (2004). Middletown Media Studies: Media multitasking . . . and how much people really use the media. *The International Digital Media and Arts Association Journal, 1*(1), 9–50. (Also available online at http://www.bsu.edu/cmd)

Pilotta, J.P., Schultz, D.E., Drenik, G., & Rist, P. (2004). "Simultaneous media usage": A critical consumer orientation to media planning. *Journal of Consumer Behavior, 3*(3), 285–292.

Stempel, G.H., III., Hargrove, T., & Bernt, J.P. (2000). Relation of growth of use of the Internet to changes in media use from 1995 to 1999. *Journalism Quarterly, 77*(1), 71–79.

CHAPTER 4

# The Converged Audience

## Receiver-Senders and Content Creators

Jeffrey S. Wilkinson
Steven R. McClung
Varsha A. Sherring

Media convergence brings about new ways to define and assess what information is and how content is presented and consumed (and at what cost). The convergence of traditional mass media, broadband Internet and World Wide Web as well as compatible digital consumer technologies (such as 3G telephony) erases past distinctions between interpersonal and mass communication.

Virtually every society is now grappling with a huge paradigm shift as audiences increasingly take on the role of content creator, contributor and consumer—sometimes simultaneously. This phenomenon is not limited to the United States. As Google, YouTube, and Wikipedia launch non-English-language variations, opportunities for social networks increase around the world as well as potential revenues associated with new forms of news, information, advertising and services.

People are increasingly connected. At the beginning of 2007, the number of Internet users worldwide was reported at around 1.09 billion (Internet World Stats, 2007). Globally, a record number of telecommunications connections—over four billion worldwide—were predicted for 2006 (Deloitte Research, 2006). This number reflects mobile, landline, VoIP (voice-over Internet protocol) and broadband connections. Due to the rise in popularity of IPTV (Internet protocol television), video-on-demand, and podcasting, convergence will become a fundamental value generator for media companies (Deloitte Research, 2006).

Understanding the shifting boundaries between consumers, contributors and content creators is important. Just as old media can quickly lose value and influence, so now new companies or individual entrepreneurs can leverage these boundary changes to quickly become a success.

This chapter is divided into two sections. The first half examines how consumers and media companies have converged in the area of content creation and dissemination. Interactivity and social networks have produced a matrix of consumer convergence roles that complement and compete with media organizations. The relationship between consumers, creators and media is dynamic and subject to a variety of changes that are addressed in the second half of the chapter.

To illustrate this process, this chapter examines how a consumer-created Web site can develop into a successful commercial site. The implications of this process for media companies are significant because the Web enables a new information franchise to develop at any time. These new competitors can come from any direction and take almost any form. In this chapter, three specific examples of new entertainment/information Web sites are explored, one each in the areas of news, sports and entertainment.

Media companies will not be going away, despite what doomsayers have sometimes predicted. Society needs institutions even as it allows for outside forces to change it. Political parties need a third choice from time to time, schools need a certain number of children to be homeschooled, and media need social networking sites and YouTube to help them change and stay relevant.

## Traditional Media Adapt

The annual report from Pew's Project for Excellence in Journalism (2007), the "State of the News Media 2007," notes that the pace of change has accelerated. Traditional media companies have flooded the world with information, news and entertainment. Since 2006, television networks, syndicators and other media corporations with content have placed entire programs (as well as clips) online to lure and keep consumers. At the same time, local news media outlets have been trying to stay relevant by offering more original content online—local news and information—for those in the community. For example, in South Carolina the *Bluffton Today* newspaper (http://www.blufftontoday.com/) devotes much of its Web site to local community news and information, actively inviting citizen contributions and participation (Storm, 2006).

Media companies watched and learned lessons about controlling content when the file-sharing phenomenon caused by Napster took the music industry by surprise.

The Recording Industry Association of America (RIAA) claimed musicians and companies lost millions (even billions) of dollars or more in lost sales due to illegal downloads and file sharing by users. While the RIAA reports piracy is still rampant in the music industry, the same problems are now being faced by audio and video news, information, and entertainment. So far, news media organizations remain profitable and generate plenty of advertising revenue, but the economics of information distribution are changing in still-unknown ways (Project for Excellence in Journalism, 2007).

To differentiate themselves from consumer-generated content sites such as MySpace and YouTube, media companies are experimenting and embracing convergence at almost every level—from general organizational structure to the types of technological applications. The result has been a great deal of effort and movement as the industry tries its best to stay relevant and—more importantly—in control.

Podcasting is a case in point. Almost every news organization now offers links to podcasts created by their employees. Others are also experimenting with including user-generated content. Most media corporations are also emphasizing new distribution channels such as mobile phones and iPods (Best, 2006). Much of this effort has been to stay competitive with online content-provider networks such as Podtrac, Podshow and Podbridge (Klaassen & Taylor, 2006). The Bridge Ratings estimate that the audience for podcasting in the United States will have grown from 820,000 in 2004 to over 60 million by the year 2010 (Shields, 2005). Media companies are trying to work with the user-generated content trend rather than combat it.

## The Matrix of Consumer Convergence

The vast number of Web sites that make up the World Wide Web is an ever-increasing, evolving jungle of interconnected pieces. New services, applications and content are constantly being discovered, created and offered in a range of forms to suit virtually every taste. As some Web sites are launched and grow in popularity, so others decline, become inactive and eventually shut down.

New technologies and convergence have brought about new forms of content such as virals, moblogs, and mashups. A viral is a marketing message that induces users to pass it along to others, usually for its entertainment value and/or creativity. Moblogs (short for "Mobile Web logs") are online spaces that allow remote uploading of pictures, captions and other information that are presented in a blog. A mashup brings together two or more elements to produce new information. An example of a mashup comes from chicagocrime.org, which combines a crime database with maps to illustrate where the crimes occurred.

These new forms of content are feeding an exponential increase in the total amount of digital material created by human beings and available from commercial, noncommercial and personal Web sites alike. As certain types become popular, so the sources of the content also become more popular.

Throughout the shifting in this "media content creation-distribution-consumption" environment, the key factors continue to be ownership and control of content and its distribution as well as the economics (costs versus revenues) to support and sustain the creation of future new content. Traditional media still exert a degree of control, but that control has considerably lessened. Similarly, the costs involved in creating and disseminating content are within the means of most people. Non–media-affiliated content creators can now exist and operate in a variety of ways to bring information, entertainment and services to others. The old dichotomy between media (content creators) and audiences (content consumers), is now a matrix, illustrated in Figure 4.1.

Some forms of content like personal Web pages give the owner considerable freedom to post any type of content, but the commercial value is low or non-existent. A Web site like The Drudge Report is internally controlled by him but has changed to commercial status and is reviewed later in this chapter. Chat rooms are monitored and controlled by the host organization but there is little commercial value in the content. Hosted photo galleries on media sites are ways for both parties to make money.

Somewhere in the center is the still-evolving phenomenon of citizen journalism. Sometimes citizen journalists are compensated for their work when it is used by a traditional media organization; sometimes the citizen journalist posts material on his or her own Web site and it is picked up by the blogosphere.

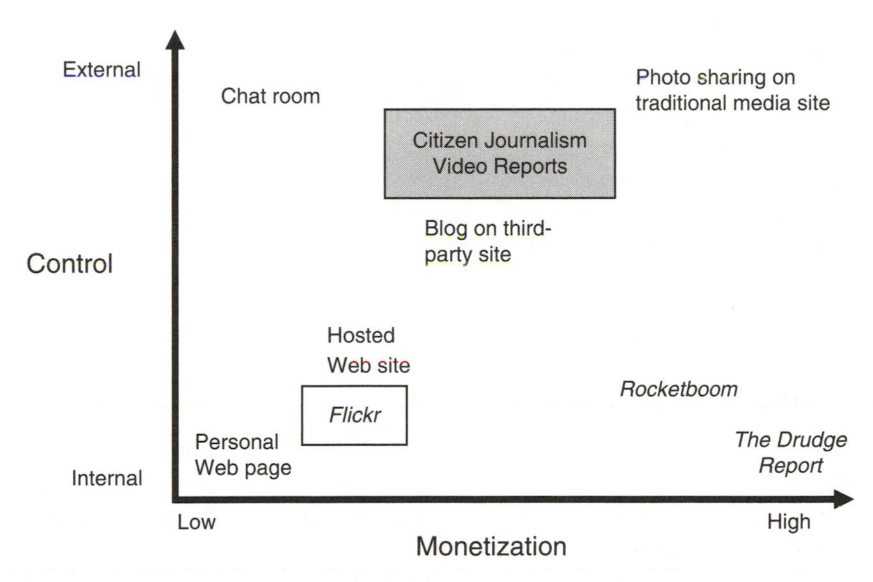


**Figure 4.1** ▰ Matrix of User-Generated Content

## From Linear to Recursive Loop

The Internet provides content creators a new resource for developing from a small niche to a brand name and the opportunity to evolve from noncommercial to commercial status. These opportunities are based on the potential of the Web to create enough buzz and bring in vast numbers of surfers and seekers of that content. Getting enough traffic to attract advertising enables such sites to change from a noncommercial status to a commercial one. These content consumers can also be termed "Receiver-Senders," because they help change the linear Source-Message-Channel-Receiver model of communication (Berlo, 1960) into a recursive loop. The receivers of mass messages are now also the sources of other messages, creating content through blogs, podcasts, and video uploads that are increasingly attended to by media organizations.

### YouTube

New economic models and means of revenue distribution have added complexity to the mix. After YouTube became widely popular in 2006, it was purchased by Google for $1.76 Billion (Sutel, 2007). As part of the agreement, Google also agreed to share revenue with the content creators. Even as some wondered if this development would herald a new age of content sharing and enjoyment, economic reality reasserted itself in March 2007 as Viacom sued YouTube owner Google, Inc. for copyright infringement. The suit claimed the video-sharing site built itself into a global brand by placing tens of thousands of video clips online without permission. A month earlier, Viacom demanded YouTube remove more than 100,000 unauthorized clips from its site, and then discovered an additional 50,000 unauthorized clips (Sutel, 2007).

These kinds of developments suggest shifts in business practices, but eventually new models will emerge that are likely to resemble the old ways of doing business. Copyright, ownership and distribution rights will still be enforced, and the bigger question may be whether—or at what point—consumers agree to pay for online content, or whether these new forms will follow traditional media, relying upon advertising for financial support.

## Repositioning Convergence

Diffusion of Innovations Theory explains how a new product or idea becomes adopted through a society (Rogers, 2005). A product or idea must have certain characteristics to impact purchasing/use decisions. These include cost-benefit advantage, compatibility with existing technologies, ease of use, presence of readily identifiable benefits and ability of those benefits to be observed (Rogers, 2005).

As a product or idea goes through the adoption process, the number of users increases over time. Those who adopt first are called innovators. They are in turn followed by early adopters, early majority, late majority and laggards. (With most innovations, there is a sixth category that is not discussed or studied as frequently: nonadopters.) Rogers predicted the s-curve of adoption closely follows a bell curve, where innovators are the first 2.5 percent of all eventual adopters to adopt an innovation, followed by 13.5 percent for early adopters, 34 percent for early majority, 34 percent for late majority, with the final 16 percent comprised of the laggards (p. 262).

But convergence is not only adoption of merged technologies. It is also a merging of functions performed by technologies. This is a fundamental shift in who can do what. The *Diffusion of Convergence* is marked by ever-increasing numbers of Receiver-Senders, following the s-curve predicted by diffusion theory. As the practice of content creation shifts from the few to the many, certain segments or groups will adopt first and others will wait or perhaps refuse altogether.

Besides *time* of adoption, *practice* must also be considered. Some people will create a lot of content, some will create occasionally, and others will prefer to primarily consume rather than create. It may be useful to consider three general categories that differ mostly in emphasis: consumers, contributors and creators.

## Consumers

Everyone enjoys being a passive consumer at times (Project for Excellence in Journalism, 2007). But diffusion theory would also predict that a number of people would choose not to create content altogether, or, if they did, would not disseminate it to others. These are the equivalent of diffusion theory's "nonadopters," who are not considered in the five categories of adopters. Thus, a person can be an adopter of one dimension of an innovation and a nonadopter (nonuser) of other dimensions.

## Contributors

Most people who have the technical skills will at least occasionally produce content. An active creator can have his or her own Web site or MySpace account, engage in social networking through posted images on Flickr or engage in being a "citizen journalist" who sends photos or provides first-hand accounts of events. Much more study is needed to identify the contexts and types of contributions these people present.

## Creators

This group represents the truly converged audience. Like those who choose to mostly or primarily consume, there will also be a significant minority who either work in media, routinely freelance and/or otherwise create as their primary work in life. These people include artists, writers, musicians, performers and others with the inspiration and skill to create.

## Characteristics of the Converged Audience

Two conditions are necessary for Receiver-Senders to exist. First, they must have access to media that allows them to reach (and be reached by) audiences. Second, they must be able to create content. Before the days of the Internet, the distribution of content was an important criterion for market entry (Wicks, Sylvie, Hollified, Lacy, and Sohn, 2004). Hard copies of the newspaper still rely on personal delivery to the door of the reader. Broadcasters could easily deliver their signal into the home of the viewer (although government licensing restrictions kept the number of operators comparatively small in number).

Contrast this with the inherently interactive World Wide Web. Anyone who receives Web content can also be a sender of content if they have (1) the hardware, (2) sufficient connectivity, and (3) a minimal set of technical skills. The first two (in the form of multimedia personal computers and broadband capability) are no longer a barrier in many places.

The deciding factor now is the third element, the skills. Schools, colleges, and universities offer countless courses, workshops, and training seminars equipping countless numbers of workers and enthusiasts to design and build an interactive presence on the Web. As authoring technology becomes more sophisticated, the skill set needed to produce content has become simpler, lowering the minimum skill level needed to create content.

The second necessary condition for a Receiver-Sender is that he or she must be able to create content. In addition, the content must have perceived value (such as information, entertainment or artistic merit), and the creator must be able to use some technical skill (by himself or herself or through another) to produce the content in a way that it will be consumed by others. Papacharissi (2002) found that some personal Web page creators were quite sophisticated and creative in their content. Personal Web pages tended to be created either as a form of self-expression or as a tool for professional advancement (p.364).

Niche theory suggests that a new medium will compete with established media for consumers (Dimmick, Chen, & Li, 2004). This competition is ongoing, as the diffusion of digital technology enables private individuals and consumers to create professional-looking content that was formerly the domain of media professionals. Dominick (1999) noted that "anybody in the audience with the right hardware and minimal computer skills can become a mass communicator" (p. 647). Any Internet user can find content on the Web—for entertainment, information, education or escape. The content can take any form, such as text, animation, audio—and increasingly—video.

## The Influence of Interactivity

One of the most talked-about features of the new media is interactivity. Interactivity has been defined as interacting with the computer (Dillon & Leonard, 1998, p. 144), as well as communication between two human beings (Williams, Rice, & Rogers, 1988, p. 10). Interactivity has typically been considered as a multidimensional construct. For example, McMillan and Downes (1998) identified six dimensions of interactivity includin, persuasion/informing, degree of control, degree of activity, one- versus two-way, set time/flexible time, and sense of place. Other research identified five dimensions of interactivity including playfulness, choice, connectedness, information collection, and reciprocal communication (Ha & James, 1998). These perspectives reflect the specific experience of interacting with the computer and/or another human being. This approach has been useful to identify a constellation of features regarding interactive behaviors.

## The Extended Model: Content Creators and Diffusion of Convergence

Interactivity can be extended conceptually as a defining characteristic of the World Wide Web and a precondition for the convergence process. This conceptualization strengthens the notion that expanded convergence represents a radical change in the existing media universe. According to Allen (2003), radical technologies "shatter the equilibrium, requiring significant learning curves and exacting high switching costs from customers" (p. 260). As the Web diffuses throughout the society, the number of participants involved in content creation has risen, bringing about such a radical change.

One form of technology diffusing through the society is original content being posted online through social-networking sites. For example, in 2006, the social-networking site MySpace boasted a membership of 90 million and ranked second among all Internet sites in individual streams viewed online. In July 2006 alone, the number of MySpace videos watched by U.S. users totaled nearly 1.5 billion streams, representing nearly 20 percent of all videos streamed by U.S. Internet users in a single month (ComScore Networks, 2006). Most of those joining MySpace and posting personal information ("original content") are young and relatively tech-savvy.

As Web site creation and design become common and simple as taking a photograph, so the number of content creators increases. Each personal Web site has the potential to transition from hobbyist or fan site to commercial status. This transition is important because, once accomplished, the Web site becomes another competitor for the traditional content creators, the mass media. Furthermore, once the sites become economically viable, their popularity may spill over into the social realm as well as the political arena. Program producers and politicians will have to learn how to captivate these computer-savvy audiences who are used to getting information from their own sources.

To prepare a new technology for commercial launch, Allen suggested three stages: (1) concept testing, (2) primitive prototyping, and (3) near-production quality (pp. 62–63). A hobbyist Web site can undergo similar stages, either on purpose or by accident. These sites may not begin with a formal business plan and budget. They can be initially created as expressions of personal tastes (such as fan sites, music), personal opinion (disguised as news or information), or the creator's own strange sense of humor.

However, that does not mean that these sites do not have management strategies. All sites expecting to attract an audience need to be managed. Flickr, a photo-sharing site, uses "tabs" to manage its user content. In its startup phase, the number of employees working for this site was less than 10, hardly the personnel one would expect to manage a site with millions of hits every day (Levy & Stone, 2006). Flickr was so successful it was purchased by Yahoo! in 2005.

## Receiver-Sender Web Site Brand Development Stages

For those sites that began as expressions of love, interest and creativity (and probably an element of entrepreneurship), organic designations are probably more useful: Genesis, Birth of the site, Adolescence, a crucial Coming-of-Age/Maturity stage, and Adulthood. The Flickr Web site demonstrates how a site develops through these stages.

Timeline for Flickr growth and development

Time 1. Genesis: approximately 2002 until launch

Time 2. Birth: February 2004

Time 3. Adolescence: February 2004–March 2005

Time 4. Coming-of-Age/Maturity: March 2005

Time 5. Adulthood: Commercial Brand: May/June 2005–present

**Figure 4.2** ▰ Example of Web Site Brand Development Stages

## Time 1: Genesis

First the creator gets the idea, develops the skills, and begins preparing for the creation and launch of content. This stage can cover a relatively short or long period of time. Time and effort are spent planning, understanding the legal implications of Web publishing, and initial investments in software and hardware. Take, for example, Flickr. As early as 2002, while working to develop an online game, Caterina Fake and husband Stewart Butterfield noticed they enjoyed a picture-sharing function more than the game itself (Graham, 2006). They pursued looking at ways to allow people to easily share digital photos online, and launched Flickr in February 2004 (Levy & Stone, 2006).

## Time 2: Birth

At this stage, the creator has launched the Web site and begins offering the content that has been created. This stage can be relatively brief and turns the creator's dream into a reality. But with the reality comes the stress of making the Web site organic; content must be routinely updated and new content created, and time and effort must be invested for critical analysis and ways to improve the content. Using the SMCR loop, feedback from friends, family, and strangers is sought and considered to improve the content.

With Flickr, the developers continued to work on improving the tools and the Web site. They solicited feedback from the users in order to help shape the direction of the company. They moved from an online to an instant-messaging service with pictures, to what it became—a way for people to upload their photos and share them with a community of users (Levy & Stone, 2006).

## Time 3: Adolescence

At this stage, the creator has established a routine and goals for the site. As a result of planning, decisions are made as to additional resources that can be invested into the project. At this stage, the site has established itself to hold an "amateur" status and may remain in this phase indefinitely. It may never attract enough visitors or capture enough media attention to change to commercial status. Many personal Web sites are in this stage, and over time the creator may lose interest, the content may decrease in value, and it may become one of the many "dead links" littering the Internet.

In the case of Flickr, the creators were innovative enough to keep the site interesting. According to Fake, "We were very small and very poor, so we built a lot of features that were deliberately viral" (Levy & Stone, 2006, p. 52). The effort paid off, and from December 2004 to December 2005, Flickr's traffic grew 448 percent, to 3.4 million, according to Internet measurement firm Nielsen/NetRatings (Graham, 2006).

## Time 4: Coming-of-Age/Maturity

The Coming-of-Age time is when the site changes in some profound way to embrace its success. The site may change from amateur to professional status, from noncommercial, to commercial. During this time, key personnel may reevaluate their involvement in the venture. Disagreements and so-called creative differences may come to a head and lead to departures and changes. For most content creators, this stage may take a relatively long time. For some, it may be reached quickly due to a *Critical Incident* that generates publicity and excitement (or cash). This can happen in conjunction with word-of-mouth buzz, other sites linking to the site and use of traditional advertising/promotional vehicles to bring even more people to the site. For Flickr, the September 2004 bombing of the Australian Embassy in Jakarta demonstrated it was more than just a place to share family photos. The first pictures of the bombing appeared on the Flickr site, demonstrating its potential as an important outlet for news content (Levy & Stone, 2006).

Another noted Web site also grew out of a disaster. Hours after the December 2004 tsunami struck South Asia, India-based writer Peter Griffin set up a blog. Soon, two other Indian bloggers joined him to help put together what became known as one of the best online resources on the tsunami. The *South-East Asia Earthquake and Tsunami blog (SEA-EAT)* had the latest news from every region, a missing persons page, and links to relief efforts. Within 48 hours, over 200 volunteer bloggers were sending in text messages, witness accounts, photographs, sounds and videos. This content was compiled long before traditional media organizations could even begin to comprehend the scale of the disaster (Srinivas, 2005, para. 7). According to *The Independent,* "Never before has there been a major international story where television news crews have been so emphatically trounced in their coverage by amateurs welding their own cameras" (Srinivas, 2005, para. 27).

As Flickr continued to generate its own headlines throughout 2004, it evolved toward the next phase of maturity, toward Adulthood. In March 2005, Yahoo! bought Flickr for $35 million so it could learn more about how to use the "audience as creators" concept and apply it to their own user base of over half a billion (Levy & Stone, 2006). After the purchase, the team was physically moved from Vancouver, Canada, to California (Graham, 2006).

## Time 5: Adulthood

The final stage for the independent content creator is becoming a full-fledged commercial business. As the Web site has evolved from noncommercial to commercial status, it increasingly adopts more formal business practices. Now the business competes for market share, new audiences, ad revenue and general financial success. Since being purchased, Flickr has undergone a number of changes and improvements. In December 2006, for example, upload limits were increased from 20 MB to 100 MB. In June 2007, a small-but-personal touch resulted in changing the tagline on the logo to read "flickr LOVES YOU" (Flickr Web site, http://www.flickr.com/explore/, July 2007).

## Media Promotion of Receiver-Senders

In order to generate interest about a site, the creators/owners must both advertise and find ways to promote what they have. Advertising must be purchased, but promotion can come from traditional mass media, other new media, and word of mouth.

All of these can play an important role when the site is on the verge of transforming from noncommercial to commercial status. Success for most Web sites is primarily determined by the number of visitors (hits). Exposure through other media produces curiosity, directing new traffic to the site. Depending on what new visitors think, they may begin to spread the word about the site to others.

New media in particular have changed the formula for promotion. Bloggers cross-link with each other, and as one reports, another may link to that report. Countless niche-news and information Web sites as well as search engines enable casual surfers to come across and be directed to the site. By building a "buzz" about a site, the Web site can—through name recognition be brought repeatedly to public awareness. This boost from being in the media spotlight can enhance the perceived value of the content. A personality who becomes "famous for being famous" enjoys certain benefits regardless of merit. The ideal is that the attention adds to the reputation of the site and helps confer a degree of credibility that enhances the new company brand.

## Media Competition with Receiver-Senders

As the pace of change accelerates, traditional media companies have had to branch ever-outward to take advantage of synergies that come from complementary businesses. Chan-Olmsted and Ha (2003) investigated which factors weighted most heavily on broadcasters' approach to incorporating the Internet into their existing business model. Revenue dependency (making money) and branding were used most heavily in the Internet strategies for broadcasters. Businesses typically focus on "the bottom line" for growth and survival.

But "personal Web sites are expositions and exhibitions of the self" (Burnett & Marshall, 2003, p. 95), and are therefore driven by factors other than revenue dependency and branding. Certainly, in the early stages, a Receiver-Sender has no brand and cannot rely on revenue because it does not exist at that point. The focus for these labors of love is the content or service provided. A prime example of producers initiating an Internet-based television program for a narrow audience who have similar tastes can be found at www.narrowstep.com. On this site, one can find television talk shows on sailing, religion, golf, babies and other topics. In the broadcast television environment, this kind of programming typically could not command big enough audiences to survive, but this type of niche programming has become a staple of the Web.

As video over the Internet has become common, increasing numbers of producers of niche programming have turned to the Web and found success. For example, the London-based Sail.tv, a sailing channel, attracted 70,000 viewers in its first month. This kind of Internet television program aims at a select audience and has been termed "slivercasting" (Hansell, 2006). As more of the niche audiences discover these specialized programs, the amount of advertising on these channels can increase, as happened with Sail.tv.

This phenomenon suggests that we should revisit what we mean by "valued content," that is, content for which people will pay. Burnett and Marshall (2003) provide some standard genres of commercial sites, including "company/corporate Web site, commercial trading site, Institutional site" (p. 94).

Media sites are also particularly popular for their streamed content, especially news clips, movie trailers, music videos and sports, in that order (Arbitron/Edison Research, 2005). In the early days of the Web, most content was free, and efforts to charge for various forms of content have typically not been successful. But more recently the trend is moving toward for-pay streaming-content services. *RealNetworks* has over 2.2 million paid subscribers to its premium digital media, and many other video content providers offer subscription services as well (RealNetworks, 2006).

Established properties on the Internet provide a range of content and services. According to Nielsen Net-Ratings, the top 10 most visited Web sites in the U.S., Germany and Hong Kong in 2005 included Microsoft, Yahoo!, and Time Warner. Other famous Internet brands that placed prominently include Google, eBay, Interactive Corp., and Amazon (A.C Nielsen, 2005). So, even as the Internet has spawned new forms of content, so also we can expect incumbent companies to experiment with new business models for marketing and distributing this content.

## Content That Has Value

A key starting point is to identify content that is competitive or compatible with what media companies typically offer (this step differentiates the professional "wanna-be" sites from amateur sites). An important characteristic is commercial value, or whether others are willing to pay for it.

If Receiver-Senders will one day become even marginal competitors to media companies, they will have to create content that is somewhat similar. Given the various categories of content creation, a certain degree of overlap will be found. Examples of amateur-to-professional Web sites can already be identified by their valued content. Various categories include *fan sites* (sports, music, celebrities), *hobbyists' sites* (photography, art, self-created music, poetry), *freelance/professional sites* (health, commentary, politics, legal, educational), *adult sites* (sex industry, erotica), and *ideology-based sites* (from mainstream religious at one end to extremist groups and fringe organizations).

## Specific Cases: News, Sports, Entertainment

Three broad areas that have been staples of media company–created content are news, sports, and entertainment. Each of these is influenced by the diffusion of convergence.

### News

Internet users are increasingly going to the Web for news and information (Project for Excellence in Journalism, 2007). While much of it is perceived as credible (Johnson & Kaye, 2000), verifying Web-based information has been a problem. Parody and hoax sites can be almost indistinguishable from news sites belonging to traditional media (such as newspaper, broadcast, and cable news organizations). The blogosphere generates discussion on numerous issues, and the line between fact and fiction is equally hard to find. For those sites developed by individuals or nonprofit organizations, the information can be quite narrow in its focus. For these organizers, there is no desire or incentive to commercialize their sites. A number of blog sites fit this category. But for others there is incentive, and they do commercialize their sites.

For example, the *Drudge Report* (www.drudgereport.com) is now a popular and commercial media outlet of great influence to media and consumers alike. The Web site was created in 1994 when founder Matt Drudge began sending out what was called "a gossipy email newsletter" (Naughton, 2006). Drudge and his report remained in the adolescent stage for roughly four years. During that time it was derided by critics and mainstream media, and on occasion his "reports" resulted in lawsuits.

But in 1998, Drudge was credited with being the first to break the news about the relationship between President Bill Clinton and Monica Lewinsky. This critical incident transformed Drudge into a household name. Traffic continues to steadily build, and in March 2007, the Web site reported that it had recorded 4.3 billion visits over the previous 12 months (Drudgereport, 2007).

### Sports

Sports Web sites are popular and plentiful. Many sports sites are created by people who simply love the sport they are involved with, whether at the level of a participant or a fan. Fan sites are spreading rapidly on the Internet, and there are many companies on the World Wide Web to help them out. For example, EZ and Invision started out as tools and programs to help amateurs gather fans and audiences in a forum known as a bulletin board. Many owners still use these boards but maintain an amateur status, some paying for the maintenance of the sites out of their own pockets and some asking for user donations to keep the site going. One of the key elements of this level of operation is that the fans generate content. The posts to the forum are the attraction, and the posts are put there by the fans. The owners simply provide the place for others to play.

Although there are many examples of this type of Internet site, there are others that still maintain amateur status, even though they move toward a more "professional" and economically profitable model. Some owners prefer to control content without allowing for the fans' input. These owners, such as Vandelay sports.com (http://www.vandelaysports.com), a site dedicated to the NCAA's Mid-American Conference, also pay to maintain the site, but at some level have personal or professional contact(s) with the team/group of which they are fans. In this instance, the owners provide a more informational model for the user, but choose to not "sell" the information. Rather, the owners provide hyperlinks to traditional, established media stories and also provide self-generated content in the form of interviews with coaches and players and "scoops."

Still, this type of site finds the owner at amateur status, but operating a site that generates interest and traffic. It is this type of site that can "graduate" to professional status and start to make money via a critical incident. In sports, it is generally a story that the owner breaks that concerns a broad range of fans, such as a coaching change, player arrest or other newsworthy event. The reporting of this critical event elevates the owner to a position that subsequent scoops are economically viable, so the option for a subscription-based site is an option.

## Entertainment

An example of a Web site that quickly caught on with the right mix of content, attitude and technology is the video blog site Rocketboom (http://www.rocketboom.com/vlog/). Launched in October 2004 by designer-producer Andrew Baron and anchor Amanda Congdon, the Web video show used a consumer video camera and a simple world map in the corner of an apartment as a background for a three-minute mock television news report.

The daily reports of news, tech news and news of the weird were an immediate hit with consumers, and by September 2005, Rocketboom was attracting over 250,000 visits per day (Green 2006). The following spring, Rocketboom made news by selling $40,000 worth of ads through eBay (Green, 2006). The meteoric rise seemed to bring the associated pressures of success, and in July 2006, Congdon left the show amidst a public airing of differences with Baron.

Since then, the new anchor has been Joanne Colan, and Rocketboom continues as a popular brand. In January 2007, Baron indicated that Rocketboom brought in $250,000 in 2006 (Gannes, 2007).

Another viable example is that of the comedian Dane Cook, who began developing his own page on MySpace in 2003. On MySpace, users put their information where friends can see it, and Dane used this concept to approve of every "be my friend" request until his network approached a million friends. By constantly promoting his CDs and talents on his page, Dane was able to land an offer to host *Saturday Night Live,* as well as cut a hit album with HBO (Levy & Stone, 2006).

These examples demonstrate that for certain types of entertainment, the adolescent stage can be extremely short. For Rocketboom, adolescence was at most 18 months before they began generating significant revenue. The T4 stage of Coming-of-Age/Maturity involved deciding whether to continue the old way or adopt a new model (or selling the show brand to a larger company). This stage resulted in the parting of the two partners. Since mid-2006, Rocketboom is in the T5 Adulthood stage and operating like a normal commercial business. These examples demonstrate how the Internet and media convergence allow one or two people with an interesting idea and entertaining manner to quickly become successful.

## Conclusion: Impact on Mass Media Companies

Because the diffusion of convergence will create as-yet untold numbers of Receiver-Senders, there are three especially important implications for mass media.

1. *Competition for Advertising and Audiences.* Because Receiver-Sender content mirrors some types of mass media content, it can attract advertising and audiences, the same as media.
2. *Choice.* Receiver-Sender content not only takes the same form as mass media content, but there is a greater likelihood of reaching underserved audiences by experimentation with new categories and forms of content.
3. *Cooperation.* Receiver-Sender content can be harvested and/or licensed to mass media for their own use. These will be shaped by legal issues that will continue to evolve and occur on a case-by-case basis.

Almost all types of speech (content) are limited by time, space and manner considerations. Web logs (blogs) and personal Web sites are under legal constraints, as are video materials that are submitted for broadcast. This content must conform to legal principles common to mass media regulation and are subject to regulations regarding libel, slander, defamation, false light, privacy, and so on.

One of the characteristics of new technology is that it raises new issues or problems before regulation has been drafted. For example, MP3 technology was created first, and widespread unauthorized music downloading followed. The capacity to take photos with a mobile phone came first, then regulations limiting where this technology could be used came afterward (for example, gymnasium locker rooms). The iPod, the latest in portable digital media players, has flooded the market, but experts say that iPods may cause hearing damage in some of its users (Allison, 2006).

For media organizations, it must be anticipated that Internet users will inevitably search for content that is fresh, new, and exciting. Companies that choose to ignore this trend will not be best positioned to take advantage of the benefits in the new "worldwide marketplace of ideas." Again, a good example of this is the Flickr site. After being bought by *Yahoo!* all Flickr users had to have a Yahoo! ID in order to access the Flickr site (Levy & Stone, 2006).

The implications for media organizations are numerous. Each site that successfully transitions from amateur to commercial status competes for audiences and types of revenue. Media companies should begin establishing protocols and practices for using and being used by these Receiver-Senders, who constitute the converged audience. For the content creators who are under their employ, greater freedom and less ownership must be considered. Encouraging employees to launch their own Web businesses should result in rewards based on the goodwill generated. To try to control or discourage such initiatives may not be effective. For content creators outside the organization, a clearer reward structure would encourage them to submit future work to that organization first. In other words, media companies can choose to compete with or co-opt these new types of content creators, but cooperating offers the best likelihood of mutual success.

A number of ethical considerations must be addressed. Contractual and licensing issues will continue to be significant because content that is first "freely given" may take on value at a later stage. Since anything appearing on the Internet constitutes publication, reprinting or rebroadcasting this content brings the potential for royalty claims and copyright infringement. As individuals create content for media organizations as well as themselves and their own private Web sites, more discussion and more litigation will result.

Former Rocketboom anchor Congdon is a case in point of the new ethical issues that arise from convergence. Congdon straddles the line between journalism and acting while working for ABC News. She acknowledges her reporting qualifies as "journalism," but she publicly calls herself a "blogger" and therefore does not choose to operate under the same restraints in terms of conflict of interests and/or taking acting roles in commercials or films (Gold, 2007).

In summary, the new millennium has ushered in the age of the converged audience. This is a new frontier of media management. The Web is unique in that there is the possibility to remain at amateur status and still have a potential worldwide audience of millions. That fact alone raises the likelihood of other new "models" of electronic media management. The important point for scholars is that, unlike traditional electronic media, the profit margin or commercial viability of the site may not be an issue in the early stages of brand development for the converged creator. Enough of these Web sites exist for us to identify these outcroppings of new content forms.

## References

A.C. Nielsen. (May 2005). *Top ten companies in Internet*. Retrieved April 7, 2006. Archived at http://www.internetworldstats.com/top10.htm

Allen, K.R. (2003). *Bringing new technology to market*. Upper Saddle River, NJ: Prentice Hall.

Allison, K. (2006, February 3). Apple faces class-action suit over ipod volume. Ft.com, Technology. Retrieved February 16, 2007, from http://www.ft.com/cms/s/7bf03be0–94de-11da-9f39–0000779e2340.html

Arbitron/Edison Research. (2005). Internet and multimedia 2005: The on-demand media consumer. Retrieved March 29, 2006. Archived at http://www.edisonresearch.com/home/archives/Internet%202005%20Summary%20Final.pdf

Berlo, D. K. (1960). *The process of communication*. New York: Holt, Rinehart and Winston.

Best, J. (2006, April 26). BBC online revamp embraces mobile, podcasts and user content. *Webwatch*. Retrieved May 26, 2006, from http://networks.silicon.com/webwatch/0,39024667,39158411,00.htm

Burnett, R., & Marshall, P.D. (2003). *Web theory*. London: Routledge.

Chan-Olmsted, S., & Ha, L. (2003). Internet business models for broadcasters: How television stations perceive and integrate the Internet. *Journal of Broadcasting and Electronic Media, 47*(4), 597–617.

ComScore Networks. (2006, September 27). *MySpace leads in number of U.S. video streams viewed online, capturing 20 percent market share; Yahoo! Ranks #1 in number of people streaming*. Retrieved February 15, 2007, from ComScore Networks Web site: http://www.comscore.com/press/release.asp?press=1015

Deloitte Research. (2006). TMT Trends: Telecommunications Predictions 2006: The media sector's top trends. Retrieved May 26, 2006 from http://www.deloitte.com/dtt/research/0,1015,sid%253D1012%2526cid%253D108166,00.html

Dillon, P.M., & Leonard, D.C. (1998). *Multimedia and the Web from A to Z* (2nd ed.) Phoenix: Oryx Press.

Dimmick, J., Chen, Y., & Li, Z. (2004). Competition between the Internet and traditional news media: The gratifications-opportunities niche dimension. *Journal of Media Economics, 17*(1), 19–33.

Dominick, J.R. (1999). Who do you think you are? Personal home pages and self-presentation on the World Wide Web. *Journalism and Mass Communication Quarterly, 76*(4), 646–658.

Drudgereport (2007). *Visits to Drudge*. Retrieved March 14, 2007, from http://www.drudgereport.com/

Gannes, L. (2007, January 20). *Rocketboom brings in $250K in '06*. Retrieved March 12, 2007, from http://newteevee.com/2007/01/20/rocketboom-brings-in-250k-in-06/

Gold, M. (2007, March 22). ABC News is ok with blogger's DuPont ads. *LA Times,* p. E-15

Graham, J. (2006, February 28). Flickr of idea on a gaming project led to photo website. *USA TODAY*. Retrieved July 30, 2007, from http://www.usatoday.com/tech/products/2006–02–27-flickr_x.htm.

Green, H. (2006, July 6). Splitsville at Rocketboom. *BusinessWeek* online. Retrieved March 13, 2007, from http://www.businessweek.com/technology/content/jul2006/tc20060706_468183.htm

Ha, L., & James, E.L. (1998). Interactivity reexamined: A baseline analysis of early business Web sites. *Journal of Broadcasting and Electronic Media, 42*(4), 457–474.

Hansell, S. (2006, March 12). As Internet TV aims at niche audiences, the slivercast in born. *New York Times*. Retrieved May 26, 2006, from http://www.nytimes.com/2006/03/12/business/yourmoney/12sliver.html?ex=1299819600en=b93a73a9426aeb16ei=5088partner=rssnytemc=rss

*Internet World Stats: Usage and population statistics*. (January 2007). Retrieved February 12, 2007, from Internet World Stats Web site: http://www.internetworldstats.com/stats.htm

Johnson, T.J., & Kaye, B. (2000). Using is believing: The influence of reliance on the credibility of online political information among politically interested Internet users. *Journalism and Mass Communication Quarterly, 77(4),* 865–879.

Klaassen, A., & Taylor, L. (2006, April 24). Few compete to settle podcasting's Wild West. *Advertising,* 77. Retrieved May 26, 2006, from the Communication and Mass Media Complete database.

Levy, S., & Stone, B. (2006, April 3). The new wisdom of the Web: Why is everyone so happy in Silicon Valley again? A new wave of start-ups are cashing in on the next stage of the Internet. And this time, it's all about... you. *Newsweek,* 47–53.

McMillan, S.J., & Downes, E.J. (1998, August). *Interactivity: A qualitative exploration of definitions and models*. Paper presented to the Association for Education in Journalism and Mass Communication, Baltimore, MD.

Naughton, J. (2006, August 13). Websites that changed the world. *The Guardian Unlimited*. Retrieved March 20, 2007, from http://observer.guardian.co.uk/review/story/0,,1843263,00.html

Papacharissi, Z. (2002). The self online: The utility of personal home pages. *Journal of Broadcasting and Electronic Media, 46*(3), 346–368.

Project for Excellence in Journalism, "*State of the American news media, 2007: Mainstream media go niche*." Retrieved March 13, 2007, from http://www.stateofthemedia.org/2007/index.asp.

RealNetworks. (2006, November 30). *Industry awards honor RealNetworks' excellence in digital entertainment*. Retrieved February 16, 2007, from http://www.realnetworks.com/company/press/releases/2006/demmx.html

Rogers, E. (2005). *Diffusion of innovations* (5th ed.). New York: Free Press.
Shields, M. (2005, November 18). Study: Podcasters to reach 62 mil. *Adweek.* Retrieved May 26, 2006, from http://www.adweek.com/aw/search/article_display.jsp?vnu_content_id=1001524777
Srinivas, S. (2005, July). *Online citizen journalists respond to south Asian disaster. A USC Annenberg online journalism review*. Retrieved November 11, 2006, from http://www.ojr.org/ojr/stories/050107srinivas/
Storm, E. J. (2006, October). *Converging the conversation: The introduction of Web-generated citizen content into newsrooms*. Paper presented to Convergence and Society: Ethics, Religion, and New Media, Columbia, SC.
Sutel, S. (2007, March 14). Viacom sues YouTube over copyrights. *Businessweek* .Retrieved March 14, 2007, from http://www.businessweek.com/ap/financialnews/D8NRNI701.htm.
Wicks, J.L., Sylvie, G., Hollified, C.A., Lacy, S., & Sohn, A.B. (2004). *Media management: A casebook approach* (3rd ed.). Mahwah, NJ: LEA.
Williams, F., Rice, R.E., & Rogers, E.M. (1988). Research methods and the new media. New York: Free Press.

CHAPTER 5

# Media Convergence and the Neo–Dark Age

Van Kornegay

In 1998 supercomputer designer Danny Hillis used the term "digital dark age" during a presentation on "Digital Continuity" at the Getty Center in Los Angeles. He made the comparison to point out that we now store much of the record of our news, art, and science on media that we know will not outlast our lifetimes (Brand, 2003). In making the seemingly incongruous connection between our current high-tech era and the Dark Age, he only meant to sound the alarm about the tenuous and changeable nature of digital storage as a medium for preserving the historical record. But information in digital form is also the foundation and catalyst for the processes behind media convergence, and if we look closely at the broader social impact of this related phenomenon, we discover other dynamics at work that make Dark Age comparisons even more plausible.

As discussed in the opening chapter of this volume, the definition of media convergence is still evolving, but generally the term refers to the process of gathering, editing, storing, transmitting, and consuming text, images, and sound in digital form with networked computers playing some mediating role. Like many of the communication innovations that have gone before it, media convergence is more than just a new system of communication. It is also an agent of change that is redefining the social, cultural, economic, and political contours of society in ways that are both subtle and overt. This process is turning the world of mass media on its head and sending ripples—or perhaps it is a tsunami—radiating out into society at large by democratizing the ways in which media messages are created, delivered, and consumed.

We become a part of the phenomenon of media convergence when we post our opinions on a blog, download a video from CNN, or create our own video and post it on YouTube. Media convergence makes it possible to download songs from our favorite artist's latest album, browse the Web site of our hometown newspaper, or a newspaper on the far side of the world. Media convergence creates the virtual spaces where we exchange opinions with like-minded friends and flame our foes. The list could go on and on. Media convergence is a driving force of modern life in the developed world, and it touches our lives just about every time we push a button.

## Media Convergence and the Age in Which We Live

It was 1844 when the first telegram sent from Washington to Baltimore famously delivered the Old Testament verse "What hath God wrought!" It was a "Wow!" of biblical proportions that announced the arrival of the Information Age, an age of electronic messages that has transformed the way humans communicate.

Now, some technology theorists assert that we have moved beyond the Information Age into yet another era known as the Intangible Economy. Rather than the simple possession of information alone, intellectual know-how and the ability to collaborate are the most important resources (Goldfinger, n.d.).

Whatever this age comes to be called, historians and social critics frequently point to media convergence as a watershed development of the times. The term "media convergence" carries with it the suggestion of an innovative process that represents a coming together, a seamless integration of information tributaries that flow into a digital river that is more malleable, more accessible, and more useful than its analog predecessors.

The marvels of media convergence are certainly worthy of a biblical "Wow!" But where is this "Wow!" taking us? If our eagerness to get our hands, eyes, and ears jacked in to the latest media technologies is any indication, then the popular vote is that media convergence is taking us somewhere good. After all, easier access to more information should surely lead to more enlightenment, and to eddy out of this digital river is to risk being left behind in a personal, cultural, and economic backwater.

But will this hunch of the herd withstand the test of time? Should our exuberance for media convergence be tempered with the rational reminder that every technology brings with it a series of unintended consequences?

In his book *Why Things Bite Back: Technology and the Revenge of Unintended Consequences* Tenner (1996) examined how every new technology creates unintended consequences that he defined as a series of revenge effects. He said revenge effects happen when a new technology reacts with real people in real situations in ways we could not foresee, and ironically, these revenge effects often subvert the original purpose of the innovation.

If Tenner is right, then we can anticipate revenge effects from the technologies of media convergence that will actually undermine some of our abilities to engage in meaningful communication. Will these revenge effects be the communication equivalent of a mild head cold or a deadly pandemic? Is media convergence ushering in an evolutionary advance in human communications or unwittingly opening the door into a Neo–Dark Age?

## The Dark and the Neo–Dark Age

Historical labels can be contentious things, and there is still much debate over what to call the roughly 1,000-year period that spanned 410 A.D., when the Visigoths sacked Rome, to the mid-1400s and the start of the Renaissance. The Italian historian Petrarch is generally credited with creating the term "Dark Age" in 1330, and other historians of that period adopted it to paint a picture of a time of lawlessness, a scarcity of written records and few cultural achievements. Most contemporary historians prefer less pejorative terms, such as Late Antiquity or Early Middle Ages, and those who do use the term Dark Ages argue that the word "dark" simply refers to our lack of knowledge about the time.

Sociologist Rodney Stark (2005), in his book *Victory of Reason,* goes even further and posits that the Dark Ages were actually a period of creative ferment that made the Enlightenment possible. He cites as evidence the numerous technological advances of the age, such as windmills, horseshoes, chimneys, waterwheels, stirrups, compasses, eyeglasses, and mechanical clocks.

However, despite the contention surrounding the term and evidence that there were significant cultural and scientific advances during this time, the concept of a Dark Age as a violent, chaotic, and mute period of cultural stagnation has endured. Perhaps it started with naked prejudice against a time when other cultures held sway but has endured because it serves as a useful deterrent against collective sloth and waywardness, the cultural equivalent of "sit up straight, or you'll ruin your posture."

Whatever the explanation for its durability, the term remains fixed in the popular imagination as a fact of history, a symbolic warning against social calamity or both.

The Dark Age did have some well-documented downsides:

- The violent invasion by outsiders that weakened an established, centralized authority in Western Europe.
- The deterioration of a general sense of order, both legal and cultural, that had given rise to an empire that spanned much of the Western world. In its

place there arose a parochial system of social organization typified by crude and chaotic legal systems and allegiance to tribe and clan as opposed to the wider body politic or governing authority.

- A decline in literacy and a dearth of the written historical record that prevented the achievements of the age from being widely circulated or passed on to future generations (Cantor, 1993).

There are intriguing parallels between these Dark Age downsides and the observable revenge effects of media convergence. Lost, or just ignored, behind all the "Wow!" factor of media convergence is the reality that the convergence trend is responsible for undermining some of the fundamental laws we have agreed to live by; it is encouraging a realignment of society around narrow, parochial interests that are leading to a lack of social cohesion and a growing cultural stasis; and in very tangible ways it is putting at risk the dissemination and preservation of the historical record.

# Evidence of the Neo–Dark Age

## Media Convergence Challenges the Law

Copyright law is one of the legal offspring of the Enlightenment. It developed after the introduction of the printing press and was meant to advance the concept that the dissemination of knowledge brings more voices to the marketplace of ideas. The rationale for one of the first copyright laws, the Statute of Anne enacted in Great Britain in 1710, was meant to support the encouragement of learning (Feather, 1980).

Since that time, ideas have been protected in most Western Democracies by laws that cover the physical (analog) expression of an idea in things such as books, recordings, photographs, films, and software, to name a few examples. But convergent media technologies transform these ideas into digital data, and, in doing so, the ideas become detached from the physical plane. This in turn makes the resulting media content easy to alter, replicate, and distribute over networked computers. Intellectual property becomes difficult to protect when it takes on this new skin.

It is an ironic revenge effect that the laws meant to advance learning and the dissemination of knowledge are being undermined by the very technologies that are making more information more accessible to more people. In the age of media convergence we have more information available to us than ever before, and yet we have less agreement on its value and status under our laws.

Barlow (1994) in his essay *The Economy of Ideas* said, "In the absence of the old containers almost everything we think we know about intellectual property is wrong. We are going to have to unlearn it. We are going to have to look at information as if we had never seen the stuff before" (p. 13).

The plight of the music industry is an example that current copyright laws are struggling to remain relevant in the age of media convergence. In the mid-1980s, music first came to the mass market in digital format via compact disks (CDs) This new technology led to a 15-year boom in music sales. But that boom evaporated and became a bust when the Internet, networked computers, and software, such as Napster, converged to enable people to compress and share digital music files for free.

Napster was released in June 1999, and by 2001, when it reached its peak of popularity, the number of CD units shipped dropped by 6.4 percent from the previous year. After a copyright lawsuit by the music industry, Napster shut down its free music-swapping service and later returned to market with a fee-based model. However, plenty of other file-sharing schemes sprang up to fill the void, and users flocked to them.

Except for a small increase in 2004, the number and retail value of CDs sold has continued to decline. Between 2000 and 2006 the retail dollar value of CD sales in the United States dropped from approximately $13.2 billion to $9.2 billio. (Recording Industry Association of America, 2007a).

More recently, the music industry has been cheered by the performance of online fee-based services such as Apple's iTunes. However, if there is any doubt that the music industry still sees media convergence as a serious threat, one need only look at the Web site of the Recording Industry Association of America. Most of its news items do not tout the creative achievements of talented artists. Instead, they sound like a call to arms against the barbarians at the gates. "Music Thieves Sentenced in Major Federal Internet Pre-Release Piracy Operation," and "Music, Movie Industries Target Theft on Internal Campus Networks," are typical of the headlines on the site (Recording Industry Association of America, 2007b).

As compression software becomes more sophisticated and access to fast, broadband Internet service grows, file sharing promises to pose a similar problem to the motion picture industry. According to the Motion Picture Association of America (2007), Internet piracy of downloaded movies cost the worldwide motion picture industry $2.3 billion in 2005 against total box office sales of $23.24 billion.

Copyright law has proved to be somewhat inelastic and impotent in the face of encroachments of media convergence. Efforts to develop new laws that protect ideas, such as the Digital Millennium Copyright Act (DCMA) enacted in 1998, have been criticized as ineffective and possibly responsible for creating their own set of revenge effects. Also in 1998, Congress passed the Sonny Bono Copyright Term Extension Act.

According to the Electronic Frontier Foundation (2006) the DCMA has "done nothing to stop 'Internet piracy.' Yet the DMCA has become a serious threat that jeopardizes fair use, impedes competition and innovation, chills free expression and scientific research, and interferes with computer intrusion laws" Copyright is just one area where media convergence has posed challenges to the legal system.

Statues related to conduct that governs protecting personal privacy, distribution of pornography, gambling, and commerce, to name a few, are becoming increasingly difficult to enforce.

As Barlow (1994) says in *The Economy of Ideas,* "Whenever there is such a profound divergence between the law and social practice, it is not society that adapts" (p. 4). An undeniable revenge effect of media convergence is that it indiscriminately levels legitimate legal barriers as easily as it erases impediments to communication. How much longer can these legitimate barriers stay down before the laws themselves become impotent?

## A Breakdown in Civic Engagement

Those who claim that media convergence will enrich and strengthen public discourse frequently point to the development of online communities, such as blogs, as forums that have the potential to make the discussion more meaningful and representative than content found in the mainstream media (It's the links, stupid, 2006). Social networking sites such as MySpace, Facebook, and YouTube have proved so popular at attracting mass audiences that corporations, nonprofit organizations, and political campaigns are now creating their own social networks to speak directly to their audiences. Several companies are now developing software tools that help people create their own social networks online. Marc Andreessen, coauthor of Mosaic, the first widely used Web browser, said this new generation of software lets people do what they want to do online, which is build and design their own worlds (Stone, 2007).

Due to their popularity, Web logs, or blogs, are an indicator of the way many of these online communities form, interact and have the potential to influence society at large. The Web site Technorati claims to track more than 70 million Web logs that comprise "the blogosphere" and states that about 120,000 Web logs are being created worldwide each day (Sifry, 2007).

Blog advocates cite these numbers as proof that there is pent-up demand for an alternative form of journalism that challenges the monopolistic and hierarchical nature of the mainstream media. They also contend that blogs offer more points of view and a more responsive self-righting process than that of the mainstream press.

But while popular in numbers, blogs struggle to live up to their promise as a better marketplace for the exchange of ideas, and many pundits have expressed reservations about their overall societal impact. On the impact of blogs, Shulevitz (2002) drew a parallel with the vision of Jorge Luis Borges, who "dreamed of a library the size of a universe, whose wealth of books would induce first delirium, then despair, then breakdown of the social order. Since we first became aware of the Web, we have ricocheted between similar feelings over a universe far more disruptive: one of unbounded, uncensorable streams of text" (para. 1).

Other critics say the narrow, parochial nature of blogs and other online communities create their own unanticipated social revenge effects. One of these is what communications theorists refer to as the "narcotizing dysfunction." People who have been narcotized by their media habits often harbor the illusion that they are actually socially engaged when in fact they only use media to surround themselves with information about social issues. As a result, these would-be agents of change become little more than well-informed couch potatoes.

*New York Times* columnist David Brooks (2005) wrote that these online communities are producing a segmentation of society that is bound to produce cultural stagnation. He said reformers and radicals no longer feel muzzled and thus compelled to challenge the status quo. Now, they host their own Web sites and cable talk shows and publish their own magazines.

> People are taking advantage of freedom and technology to create new groups and cultural zones.... People are moving into self-segregating communities with people like themselves, and building invisible and sometimes visible barriers to keep strangers out... far from converging into some homogeneous culture, we are actually diverging into lifestyle segments. The music, news, magazine, and television markets have all segmented, so there are fewer cultural unifiers like *Life* magazine or Walter Cronkite. (para. 6)

Paul Saffo, a futurologist with the Institute of the Future, offers a warning that sounds even more like a Neo–Dark Age prophecy. "Each of us can create our own personal-media walled garden that surrounds us with comforting, confirming information that utterly shuts out anything that conflicts with our world view," he says. "This is social dynamite" and could lead to "the erosion of the intellectual commons holding society together.... We risk huddling into tribes defined by shared prejudices" (What sort of revolution? 2006, para 6).

Instead of building a sense of community, the lack of hierarchy and editorial control that is a hallmark of many news and political blogs seems to unleash a Dark Age dynamic that is hostile to social interaction and similar in structure to a feudalistic culture. Peter Kolock and Marc Smith have observed the effects of the Internet on community interaction and found many online groups have the structure of either an anarchy (if unmoderated) or a dictatorship (if moderated) (Putnam, 2000).

The mainstream media is gamely trying to figure out where it fits in this "no rules" world of blogs and discussion groups. They have even coined the term "citizen journalist," in an effort to corral their audiences into the culture and cause of news gathering. Mainstream journalists seem to be saying that if we cannot be the gatekeepers, we will agree to be the gateway. But their efforts to fit in, collaborate, or co-opt often seem to miss the point that the ground has shifted underneath everyone's feet. This is not your father's journalism, and the marketplace of ideas has turned into something more akin to a demolition derby where the audience is driving their own cars with their hands firmly on the wheel.

Gil Thelen is a former editor and publisher of *The Tampa Tribune* and was a leader in developing the innovative News Center, a multimedia news effort that reaches the Tampa area audience via the newspaper, a Web site, and radio and

TV broadcasts. Speaking at a conference on convergence, he defined his organization's approach to media convergence as using multiple media on multiple platforms to reach an ever-changing audience to help them live their public life together (Thelen, 2003).

That is a noble and worthy aim of a civic-minded journalist. But, as it turns out, the audience gets an even bigger say in defining media convergence, and here is what it more often than not might mean to them: "Me creating or accessing multiple media on multiple platforms to live my narrowly focused, increasingly private life any way I darn well please."

An effort to integrate blogs into the online site of *The State* newspaper in Columbia, SC, ran into problems that are illustrative of dealing with online culture. The blog editor, Brad Warthan, had to impose rules on the newspaper's editorial blog due to the growing hostile nature of many participants' posts. "My less mature correspondents are running off the serious, thoughtful people," he said, adding that those were the people "who came to the blog hoping for the very thing I'd like that venue to be—a place to exchange sincere, constructive ideas about the challenges facing South Carolina and the rest of the world" (Warthan, 2006, p. D2).

Warthan began requiring people using the blog to sign their names, and he began deleting posts that resorted to personal attacks. It is ironic that his new rules to make the blog more hospitable were in effect reshaping the site in the image of an older, mainstream media institution: the op/ed page, a space where editors ensure brevity, spelling, grammar, salience, and good manners.

The experience at *The State* is not an isolated event among newspapers that have tried to open the doors to this form of citizen journalism. *The Washington Post* shut one of its blogs down in January 2006 because of the ferocity of personal attacks, profanity, and hate mail directed at the paper's ombudsman. A *Los Angeles Times* experiment in letting readers rewrite the paper's editorials, or "Wikitorials" as they were called, only lasted three days, after users flooded the site with foul language and pornographic photos ("Los Angeles Times suspends 'Wikitorials,'" 2003).

There are media outlets that have been successful with experiments in citizen journalism. But it is worth noting that these publications have imposed some traditional, mainstream media controls on content. The Korean news site *Ohmy New,* is one such example. *Ohmy News* encourages readers to submit content, but it uses feedback, rating systems, and a "tip-jar" that invites readers to pay writers for good work. This system creates a hierarchy of stories so that the better articles are featured more prominently on the site (Lee, 2007).

## The Digital Divide

If media convergence is casting the shadows of a Neo–Dark Age on those who have access to the main channel of convergence, the Internet, those who do not or cannot tune in to this channel are at risk of living in a total media eclipse.

Early U.S. government studies found considerable discrepancies in Internet access among income and ethnic groups. Whites and Asian Americans had the highest access and African-Americans and Hispanics had much lower rates of access (NTIA, 1999). More recent studies show there continues to be significant gaps along race, age, income, and gender lines among those who use information and communication technologies (ICT). A Pew Internet and American Life study (Horrigan, 2007) found males and whites continue to dominate the "high-tech" categories of ICT use, while those in the "Off the Network" category had the oldest median age and reported lower levels of income. Members of this group were also more likely to be female and were more ethnically diverse than users who reported heavy ICT use.

There are also considerable global differences in Internet access. The United Nations' International Telecommunication Agency reports that the total number of Internet users in the G8 countries (Canada, France, Germany, Italy, Japan, Russia, the UK and the US) is roughly the same as in the rest of the world combined. In the United States alone there is eight times the number of Internet users as in all the countries of Africa (International Telecommunications Union, 2004). Castells (2001) says that "the differentiation between Internet-haves and have-nots adds a fundamental cleavage to existing sources of inequality and social exclusion in a complex interaction that appears to increase the gap between the promise of the Information Age and its bleak reality for many people around the world" (p. 247).

A central contention of this chapter is that media convergence, a phenomenon that goes hand-in-hand with Internet access, tends to reinforce more narrow social, economic, political, and cultural ties at the expense of wider societal cohesion. But if media convergence has negative consequences for the Internet-haves, it has even greater potential to inflict collateral damage on the Internet-have-nots. With traditional, mainstream media there were at least tenuous connections between different social and economic groups. Newspapers are inexpensive and readily available to all. Broadcast signals reach areas not wired for Internet access, and even illiterate audiences can see, hear, and understand an event on television and radio.

As these media migrate to the Internet, or disappear due to competition from Internet-based media, the opportunities for wide segments of society to share experiences and have a part, however small, in the public discourse will diminish. Thus, as media convergence undermines more traditional and accessible forms of media, it threatens to turn the digital divide into a chasm.

## The Personal and Social Costs—the Neo–Dark Age Begins at Home

If media convergence undermines the role of the mass media to serve as a binding agent in society, there is also evidence that a growing dependence on computer-mediated communication has a corrosive effect on some of the most intimate institutions of community and personal life.

McPherson, Smith-Lovin, and Brashears (2006) found that Americans have fewer confidants and community-based connections and are retreating more into the nuclear family. The study cited increased reliance on Internet-based communications as one possible reason for the disconnect. In contrast, researchers said a broader network of friends creates a safety net, more civic engagement, and local political action.

Kraut, Patterson, and Lundmark (1998) examined the social and psychological impact of Internet use on 169 people in 73 households during their first one to two years online. Even though participants in the study used the Internet primarily for communication, the authors found that greater use of the Internet was associated with declines in participants' communication with household family members, declines in the size of their social circle, and increases in their depression and loneliness.

In addition, there is ample research on the social revenge effects from earlier forms of electronic mediums of communication, such as the telephone and television, which that show these electronic cousins of the computer play some role in reducing face-to-face communication and civic involvement (Putnam, 2000). Add the interactive and anonymous nature of online groups into the mix, and it is not too much of a stretch to predict that a computer connected to the Internet could more readily turn out to be a hybrid of the television and telephone with amplified negative social consequences.

## Lost or Inaccessible Historical Record

Of all the revenge effects of media convergence that have introduced a Neo-Dark Age dynamic, none is more observable or chilling than what is happening to the historical record. The record of our culture found in mass media such as newspapers, magazines, photography, videography, recordings, and Web pages are now in danger of evaporating forever into the digital ether. This is because much of the information stored digitally using current methods of professional archivists lacks the digital "hooks" that over time will enable future users to access stored files.

The term "digital dark age" now regularly appears in news reports and the conference proceedings of archivists and librarians who are concerned with preserving the record of our culture that appears in the mass media. Yet there is still no solution to the problem of preserving digital records.

McCargar (2003), a senior editor at the *Los Angeles Times* with a master's degree in library science, says the data may today be sitting on a perfectly functional storage medium with an up-to-date computer on top. But access to that data will eventually slip away without some technological breakthrough that allows that data to be translated into as-yet unknown formats and read by yet-to-be-developed operating systems.

The problem is that when media, such as a photograph or recording, are removed from the physical plane and converted into a digital format they become just a string of ones and zeros, known as a bit stream. Software has the hooks to decipher the bit stream and tell a monitor or printer what the photograph looks like or send a digital signal to the speakers to tell them how the recording sounds. But software is always changing and formats are in a constant state of flux. With each iteration of new software some of these hooks are lost. At some point earlier formats become unreadable by newer software, and the result is that the bit stream starts to look like hieroglyphics without a Rosetta stone.

It is not only a problem of software; the medium on which digital information is stored also has a short life span. McCargar (2003) reports that Paul Conway, a researcher at Yale, has analyzed the life span and carrying capacity of a variety of storage media used throughout history. He found the most durable storage medium was the Sumerian Clay tablet, which has a capacity of only 75 characters per square inch but a life span of 10,000 years.

In contrast, the now-outdated floppy disk held the equivalent of 100,000 characters per square inch but only had a life span of about six years. The medium itself did not deteriorate, but the hardware and software to read it has been relegated to the recycling bin.

In addition to the ephemeral nature of digital storage mediums, there are no set standard formats for preserving things such as sound, video, animation, and Web pages that are here today and gone tomorrow. It is quite possible that future historians who would like to look at the very first pages of *Yahoo!* would have more luck finding an original newspaper that covered the Civil War.

There are efforts underway to make an archival record of much of the content that appears on the Web. The most notable of these is the Internet Archive found at http://www.archive.org. The site is building an online library of Internet sites and other artifacts in digital form such as moving images, audio files, and texts. However, the site only archives periodic snapshots of Web sites, leaving large gaps in the record, and many of the archived pages on the site are missing links and navigation graphics.

Ironically, efforts such as the Internet Archive can inadvertently contribute to the myopic belief that the entire world's knowledge is stored on the Web. The fact is that most of the historical record, including the cultural record from the mass media, is still not in digital form. At the Library of Congress only 10 percent of its 132 million objects will be digitized in the near future. The library has an archive of more than 7 million photo prints from newspapers and magazines, including more than 5 million from *Look* magazine, yet only 6,086 of these images have been digitized (Hafner, 2007). Historians worry that researchers looking only at the digital record of our past will overlook or ignore undigitized physical collections that have much to say but that are not searchable in digital form.

## Conclusion

The title of journalist Thomas Friedman's (2005) best-selling book, *The World is Flat,* turns the Dark Age geographical description of the world on its head to describe how media convergence—along with a host of other technological and economic trends—is overcoming barriers to the exchange of information.

Friedman points out in exhaustive detail how these digital channels are leveling the barriers of time and space. In his view there are plenty of risks, challenges, and human casualties lurking in this period of change, but in the end this trend is creating a metaphysical flat world where information and opportunity are more accessible to us all. This is the default view of technological determinism, and it certainly has momentum on its side. It is decidedly more optimistic than the alternative view that we are rushing headlong over a cliff.

As the technologies of media convergence continue to flatten barriers to communication it is helpful to remember that some barriers, hierarchies, and organizing authority can set useful limits, pave channels for productive creative energies and nurture cultural progress. The concept of a Neo–Dark Age can provoke us to consider the possibility that the travails we associate with the first Dark Age—a lack of social cohesion, breakdown in the rule of law, a lost historical record—could be triggered by a flood of information, not just a drought.

Using the lens of a previous period in history as a prophetic vision of our own times is uncertain business. But most certainly we will one day look back on this sorting-out period and recast the labels that define an era in which media convergence played such a major role. Perhaps it will be clearer to us then than it is now whether the uncertainty of this new era, its darkness, was a darkness of the tomb or of the womb.

## References

Barlow, J. P. (1994). The economy of ideas [Electronic version]. *Wired, 2.03*. Retrieved June 22, 2007, from http://www.wired.com/wired/archive/2.03/economy.ideas.html

Brand, S. (2003). Escaping the digital dark age. *Library Journal, 124,* 46–49.

Brooks, D. (2005, August 11). All cultures are not equal [Electronic version]. *The New York Times*. Retrieved June 10, 2007, from http://nytimes.com.

Cantor, N. S., (1993). *The civilization of the Middle Ages*. New York: HarperCollins.

Castells, M. (2001). *The Internet galaxy: Reflections on the Internet, business, and society*. New York: Oxford University Press.

Electronic Frontier Foundation, (2006) *Digital Millenium Copyright Act (DCMA)*. Retrieved September 2, 2006 from http://www.eff.org/IP/DMCA/

Feather, J. (1980). The book trade in politics: The making of the Copyright Act of 1710. *Publishing History, 8*, 19–44.

Friedman, T. L. (2005). *The world is flat: A brief history of the twenty-first century*. New York: Farrar, Straus and Giroux.

Goldfinger, C. (n.d.). What is the new economy? *GEF CEO Electronic Forum*. Retrieved June 22, 2007, from http://www.gefma.com/Intangible.htm

Hafner, K. (2007, March 11). History, digitized (and abridged). *The New York Times,* p. 1, Section 3.

Horrigan, J. B. (2007, May &) *A Typology of information and communication technology users*. Retrieved July 16, 2007, from http://www.pewinternet.org/PPF/r/213/report_display.asp

International Telecommunications Union (2004). *The digital divide at a glance*. Retrieved June 23, 2007 from http://www.itu.int/wsis/tunis/newsroom/stats/

It's the links stupid. (2006, April 20). *The Economist* [Electronic version]. Retrieved June 22, 2007, from http://www.economist.com/surveys/displaystory.cfm?story_id=6794256

Kraut, R., Patterson, M., Lundmark,V., Kiesler, S., Mukhopadhyay, T., & Scherlis, W. (1998). Internet paradox: A social technology that reduces social involvement and psychological well-being? *American Psychologist, 53*(9), 1014–1016.

Lee, D. (2007, June 18). Citizens are the media is S. Korea. *Los Angeles Times*. Retrieved July 14, 2008 from http://articles.latimes.com/2007/jun/18/business/fi-ohmynews18

Los Angeles Times suspends 'Wikitorials,' (2003, June 21). Retrieved June 23, 2007, from http://msnbc.msn.com/id/8300420/

McCargar, V. (2003, November 15). *Losing the first draft of history*. Panel discussion at the opening of Newsplex, University of South Carolina, Columbia, SC.

McPherson, M., Smith-Lovin, L., & Brashears, M. E. (2006). Social isolation in America: Changes in core discussion networks over two decades. *American Sociological Review, 71*(3), 353–375.

Motion Picture Association of America. (2007). *Internet piracy,* Retrieved March 15, 2007, from http://www.mpaa.org/piracy_internet.asp

NTIA. (1999). *Falling through the Net: Defining the digital divide. A report on the telecommunications and information technology gap in America*. Washington, DC: U.S. Department of Commerce.

Putnam, R. D. (2000). *Bowling alone: The collapse and revival of American community*. New York: Simon & Schuster.

Recording Industry Association of America. (2007a). *2006 year-end statistics*. PDF retrieved June 22, 2007, from http://www.riaa.com/keystatistics.php

Recording Industry Association of America. (2007b). Retrieved August 30, 2006, from http://www.riaa.com/default.asp

Shulevitz, J. (2002, May 5). At large in the Blogosphere [Electronic version]. *The New York Times*. Retrieved June 22, 2007, from http://www.nytimes.com

Sifry, D. (2007, April 5). Technorati, *The state of the live Web, April 2007,* Retrieved from http://technorati.com/weblog/2007/04/328.html

Stark, R. (2005). *The victory of reason: How Christianity led to freedom, capitalism, and Western success*. New York: Random House.

Stone, B. (2007, March 3). Social networking's next phase. *The New York Times*, p. A1.

Tenner, E. (1996). *Why things bite back: Technology and the revenge of unintended consequences*. New York: Knopf.

Thelen, G. (2003, November 14). *The dynamics of convergent media*. Panel discussion at the opening of Newsplex, University of South Carolina, Columbia, SC.

Warthan, B. (2006, July 30). Making the blogosphere safe for decent folk. *The State*, p. D2.

What sort of revolution? (2006, April 20). *The Economist* [Electronic version]. Retrieved June 22, 2007, from http://www.economist.com/surveys/displaystory.cfm?story_id=6794256

CHAPTER 6

# Converging Communication, Colliding Cultures

## Shifting Boundaries and the Meaning of "Our Field"

Jeffrey S. Wilkinson

As discussed by Grant in the Introduction to this volume, the term "convergence" is used to mean a number of things, but essentially it is a short-form descriptor used by media scholars to refer to the huge shifts and changes taking place among media companies over the past decade. More specifically, media convergence attempts to capture the process by which traditional content creators (journalism and mass media) adapt to an on-demand society where the consumer wields control. As a result, the party able to decide who gets to read, hear, or view what, when, and how, has shifted away from the source and toward the receiver.

But this shift in control only touches upon the surface of several other deep and profound changes occurring in modern society. Convergence is not limited to media, and catching a glimpse of the big picture can help scholars and practitioners anticipate future trends, changes, and developments.

Journalism programs have long prided themselves for teaching writing and reporting and occupying the unique role of the "fourth estate," the watchdog of government and protector of—and protected by—the First Amendment. For generations "journalism" was synonymous with newspapers, and academic programs positioned themselves by training students to work for newspapers as reporters, editors, and photographers.

With the advent of electronic media, certain other skills became part of that trade. Learning the equipment was part of the craft (and the art) of "doing radio" or "making television." The result was that media workers, and journalism and mass communication educators considered these skills and practices as uniquely their own. For years, the economics of the technology reinforced this view. When television cameras were first invented in the 1920s they were expensive, large, and temperamental. Special facilities, parts, and technical support were needed. A certain amount of "insider knowledge" was needed to use them, and people working in radio and television developed their own language and unique sets of practices (as had newspapers). These types of elements became the core for college programs in journalism, media, and mass communication (for more, see Bajkiewicz, Chapter 16, this volume).

But fast-forward (an old broadcast term) to today. Parents give their children camera phones, and video is a built-in feature on many home computers. Teenagers with little training routinely engage in "live webcasts" from their bedrooms and produce podcasts or video blogs that are uploaded to the Internet for others to watch on their broadband cell phones.

The diffusion of media technology into all sectors of society creates new ways of using and disseminating communication content. To best understand the implications for mass media and communication, it is important to be aware of the so-called "killer applications" in unrelated fields. In May 2005, John Podhoretz of the *New York Post* noted the inexorable trend:

> But it can't be a coincidence that the five major pillars of the American media—movies, television, radio, recorded music and newspapers—are all suffering at the same time. And it isn't. Something major has changed over the past year, as the availability of alternative sources of information and entertainment has finally reached critical mass. Newly empowered consumers are letting the producers, creators and managers of the nation's creative and news content know that they are dissatisfied with the product they're being peddled. (par 3–4)

Much has already been written on the convergent synergy enjoyed by media and media-related areas (advertising, public relations, journalism, and entertainment—music, film, sports, theater, and even gaming). But the same developments are happening across the spectrum—in the fields of medicine, political science, education, law, and elsewhere. Practitioners in these fields are also creating messages for mass audiences. By creating content with entertainment and information value, using digital technologies and delivery systems, we have entered a period of social and economic Darwinism, a survival of the fittest.

This chapter presents a discussion of convergence in non-media–related fields and some means for interpreting the implications of such convergence. This discussion is designed to help scholars better anticipate future developments related to convergent media technologies.

## Convergence and Content

Although definitions of media convergence vary according to emphasis, it is essentially the "blending of the media, telecommunications and computer industries, and the coming together of all forms of mediated communication in digital form" (Burnett & Marshall, 2003, p. 1).

Many dimensions of convergence are enabled and realized through the development of the World Wide Web and high-speed computer networks. The tools for creating digital content are rapidly spreading into the hands of consumers (Pew Internet and American Life Project, 2004). This trend means consumers are increasingly producing material for others to enjoy.

Since convergence is the coming together of technologies and their applications, we need to consider how all the content creators and all the users employ it. Here, "content" can be brought out to the widest application of meaning to include any message or communication that is attended by another. This perspective transcends journalism and mass communication.

## The Context of Media Convergence

For a century communication scholars and theorists have created as well as borrowed and adapted models from other disciplines to better explain communication processes and effects. For example, Berlo's (1960) SMCR model provided a useful

| Source (originator) | Message (content) | Channel (mode) | Receiver (consumer) |
|---|---|---|---|
| Professional | Informative | Visual | One |
| Amateur | Entertaining | Aural | Many |
| Consumer | Persuasive | Text | One-time |
| Corporate | Selling | Animation | Repeated |
| Educator | Educating | Still or Moving | "Source/originator" |
| Marketer | Hypes | Graphics | Other |
| Politician | Shares Opinion | Other | |
| Lawyer | Other | | |
| Retiree | | | |
| Other | | | |

**Figure 6.1** ▰ SMCR Revisited

framework for conceptualizing the process of communication into four components: Source, Message, Channel, and Receiver. Source was represented as media companies and their workers, which typically meant television companies, radio stations, newspapers. But now, both the Source and Receiver can be represented by almost anyone (see Figure 6.1). Any individual with the right skills can create content (Message) and upload it to the Internet (Channel) to any number of consumers (Receivers). Rather than a linear model, the process has now become recursive in that receivers now influence traditional sources (media, politicians), who in turn take on the role of Receiver (*The YouTube-ification of politics,* 2007).

## Convergent Journalism Training

Each year a nationwide survey of university graduates entering the field/discipline of mass media/communication is conducted by scholars at the University of Georgia. In 2006, it was reported that roughly 49,100 students earned bachelor's degrees and 4,400 students earned master's degrees from the 472 colleges and universities across the United States and Puerto Rico offering programs in journalism and mass communication (Becker, Vlad, & McLean, 2007a). Curricular specializations include print (news editorial) journalism, telecommunications (broadcasting, broadcast journalism), advertising, public relations, and related areas such as magazine, radio, or video production.

A growing area is Internet and multimedia skills. In 2007, "four in 10 of the 2006 journalism and mass communication bachelor's degree recipients with a job in communication reported that at least part of their assignment involved writing and editing for the Web" (Becker, Vlad, & McLean, 2007b, p. 7). The year before, the number was "three in 10" (Becker, Vlad, Tucker, & Pelton, 2006, p. 5), and in 2004 the number was roughly two-in-10. These particular skills include designing and creating computer graphics, nonlinear editing, designing Web pages, and photo imaging. Many or most of these graduates in journalism and communication have been trained to do these tasks, and even if they do not go into a related job, the chances are that they may be called upon to use them wherever they find themselves at work.

## Diffusion of Journalism and Media Skill Sets

But even as Web skills and multimedia training are now common in journalism and mass communication programs, these skills are not limited to our field. A number of other opportunities enable virtually anyone access to multimedia training, and journalism is often a part of this training, which is provided through degree and certificate programs, on campus or via distance education, in art, design, gaming, and/or continuing education. The result is a rapid spread of the ability to use multimedia tools and basic journalism skills. In the midst of this convergence, it is helpful to consider the two separately.

Pierce and Miller (2006) studied editors' perceptions of what skills journalism majors needed. A questionnaire identified 20-journalism-specific skills and was based on a survey from the American Society of Newspaper Editors (ASNE):

> The 20 traditional and emerging journalism-specific skills are: How to write efficiently, How to interview, Objectivity, How to develop sources, how to write a lead, how to write a nut paragraph, basics about libel law, how to write about government meetings, how to write about crime and civil trials, how to write a profile, how to copy edit, how to use computer assisted reporting, how to write for a web site, how to write headlines, how to shoot news photos, how to design pages, how to write obits, how to write narratives, how to work with TV on packages, how to report for live TV. (p. 10)

Essentially, scholars report that the most important journalism skills are (1) writing, (2) reporting, and (3) critical thinking (Pierce & Miller, 2006). Somewhere between the list of three and the list of 20, perhaps the following list of seven might cover needed convergent journalism skills:

1. Able to write stories well across platforms (newspaper, television, the Internet) about a variety of topics (government, law, individuals living or dead)
2. Able to write specific journalistic elements (lead, lead-in, headline, tease, etc.)
3. Being objective enough to overcome personal bias, being able to talk to strangers (interview) and cultivate trust (so they can become a source)
4. Being knowledgeable about laws relating to journalism (and avoid lawsuits)
5. Able to capture images (photos as well as video)
6. Able to speak and perform "live" as well as provide narration
7. Able to use computer software to write, edit and produce for print, electronic, and Web pages.

Other studies suggest more emphasis is needed on technical and multimedia skills and that journalists must have a basic working knowledge of the technology (Dupagne & Garrison, 2006; Tanner & Smith, 2007). Additional skills include Web site design, audio/video production, and animation and special effects (Becker et al., 2007a). Courses that cover these skills can be found in any number of programs at universities, colleges, community colleges, and high schools. What was once special is now common.

To stay relevant, textbooks on convergent journalism include exercises to develop technical skills as well as writing and being a good storyteller (Kolodzy, 2006; Quinn & Filak, 2005; Wilkinson, Grant, & Fisher, 2008). But the list of technical skills is becoming lengthy, and can include the following:

1. writing stories of varying lengths,
2. typing (writing with a keyboard),
3. photo taking (and editing them),
4. recording and editing audio,
5. recording and editing video,
6. using a telephone,
7. taking pictures with a telephone,
8. uploading various forms of content to the Internet,
9. downloading various forms of content from the Internet,
10. designing a Web site, and
11. sending/receiving "live" text, audio, or video through the Web (SMS, instant messaging (IM), podcasting, webcam).

This list of needed skills has changed significantly from a generation ago. For example, the demand for peopled skilled in working with ¼-inch audiotape or ¾-inch videotape have all but disappeared since the technology was replaced by digital recording and editing equipment. As new technologies and applications emerge, associated skills will also be added to the list.

## Issues of Identity, Relevance, and Competition from Other Fields

Expanded skill sets are also the trend in nonjournalism and nonmedia fields, resulting in several areas of overlap. Just as convergence brought competition between media, computer, and telephony firms, so now there are competitive opportunities for content creators in almost every field. For example, politics, medicine, law, and education have positions and job titles for workers who create, produce, and disseminate information in an interesting and entertaining manner. With so much overlap, scholars and practitioners of mass media and communication must consider the related issues of identity and relevance.

The first point, identity, deals with the broad function of journalism and mass media. To stay viable, most if not all professionals must identify their "unique selling point"—what they do better than anyone else. Doctors practice medicine. Lawyers represent people in court. Pilots fly planes. Politicians govern and provide leadership. Actors act. Journalists write, report, and tell stories.

But because the tools of the media trade are now common, there has been an explosion in the amount of content that has been created. Content produced by almost anyone can be viewed as a "news story" and can be found almost anywhere. The audiences of the blogosphere, YouTube, and search engines continue to increase while newspaper readership continues to decrease. Younger people in particular report getting their news from nontraditional sources such as comedy programs (Hollander,

2005). Until and unless journalists are recognized—not by themselves, but by the audience—as performing a unique and special role, the issue of identity will remain.

Similarly, because the debate continues over who is or is not a journalist, the relevance or perceived value of journalism and mass media is also unresolved. News media continue to argue that "quality matters," and sometimes this seems to be true. During emergencies and times of crisis, television and cable news networks in particular experience spikes in viewership (State of the News Media, 2004). But such events are relatively rare, and the overall trend for TV and cable news for audiences has been downward (State of the News Media, 2007).

The degree to which audiences care about professionally written and produced journalism is difficult to ascertain. Media owners and managers measure it by readership or viewership. If a story (or a show or a Web site) is not compelling, it will not generate or maintain an audience. The practice of using such imperfect ratings holds true for journalism, advertising, public relations, entertainment, sports, music, government, and a host of other fields that routinely "generate content." If a Web site, news story, or program creates a buzz and brings enough people to it, it is perceived as relevant. Therefore, there is a continuous search for those who can take mundane content and make it "special."

As the demand for those who can take information and make it compelling increases, would-be employees run the risk of becoming undervalued through price wars. Working in television and radio has always been highly competitive because too many people apply for too few jobs. The required skills have always been relatively undefined, enabling journalists to do the weather, athletes to join the announcer's booth, and entertainers to become news anchors.

If specialists in the field of education create multimedia, medical professionals create literature, government employees produce journalism (or "faux journalism"), and legal services include animation and video documentaries, will these present threats or opportunities for media professionals to remain relevant?

One possibility is that this type of "skill set" convergence—blurring the lines between formerly disparate fields—brings some welcome opportunity for those with journalism and mass communication training. The idea that journalism students can compete for jobs in other areas besides media can mean that the field will rise in status on college campuses and in society.

## Convergence Brings Collisions

Of course, there are other possible outcomes. The widening application of journalism skill sets can also mean the professions enjoying higher status in society can co-opt the training of these skills sets. For example, a generation ago it did not appear that media production was part of a career in government, law, or even medicine. Although business schools have long offered courses in "business communication," the emphasis was more toward creating effective business-related correspondence. But business schools have now expanded the range of course offerings to include multimedia and Web design. Since resources on campuses tend to follow enrollments, business schools (which tend to have large numbers of students) may be in a more advantaged position to teach these courses than journalism and communication programs.

Journalism educators must anticipate possible areas of overlap and points of potential collisions between programs brought about by this type of convergence of skill set applications. Five areas of overlap are presented here involving journalism and mass communication and the fields of architecture, law, medicine, education, and government. The reader is encouraged to search and find additional areas that emerge due to the constant shifts in society and its institutions.

## Collision 1: Art and Architecture

Architecture is concerned with the design of buildings, but this field has also expanded into the application of the arts, and visual arts in particular. Architecture firms are routinely providing a wide array of multimedia services to present novel layouts and designs. There is tremendous demand for being able to create and disseminate high-quality images to specification. To meet this need, workers in art and architecture employ high-quality computer and media technology to create valuable content. No longer the domain of media companies, architecture firms provide professionally designed content using animation, still image, video, and QTVR (Quicktime-virtual reality) displays that are available online for potential clients.

For example, architectural firms expect to offer computer-generated images and mock-ups of projects involving cityscapes, skylines, office buildings, the insides of homes. Virtual tours are available online that take the prospective customer inside homes or offices as well as enable a birds-eye view of the outside area, whether it be a home, neighborhood, downtown office, or park and recreational area. This category of content creation, called "VR photography" is aptly demonstrated on a German Web site called "Panoramas.dk." This Web site offers a variety of stunning images in art, architecture, museums, and even the moon and planets.

## Collision 2: Law

In the legal field, there has been tremendous growth in media-related services offered to clients. Law firms commonly use video for supporting evidence and testimony. There is increasing use of three-dimensional (3D) animation to re-create alleged activity at a crime scene. Throughout the United States there are businesses available to do multimedia postproduction work specifically for law firms (see, for example, http://dmoz.org/ and continue to "evidence and presentations.").

The demand for animation is also increasing. To assist in legal work, a number of businesses are specializing in 2D and 3D renditions of case evidence. The range and price for this work can run as low as $200 (very simple indeed), but added complexity and realism can easily run into tens of thousands of dollars and more. Some businesses focus on animation or video legal work, but others offer consolidated full-service video and animation production and postproduction services.

Probably the biggest co-optation of journalism skills and applications is in the area of video production. Video has become an integral part of legal casework and proceedings, used for depositions, settlement documentaries, video wills, "day-in-the-life documentaries," mock trials, insurance fraud, pre-construction video surveys, scenes of incidents, courtroom presentations, and other applications.

As mentioned earlier, a staple of broadcast journalism, the news documentary, is now a common service offered in the arsenal of legal services. Former network news producers and photographers advertise their skills and experience to produce content that uses journalistic styles and techniques.

But this type of work is not open to any ordinary video journalist. While traditional journalists and production specialists have long resisted any type of formal certification, the legal profession has adopted a different approach, resulting in The American Guild of Court Videographers. The AGCV has been established with set rules and guidelines for recording testimony that can be used in legal proceedings. To be a member, videographers must demonstrate knowledge of various rules governing legal evidence and procedures for getting that evidence. Guild members are skilled in multimedia and video production skills as well as the basic interviewing skills needed to help clients produce "video testimonies" that are used in court depositions. According to the guild Web site, members are located in almost every state in the United States as well as Canada and Europe (http://www.agcv.com).

Certification does not end there. The AGCV Web site notes "the vast Majority of our members have gone on to successfully complete all the requirements allowing the AGCV to award them higher designation of "Certified Court Videographer" (CCV)."

Providing ever-increasing levels of recognition and certification is one way to help practitioners justify their relevance and serves as an effective barrier to prevent outsiders or nonmembers from competing. This type of certification and justification has long been missing from journalism-related fields, and it is perhaps not surprising that a number of AGCV members advertise they once worked in broadcast journalism.

Another key point of departure is the degree of specificity in the types of video content that these specialists can capture for clients. According to the "Legal Video" Web site (www.legalvideo.com), "A Legal Videographer produces legal video and is also known as Forensic Videographer, Court Videographer, or Video Court Reporter. They produce Legal Video, Forensic Video, Courtroom Video, or Visual Evidence. The process is known as Legal Videography, Court Videography, or Forensic Videography." These certified video specialists may also set up Web pages (photos, services offered, PayPal setup) and even Web logs.

## Collision 3: Medicine

The medical field has also quietly adopted almost all the skills inherent to journalism and mass media and is now using them in various forms. Generally speaking, there are four areas of application:

**a.** One-way visual images (including photos, video)
**b.** Two-way visual images (doctor-patient video and virtual surgery)
**c.** Computer animation (3D animation and diagrams)
**d.** Medical "news" and information services

**a. One-way visual images.** Medical applications of high-resolution photography have long been established. Microcameras are commonly used with certain types of tests and surgeries (for example, the colon or the larynx).

**b. Two-way visual images.** Interactive video is so common in medicine that Britain's Royal Society of Medicine now publishes the *Journal of Telemedicine and Telecare* (http://www.rsmpress.co.uk/jtt.htm). Medical researchers continue to examine all aspects of the effectiveness of doctor-patient videoconferences. Virtual surgery is becoming increasingly common as specialists are not always able to be in the same geographical location as the patient.

**c. Computer animation.** The medical field also employs computer animation to show various processes using 3D animation. This tool helps specialists better understand how cancers work or the effect of various drugs on the system.

**d. Medical news and information.** Finally, the field of medicine is not just employing expensive multimedia tools and equipment. Medical journalism is a growing subfield within journalism that includes reporting on health or medical-related stories, often published through nationally available health news services. As another example, health maintenance organizations (HMOs) will continue to report on new techniques and services using in-house media, to be distributed to traditional media and the public.

Then there is one final intriguing example of a skill formerly belonging to the media that is fast becoming routine for one group of medical practitioners. It is becoming common practice among anesthesiologists to talk to patients before a surgery to compile a list of their favorite songs. These are prepared in advance and played for the patient through headphones during the surgery—a personal "all-request music marathon" (Music to operate to becomes specialty, 2005).

## Collision 4: Education

Mass media practitioners and academics have long considered educational media as something separate and far removed from what they do. But educational media has proven itself as an innovative and dynamic field for multimedia applications and media convergence, too. Primary and secondary schools have significant resources invested in computer and video technology, and teachers are increasingly urged to use these technologies in the development and presentation of class lessons.

In particular, Teachers of English to Speakers of Other Languages (TESOL) borrow heavily from media to train students to learn English. Clips from Hollywood films and network television programs are digitized and placed on the Web and made part of the course lesson. An online search using keywords "TESOL movie clips" produced over 66,000 links, many discussing the benefits of using these to help non-English speakers learn the language.

Education has also forged ahead in implementing interactive video over the Internet. At all levels, educators are using the technology to facilitate cross-cultural education and learning (Thurston, 2004; Wang, 2004). The technology has been especially effective for language learners (Kinginger, 1998; Wilkinson & Wang, 2007).

Administrators of Internet 2 (a faster, experimental successor to the Internet) have recognized the annual "megaconference" initiatives as a means of demonstrating novel ways of using high-speed videoconferencing to simultaneously link hundreds of schools, colleges, and universities around the world ("Internet2 Announces Winners of the First Annual Internet2 IDEA Awards Program," 2006). Organized by Ohio State University, the day-long event features innovative uses and presentations from kindergarten classrooms to advanced computing research institutions.

One other place where education scholars are adopting media proficiency is in the area of live, on-air interviews. Once almost exclusively the domain of trained professionals like broadcast journalists,live, unedited interviews are now routinely conducted by nonmedia people. Asking whether the interviews are "professionally" done is moot because the Web is now the repository of countless numbers of such interviews, and the only thing that matters is whether there is an audience for it. The evidence seems to be that "broadcast quality" or having a "trained interviewer" does not matter so much anymore.

## Collision 5: Government

A final area to consider that is challenging the media's hold on convergent media skills is government. In the past decade, there have been increasing instances of government-sponsored or -created media. More disturbing is that the content of these media releases is often not identified as such. The blurring of lines between public relations, journalism, and government propaganda has entered a potentially dangerous area. In the past few years, policies such as No Child Left Behind and the Medicaid bill were pushed on the public through professionally crafted content that was thought to be impartial and unbiased journalism. The budget for public relations contracts (which resulted in media releases and reports) during George W. Bush's first term was over $250 million. This was double the amount spent in the previous term by the Clinton administration (Barstow & Stein, 2005).

In Great Britain, there is now the Government News Network (GNN) that works specifically with government personnel to manage news and issue campaigns. The GNN is staffed by people trained in journalism and public relations, and they work specifically to help the government "sell" itself to the people. As the Web site acknowledges (www.gnn.gov.uk), "as part of the Central Office of Information (COI), our teams work with specialists in the publicity, marketing and advertising sectors—to provide you with an all-round communications service" (para. 5).

Perhaps the most telling example of how institutions of government are developing resources to persuade the public is to visit the official U.S. government Web site. As children tend to be less sophisticated and more trusting of authority figures, the U.S. government includes specific information campaigns to present its views directly to children (http://www.whitehouse.gov/kids/).

Government efforts to bypass journalism and media organizations are not limited to the federal government. Local institutions such as police departments have also changed the dynamic, using technology and in-house media skills. A generation ago, the staple of novice reporters was to be assigned the "police beat," where the rookie reporter went to the police station to look through stacks of filed reports. The power of decision making was in the hands of the reporter, who could follow up on anything that seemed interesting. Today that system has changed, ostensibly to make the reporter's job easier, but it may result in a slightly less-informed public.

Recorded police announcements are provided to media with what the police department has determined to be newsworthy. As news organizations become smaller and leaner, time becomes important. The temptation to accept being spoon-fed news items may be too great to resist. But this capability essentially gives editorial control to the police. This practice does not become an issue until something happens the police do not want publicized. If the newsroom deems it to be newsworthy, it might become difficult for the journalist to seize back the role of gatekeeper and watchdog.

The technologies are also being used to create new forms of police news and information. Many residents in New York City no longer need rely on the media for crime news. The NYPD Web site regularly provides podcasts with the latest local police news. Such a service may sound like journalism, but it is still sponsored, controlled, and presented by the public relations arm of a government institution.

## Areas of Overlap, Areas of Divergence

The core issue is whether media skills are an art, a craft, or simply a set of tools. While the field may have begun as a set of simple practices with relatively simple tools—a vocation—perhaps it has evolved to become a hybrid of high-minded idealism and low-minded sensationalism. Journalists have long been criticized for a seeming fondness for tragedy and scandal, and at the same time (sometimes by the same person) lauded for making a positive difference in the community. Due to proximity, journalists have strained to be part of the dialogue between the powerful and the weak, the elected and the electorate. Journalism—of all fields—is forever torn between upholding the ideals of the First Amendment freedoms and doing whatever is needed to grab the largest possible audience in order to pay for itself.

The next set of distribution technologies will influence how this content is to be consumed because it will introduce new options regarding the economics of content consumption. The possibilities include a pay-per-click model, pay-per-view, a weekly or monthly fee, copyright royalty charge, or some other hybrid involving advertising. Although it is difficult to predict which form will become dominant, many key elements are in the hands of Web developers. Oversight organizations and governments will help decide the issue, either by policy and regulation, or by simply looking the other way and allowing market forces to exert themselves.

## Media Skills and Tools

As mentioned earlier in this chapter, journalism and mass media have to know the business basic "unique selling point." As the tools and skills are increasingly used in other areas of society, the argument of ownership becomes more difficult to make. The question may return to that of Identity. Journalists—and pretty much everyone else—should know how to "do" multimedia. They may not need to know everything unless they wish to specialize, but it is advisable to pick these up as soon as possible:

Multimedia skills: writing, typing (writing with a keyboard), photo taking (and editing them), recording and editing audio, recording and editing video, using a telephone, taking pictures with a telephone, 2D animation, 3D animation, uploading various forms of content to the Internet, downloading various forms of content from the Internet, designing a Web site, and sending/receiving "live" text, audio, or video through the Web (SMS, instant messaging, podcasting, webcam).

As the tools become second nature, they take their appropriate place as a foundation laid before undertaking journalism practices. While multimedia training can occur at earlier ages, there are few substitutes for the maturity and wisdom that comes with life experience. The tools of journalism are the critical thinking and interpersonal skills—dealing with people—that journalism educators have championed since the beginning.

Journalism skills: reporting, editing, critical thinking, listening, asking questions, fact finding, fact checking, storytelling, distilling a set of facts into a logical and coherent story, quick thinking, resourcefulness, dogmatism, politeness-yet-tenaciousness, maturity, being well-read, being ethical/moral, awareness of legal protections and First Amendment goals, public speaking, performing, and presenting information.

Traditionally the press has not done well embodying these practices, resulting in ongoing criticisms by the public. Outrage over "treasonous" reporting has been expressed from the American Revolution to the invasion of Iraq (Pember & Calvert, 2007/2008). The matter has become further complicated with the rise of citizen journalism, the blogosphere, media pundits, and talk radio. For every "Dan Rather incident" there are attempts to "swift-boat" political candidates. When bloggers and columnists are secretly hired by government to push unpopular programs, all media become tainted. Discovering that media play favorites and make mistakes has empowered everyone to think any story can be a sham or matter of personal opinion.

## Future of Journalism and Media?

Convergent communication technologies are blurring the boundaries among a number of fields, vocations, and professions, including journalism and mass media. In the 1990s, a number of scholars trumpeted that "content is king!" yet few anticipated the coming of YouTube. Whether it is idle doodling, juvenile poetry or the serendipitous recording of a cat striking piano keys, all can be considered "content" and even "news" for hundreds of thousands of Web users.

Supporting this trend, media education has become ubiquitous. High schools and even elementary schools provide broadcast and journalism training, and there seems no limit to how young a child can be to get on-air experience (Kohlenberg, 2003). For journalism and mass communication to survive as professions, crafts, or vocations, it seems some setting of standards is needed. But credentialing and enforcing those standards has been an insurmountable problem to date.

It is vital that citizen journalism, the blogosphere, and traditional media and journalism programs come together to sort this issue out. But it may take some doing, and battle lines have already been drawn. Calls for standards by the industry are seen as mere attempts at control. A December 2007 op/ed piece in the *New York Times* by several deans of noted university journalism programs expressed concerns about "opinion journalism" and the need for standards in the new media environment. They stated "we do not believe that the market can be absolutely trusted to provide the local news gathering that the American system needs to function at its best" (Hart et al., 2007). But a response by blogger Steve Boriss (2007) likened the views of the "so-called scholars" to that of religious cults and Karl Marx.

Two weeks earlier there was similar outrage among citizen journalists when University of Georgia journalism professor and former network correspondent David Hazinski called for some form of licensing and standardization.[1] An aggressive

1. The article titled "Unfettered 'citizen journalism' too risky," was an op-ed in the Atlanta Journal-Constitution, published in December, 2007, and is no longer available online. As discussed in Chapter 5, losing information in this manner is one of the weaknesses of Internet distribution.

response by Dan Gillmor at the Center for Citizen Journalism (and the posts by other citizen journalists) attacked the notion and argued the traditional view long held by big media and journalism professors alike, that of self-regulation:

> The media industry and journalism educators do have a valuable role to play in all this. It's to teach media literacy for a media-saturated world. That is not about regulation or do-it-this-way standards. It's about helping media audiences and creators alike to understand how media and persuasion work. For journalists, citizen or otherwise, it is very much about principles and ultimately honor. For the audiences, we need to instill deep, critical thinking and a solid grasp of media techniques.
>
> Let's regulate ourselves to end up with a diverse, vibrant journalistic ecosystem that serves and informs us. (Gillmor, 2007).

It is past time for those of us formally trained in "the business" to expand our scope beyond feeding entry-level workers to newspapers and broadcast companies. Those of us who teach the skills listed earlier can take comfort that we have essentially accomplished the "big goal." Instead of training several thousand workers for the media, we have helped in the transfer of reporting and multimedia skills to millions who are transforming every sector of every society around the world. As we have so often said to our students, our ultimate goal is to bring free and open debate to society. Through convergent media technology, that goal may now be accomplished globally.

As we implement this type of national (and global) program, we will begin to better understand what professional communicators have in common and what sets them apart from the rank and file. If we do not, we may find that medicine, business, or computer science does a better job training multimedia specialists than we do, and they will shift funding accordingly.

## Implications of the Diffusion of Convergence for Journalists

Convergence is the tangible result of a postmodern age where formerly diverse fields and disciplines now collide and intersect with each other. Over time this trend can be expected to continue, expanding the opportunities for researchers in communication and media. While nonmedia fields are only now employing the tools of communication in a systematic way, media and communication workers have a century head start.

Message creation strategies include the use of language, story structure, and organization. Journalists are experts in identifying puffery and jargon and trained in the art of weaving statistics, personal observations, and the experiences of others to create dramatic and interesting accounts that can have deep impacts upon readers.

People trained in media practices are also professionals in the visual aesthetics of message design. For print messages, this includes use of color, framing, white space, fonts (lettering), spacing, and symbols. For video, principles of continuity, framing, lighting, visual sequences, editing, and pacing are standard.

Students trained in journalism and mass communication should also be best at public speaking and comfortable being in "the public eye." Specific skills include knowing how to dress, how to act, how to gesture, and how to speak in front of groups or on camera.

Finally, media professionals should be expert at asking questions that do not insult or anger people. They should be personable and intuitive, knowing principles of psychology and interpersonal communication to be able to interpret nonverbal communication (such as when a person is unwilling or unable to speak about a matter). Furthermore,they should be knowledgeable about laws and ethics regarding libel, privacy, access, and other relevant areas of press freedom.

## Conclusion

The diffusion of convergence has tremendous implications for scholars and practitioners in media and mass communication. As the amount of content creation increases across disciplines, fields, and all of society, the number of people who are trained in convergent journalism practices also increases.

In addition, the diffusion of convergent content creation tools results in the analogy that "a rising tide lifts all boats." The result of this proliferation of multimedia content will be seen in a number of ongoing trends—law firms will increasingly employ skilled workers in multimedia, government will increasingly hire and use experts in public relations, and educators will use media to facilitate all areas of learning.

Music will continue its shift from art to commerce. For example, the craft of jingle writing has shrunk to a small but lucrative niche because it is easier (and cheaper) to use an existing, familiar pop song than to write your own.

Performers already have to think in terms of repurposing themselves. Actors/actresses who are young and attractive must find ways to stay current and popular as they age and start to lose their youth appeal. Options may be to write children's stories (Madonna, Katie Couric, Billy Crystal, Jerry Seinfeld, Whoopi Goldberg), launch fashion products (Britney Spears, Naomi Campbell), own restaurants (Alice Cooper, Dan Akroyd) or engage in other ventures (Ozzy Osbourne, though still involved in music, has made another career through film, television, and theater). For most of these performers, their celebrity opens the possibility of a second career in politics or media. Likewise, media personalities, politicians, and athletes can benefit by appearing in entertainment programs (playing themselves on television or in a film).

Journalists individually and as a group must determine how the field may remain relevant in the 21st century as its primary product—the "news story"—

competes with similar-looking messages from other fields. Journalists may increasingly experiment with new forms of storytelling (as they are), injecting principles of entertainment to boost the informational aspects. In addition, the economics of journalism continue to change, and the value of a multiskilled journalist remains in flux.

Experiments in converged journalism teaching and practice are still tentative; "employee reductionism" (one-person-does-everything-across-all-media-platforms) may not result in the best journalism product. Practitioners and scholars alike have to better position what they do in order to survive in an age of converged media. Simply narrowing the focus to "hard news" is not a solution. Expanding the definition of news to include almost everything has also not been helpful (celebrity gossip is the crack cocaine of journalism).

Perhaps we can take a page from history and avoid the mistake of the railroads and the telegraph. The railroad companies did not see themselves as providers of transportation and did not embrace later innovations in transportation (Beniger, 1986). Companies skilled only at railroad operations found themselves increasingly irrelevant and marginalized. Similarly, the telegraph firms (in the United States, and especially in Europe) did not see themselves in the consumer communication field and rejected the fledgling telephone device (Czitrom, 1982).

Therefore, journalism scholars and practitioners must reinvent themselves in the broadest terms professionally, and aggressively move into those areas. It would be helpful for us to distinguish between journalism process (unique) and multimedia skills (common), even as the two become intertwined in daily practice. To do so would help in setting protocols, practices, and standards by shifting the emphasis toward the principles of message creation. These principles should result in the creation of compelling information that is effective and popular. Journalists should be best able to specifically tailor content for each platform. In this manner journalists can remain relevant, viable, and important in the age of media convergence.

## References

Barstow, D., & Stein, R. (2005, March 13). Under Bush, a new age of prepackaged television news [Electronic version]. *The New York Times*. Retrieved December 11, 2007 from http://www.nytimes.com/2005/03/13/politics/13covert.html?_r=1&oref=slogin.

Becker, L., Vlad, T., & McLean, J.D. (2007a). 2006 Enrollment report: Enrollments level off: Online instruction now routine. *Journalism and Mass Communication Educator, 62*(3), 263–288.

Becker, L., Vlad, T., & McLean, J.D. (2007b, November). Annual survey of journalism and mass communication graduates. *AEJMC News, 41*(1), 1, 6–9.

Becker, L., Vlad, T., Tucker, M., & Pelton, R. (2006, November). 2005 Annual survey of journalism and mass communication graduates. *AEJMC News, 40*(1), 1, 4–7.

Beniger, J. R. (1986). *The control revolution: Technological and economic origins of the information society*. Cambridge, MA: Harvard University Press.

Berlo, D. K. (1960). *The process of communication*. New York: Holt, Rinehart and Winston.

Boriss, S. (2007, December 23). *New group forms: "J-School Deans Against a Free Press." The future of news: A vision of the future + commentary on developments*. Retrieved December 28, 2007, from http://thefutureofnews.com/2007/12/23/new-group-forms-j-school-deans-against-a-free-press/

Burnett, R., & Marshall, P.D. (2003). *Web theory: An introduction*. London: Routledge.

Czitrom, D.J. (1982). *Media and the American mind: From Morse to McLuhan*. Chapel Hill, NC: University of North Carolina Press.

Dupagne, M., & Garrison B. (2006). The meaning and influence of convergence: A qualitative case study of newsroom work at the Tampa News Center. *Journalism Studies, 7*(2), 237–255.

Gillmor, D. (2007, December 13). Needed: Regulation to prevent journalists-turned-professors from embarrassing themselves. *Center for citizen media blog*. Retrieved December 29, 2007 from http://citmedia.org/blog/2007/12/13/needed-regulation-to-prevent-journalists-turned-professors-from-embarrassing-themselves/

Hart, R.P., Jones, A.S., Kunkel, T., Lemann, N., Lavine, J., Mills, D., Rubin, D.M., & Wilson, E. (2007, December 22). A license for local reporting [Op-ed contributors]. *New York Times Online*. Retrieved December 28, 2007 from http://www.nytimes.com/2007/12/22/opinion/22lemann.html?ex=1356066000&en=91281da561ae3cbf&ei=5124&partner=permalink&exprod=permalink

Hollander, B. (2005). Late-night learning: Do entertainment programs increase political campaign knowledge for young viewers? *Journal of Broadcasting & Electronic Media, 49*(4), 402–415.

*Internet2 announces winners of the First Annual Internet2 IDEA Awards Program* (2006, April 20). [Awards recognize innovators of advanced network applications in the research and education community, Ann Arbor, MI]. Retrieved December 11, 2006 from http://digitalunion.osu.edu/megaconference/IDEA.html

Kinginger, C. (1998). Videoconferencing as access to spoken French. *The Modern Language Journal, 82*(4), 502–513.

Kohlenberg, L. (2003, Winter). *Practicing journalism in elementary classrooms: 'Could eight-, nine- and 10-year-olds, who had trouble sitting still for more than 10 minutes at a time, develop the skills to become reporters?'* Nieman Reports. Retrieved February 2, 2008, from http://www.nieman.harvard.edu/reports/03–4NRwinter/39–42V57N4.pdf

Kolodzy, J. (2006). *Convergence journalism: Writing and reporting across the news media*. Lanham, MD: Rowman & Littlefield.

*Music to operate to becomes specialty*. (2005, October 4). *CNN.com*. Retrieved October 6, 2005 from http://www.cnn.com/2005/HEALTH/10/04/djs.of.the.or.ap/index.html

Pember, D.R., & Calvert, C. (2007/2008). *Mass media law* (15th ed.). New York: McGraw-Hill.

Pew Internet and American Life Project (2004, February 29). *Content creation online*. Retrieved March 3, 2004, from http://www.pewinternet.org/reports.toc.asp?Report=113

Pierce, T., & Miller, T. (2006, October 19–21). *What editors want from new journalists: Converging skills or old school standards?* Paper presented at the conference Convergence and Society: Ethics, Religion, and New Media, Columbia, SC.

Podhoretz, J. (2005, May 12). Mass-media meltdown. *New York Post*. Reprinted in *Shoptalk* newsletter at www.tvspy.com

Quinn, S., & Filak, V. (2005). *Convergent journalism*. Boston: Focal Press., and *State of the news media, the: An annual report on American Journalism*. (2007). Retrieved December 17, 2007, from http://www.journalism.org

*State of the news media: An annual report on American Journalism (2004)*. Retrieved December 17, 2007 from http://www.journalism.org

Tanner, A., & Smith, L. (2007). Training tomorrow's television journalists: In the trenches with media convergence. *Electronic News, 1*(4), 211–225.

Thurston, A. (2004, August-September). Promoting multicultural education in the primary classroom: Broadband videoconferencing facilities and digital video. *Computers and Education, 43*(1–2), 165–177.

Wang, Y. (2004). *Supporting synchronous distance learning with desktop videoconferencing, language learning and technology, 8*(3), 90–121. Accessed on July 11, 2007, from http://llt.msu.edu/vol8num3/wang

Wilkinson, J.S., Grant, A.E., & Fisher, D. (2008). *Principles of convergent journalism*. New York: Oxford University Press.

Wilkinson, J.S., & Wang, A.L. (2007). Crossing borders: How cross-cultural videoconferencing can satisfy course goals in dissimilar subjects. In P. Tsang, R. Kwan, & R., and Fox (Eds.), *Enhancing learning through technology*. London: World Scientific Publishing.

*YouTube-ification of politics, the: Candidates losing control* (2007, July 18). CNN.com. Retrieved July 18, 2007, from http://www.cnn.com/2007/POLITICS/07/18/youtube.effect/index.html

CHAPTER 7

# Culture, Conflict, and Convergence

## A Theoretical Discussion of Group-based Identity and Bias Reduction in a Converged Newsroom

Vincent F. Filak

At the first South Carolina Conference on Convergence in 2002, some concerns regarding convergence included how to best to define convergence, how to technologically deliver the content, and how to create cooperative partnerships between television and print journalists who had long viewed each other as competition. In the years that followed, the technology has improved and definitions of convergence are taking shape (Lowrey, Daniels, & Becker, 2005). One of the more difficult challenges remains getting journalists from varied backgrounds to work well together.

Reports from the industry in the early part of the decade would indicate that much of the cultural infighting has been ameliorated (Callinan, 2001; Murphy, 2002; Robins, 2000). However, some critics note that while much of the positive spin about convergence is coming from the top managers at converged news operations, problems still exist between frontline journalists (Glaser, 2004). Runett (2004) noted that even for some media executives, full convergence remains "a mythical Emerald City." If convergence is to move beyond some simple cross-promotion and an uneasy peace, he states, executives must be willing to conduct "a sandblasting of prevailing newsroom culture" (p. 10).

While this choice of words is an abrasive one, both literally and figuratively, it makes the point of this piece. Culture has long been the bedrock of newsrooms and in most cases an impediment to large-scale changes. A 2001 study from the Media Management Center at Northwestern University found that 80 percent of newsrooms had a defensive culture (Nesbitt, 2001). Defensive cultures resist change and eschew collaboration outside of their cultural group. Thus, early attempts to put print and broadcast journalists in the same newsroom in hopes that a cultural osmosis would occur met with resistance and left some scholars wondering if this barrier could ever be overcome (Killebrew, 2001).

Ongoing academic and professional discussions highlight the ways in which culture has become an impediment to newsroom convergence (Silcock & Keith, 2006; Singer, 2004a; Tompkins, 2001). Still, there has been little discussion about how to ameliorate tension between journalists and even less that has attempted to provide a larger theoretical perspective regarding this phenomenon (see Singer, 2004b, for an exception). Some of the professional literature stresses the importance of teamwork (Gabettas, 2000) or seeking out "wins" (Callinan, 2001). These approaches are important in seeking ways to make convergence work, but without understanding precisely what makes for a "win" or successful "team," each new convergence operation will press forward into the unknown, uncertain whether its model of convergence will work. Without a larger theoretical framework to guide convergence practitioners, there is no way to know why one group may peacefully merge newsroom cultures (Murphy, 2002) while another ends up with a war (Hammond, Petersen, & Thomsen, 2000).

To add a framework that helps our consideration of convergence, this study applied group-based behavior theories to the newsroom workplace (Gaertner & Dovidio, 2000; Hewstone & Brown, 1986; Tajfel & Turner, 1986). Prior work on newsroom culture has found that journalists (Filak, 2004; Singer, 2004a) and journalism students (Filak, 2006; Filak & Thorson, 2002; Hammond et al., 2000) have group-based social identities that can create barriers to convergence efforts. While social psychologists have found numerous ways to decrease bias between diverse groups, such work needs to be applied to newsrooms.

The issue of newsroom culture and its impact on convergence will be explored by analyzing relevant theoretical perspectives on group-based behavior. Theories in the intergroup relations field can provide an understanding of how newsroom culture works and offer practical ways to diminish culture clashes.

## Culture and Convergence

Journalists take pride in being able to report freely and without fear. Therefore, it is perplexing that culture and control are among the key hindrances to embracing convergence. News practitioners studied throughout the years have rebuffed claims that they were controlled within the newsroom (Schlesinger, 1978). But most studies on this issue report that journalists have been inculcated with a series of norms and values that dictate their behavior (Fee, 2002; Sigelman, 1973; Westin, 2001). Erickson, Baranek, and Chan (1987) succinctly noted that newsroom culture is embedded in the day-to-day actions of its members, as opposed to being overtly impressed upon journalists through rules and regulations.

The seminal study of newsrooms and social norms was Warren Breed's 1955 study, titled "Social Control in the Newsroom." Breed documented how staffers policed one another, how editors reshaped writers' attitudes and how newsrooms perpetuated a culture among their members. A former reporter himself, Breed remarked almost 50 years after his original work was published that he recognized much of the social policing that he studied in 1955 had been present in his own newsroom back in Oakland (Reese & Ballinger, 2001). For example, he noted how he was advised by an older reporter to cover a patriotic event in a certain fashion. A year later, he in turn advised a new reporter to do the same.

As Breed (1955) noted, newcomers are more likely to become part of the newsroom culture than they are to change it. Those who already work within a system tend to resist new ideas for a variety of reasons (Giles, 1991), including a loss of control, excessive uncertainty, and fear of future competence. Fee's (2002) study of newsroom culture and values found that over a three-year period, more than one-fourth of a newsroom's staff had turned over. However, his work found no significant differences in the cultural norms from the time of the first set of data to the second set, collected three years later. The culture, he argued, had subsumed the individuals who joined the newsroom.

With culture being an entrenched aspect of both print and broadcast newsrooms, the importance of addressing its role in convergence cannot be understated. (For the purpose of this piece, convergence will be defined as Quinn [2005] defined it: "doing journalism and telling stories using the most appropriate media." [p. 7].) Through interviews with convergence practitioners, Shearer (2003) found that the most common barriers to convergence are related to personal interactions rather than technology or other, more practically oriented variables. In her examination of the sociology of convergence, Singer (2004a) found that journalists have resisted change from traditional newsroom settings to converged news operations. She notes that cultural barriers exist between print and broadcast journalists and that the competitive nature of the news business can get in the way of even the simplest convergence efforts. Silcock and Keith (2006) also found that convergent media operations should brace for conflicts, as differences in language, culture and values are likely to impede efforts to work with others from different media outlets.

Hammond and Peterson (Hammond et al., 2000; Petersen & Hammond, 1999) found similar issues with their analyses of convergence efforts at Brigham Young University. Multiple-year projects found that cultural differences were the largest determining factor in students' distaste for the operation. Students noted in both cases that there was a good amount of distrust, as well as a general dislike for each other and the convergent experience. Filak and Thorson (2002) provided the theoretical underpinning for this culture clash in their study of print and broadcast students, finding that college students had been taught to operate in medium-based groups, which led to the formation of intergroup bias.

Filak (2004) found that professional print and broadcast journalists also held an intergroup bias. Journalists from each medium were more likely to show favor to their own career choice and medium while failing to recognize the value of the other's career or medium. The presence of bias was further confirmed through an experiment in which journalists from each medium were asked to rate a convergence plan. The source of the message was manipulated to be either a group of print journalists, a group of broadcast journalists or a mixture of both print and broadcast journalists. The results suggest that when the individual thought the plan was the work of the outgroup (e.g., when print journalists were presented with a plan they believed to have been created by broadcast journalists), they were significantly less positive in their responses.

Collectively, these studies indicate journalists at all levels of development are attached to the culture of their medium, and that intergroup bias is at the root of at least some of the intercultural conflicts researchers have noted. Thus, possible solutions to these problems might be found by mining the intergroup literature, beginning with an examination of Social Identity Theory.

## Social Identity Theory

Social Identity Theory (SIT) posits that individuals develop and retain multiple representations of themselves and define themselves primarily based on their group memberships. According to Tajfel (1978), social identity is "that part of an individual's self concept which derives from his knowledge of his membership of a social group (or groups) together with the value and emotional significance attached to group membership" (p. 255). The cognitive (knowledge), evaluative (value) and affective (emotional significance) components of this definition provide a basis for understanding an individual's desire to cling to group attitudes and espouse group beliefs.

Scholars posit that the reliance on group-based identities drives individuals to be biased toward fellow group members and biased against those in a competing social group (Tajfel & Turner, 1986). Studies of intergroup bias have reported that individuals are more likely to cooperate with ingroup members (Kramer & Brewer, 1984), allocate more resources to ingroup members (Tajfel, 1969), explain away mistakes made by ingroup members (Pettigrew, 1979) and engage in behavior that reinforces notions of ingroup superiority. They are also more likely to negatively stereotype outgroup members (Rothbart, 2001).

At the root of these conflicts and the ability to diminish them has been the view that contact will allow for improved relations among competing groups (Allport, 1954). Contact also has been viewed as a panacea for improved relationships between print and broadcast journalists in a converged newsroom (Silcock & Keith, 2006). However, the mere presence of journalists from another medium has been linked to bias, distrust and dislike (Hammond et al., 2000). The issue of contact appears to be a valuable one, in the sense that contact fosters familiarity. Familiarity, in turn, moves the players from a group-on-group dynamic to one in which they operate as individuals (Miller, Brewer, & Edwards, 1985). However, contact alone will not establish new social norms, nor will they foster a more collective sense of identity (Hewstone & Brown, 1986).

What will aid in the improved relations among group members, such as print and broadcast journalists, is high-quality contact (Gaertner & Dovidio, 2000). Quality contact includes events in which participants offer personal disclosure, demonstrate skill that is valued by the larger collective or provide information, either formally or informally, that disconfirms previously held stereotypes (Ensari & Miller, 2002). Singer (2004a) noted that journalists in converged newsrooms who got to know each other well through social contact and news-based activities garnered more respect for each other. One newspaper reporter in Singer's study noted that he had previously seen television journalists as "hair spray, bow ties, vapid airheads" (p. 848). Activities that allowed the newspaper journalist to see that broadcast journalists held professional skills that were different from his own and yet valuable to news consumers gave him a chance to see that his view of broadcasters was incorrect. Other converged operations reporting success noted that activities promoting contact within the newsroom were helpful in tension reduction (Dyer, 2002; Murphy, 2002). The activities included working together on large projects as well as personal activities where people could discuss their concerns about converging operations.

## Allport's Three Points

Bias also can be eliminated or at least reduced if several other preconditions are met. Allport's (1954) seminal work on intergroup relations found three keys to diminishing bias that remain almost unaltered 50 years later.

First, the groups must be of equal status, both in practice as well as perception. If there is an imbalance when two groups become one, the lower-status members will tend to cling to their group identity (Terry, Carey, & Callan, 2001). Newsrooms attempting convergence are already finding this point to be a challenge. For example, Brahm Resnik, a member of the convergence team in Arizona, noted that print people were required to learn broadcast terms because it was more likely that newspaper reporters would be required to provide material for television. However, broadcast reporters were not required to learn print terminology because pulling broadcast journalists from their television assignments was considered a waste of resources (Silcock & Keith, 2006).

This approach, in which one organization is required to socially acquiesce to another, diminishes the likelihood of equal status among the groups and can make one group feel disadvantaged in the process. Singer's (2004b) study of converged news outlets reinforces the notion that equality is a key to a positive convergence environment. Several subjects in her study noted "it has to go both ways" between print and broadcast for convergence to work. Newspaper journalists in particular said they felt slighted because the television journalists "don't do a damned thing for the newspaper" (p. 10). Pay disparity between print and broadcast reporters can also create tension in the newsroom (Tompkins & Colon, 2000). While these items may be resolved in specific convergent news operations, convergence practitioners should remain aware that these social, financial, and cultural imbalances can undermine convergent efforts.

Second, the groups must be engaged in cooperative rather than competitive activities. Miller, Brewer, and Edwards (1985) found many efforts to promote intergroup activities failed because competition undermined the participants' willingness to work together. Cooperative activities that allow individuals to demonstrate their competence on a valued portion of the task has been shown to diminish bias and promote enjoyment of that activity (Dovidio, Gaertner, & Validzic, 1998).

Competition has often been cited as a primary reason print and broadcast journalists react unfavorably to convergence attempts (Gabettas, 2000; Singer, 2004a). The prospect of "scooping" the competition, or getting the news out first, has long been a measure by which journalists assess their work (Downie & Kaiser, 2002). When the *Dallas Morning News* and WFAA began convergence discussions in early 1997, the mood in the newsroom was one of distrust, mainly because each looked at the other as their competition (Murphy, 2002). Gabettas (2000) recalled an incident at a converged news operation where a broadcaster was tipped on a story. She was asked to write it up for the newspaper, but expressed reservations about being asked to "scoop" herself. Bulla (2002) also found this concern among journalists involved in converged environments, who said that by allowing their stories to run on another medium first, they were scooping themselves. The concern about which medium gets the story out first remains a key sticking point among print, broadcast, and online journalists in converged news operations. In many cases, it is the newspaper journalists whose voice the loudest concerns because they have the fewest opportunities to break a story in their medium.

Finally, the support of institutional authorities must be present if group collaboration is to occur at an optimum level (Gonzalez & Brown, 2003). Failing to have an authority supporting the collaborative effort signals to participants that their work has no meaning or is frowned upon. While it is not suggested that authorities impose their will upon the groups in order to perpetuate collaboration, a vested interest and supportive environment helps further the cause. George Rodriguez of the Belo Corporation's news division reinforced the important role leadership plays in supporting convergence, noting the value of repeated recognition of convergence successes by publishers and top editors (Shearer, 2003). Celebrating any incremental movement toward convergence will demonstrate that individual efforts are valued and that authorities will continue to support those activities.

## The Three Main Models

Although this set of precursors is valuable in helping to reduce bias, it is not a panacea for change. From the massive number of studies conducted on intergroup bias and social identity, three dominant paradigms have emerged to diminish the negative effects of these social linkages: the Decategorized Contact Model (DCM), the Common Ingroup Identity Model (CIIM) and the Mutual Intergroup Differentiation Model (MIDM). Each of these approaches will be considered in turn and evaluated in terms of media convergence.

### Decategorized Contact Model (DCM)

The Decategorized Contact Model (DCM) proposed by Brewer and Miller (1984) argues that individuals' awareness of their own group identity is enough to foster group-based biases. Support for this approach has been found in studies of how groups allocate resources (Billig & Tajfel, 1973) and the placement of positive and negative attributes on groups (Ashburn-Nardo, Voils, & Monteith, 2001). In each of these cases, simply being aware that the individual was parsing out a positive and negative outcome to one's own group and any other group was enough to foster bias. The perception of bias held even when it was unclear as to the status and competitive nature of the other group involved.

This model suggests that the best way to ameliorate intergroup bias is to break down the group identities and form more personalized relationships. Disclosure has been found to be a strong predictor of bias reduction (Ensari & Miller, 2002) and has been demonstrated to be far more valuable than contact alone in reducing bias. By offering an insight into important individualizing actions and beliefs, members from opposing groups are allowed the opportunity to disconfirm previously held stereotypes. Seeing each other as individuals, rather than representatives of groups, will facilitate the de-homogenization of the outgroup and thus help eliminate stereotyping behavior (Brewer & Miller 1984).

This model also suggests that bias will be reduced when task-specific qualities are emphasized rather than group-based identities (Miller, Brewer, & Edwards, 1985). To that end, the valued quality each member brings to the group ("Tommy is here because you need someone who is good at math to complete this assignment") will likely be more conducive to bias reduction than what group that individual represents ("Tommy is here because you need another boy in this group"). Research in this area suggests that interpersonal interactions among competing group members will prevent social categorization by participants and thus decrease identity salience (Bettencourt et al., 1992).

The *Arizona Republic* and KPNX-TV have forged a convergence partnership that reflects this model. The two news operations have contributed individuals to form a convergence team that reports for both the newspaper and the television station (Callinan, 2001). This has become a convergence "beat" of sorts, comprised of four people committed to improving the news operations' convergence efforts. America (2002) reported that initial convergence efforts between the *Quad City Dispatch* and WQAD-TV were based on using print "experts" to offer commentary as well as sharing a few individuals between the two news outlets. Other operations have also cited small teams of specific individuals cooperating as their main entry into convergent journalism (Dyer, 2002).

While this approach has promise in small-group interactions, several key drawbacks must be considered. First, it is almost impossible to fully remove group-based identities, especially those that are deeply entrenched in the individual. Even when the categories are not so inextricably intertwined with the individual's identity, this approach fails to acknowledge the cognitive value of using categorizations as mental shortcuts (Gonzalez & Brown, 2003). Years of research in schema theory have demonstrated that individuals seek ways to examine consistent patterns of behavior and draw conclusions from them (Schank & Abelson, 1977). This prevents individuals from having to relearn everything each time they engage in an action, ranging from turning on a light to driving a car. It is impossible to strip entirely from the human mind the desire to place items in categories. Even more, if decategorization does take place, it limits the ability to generalize beyond the individually based experiences (Hornsey & Hogg, 2000).

The last point illustrates one of the potential hazards for convergent news operations that take the DCM approach. In creating a small group of convergent journalists, it is possible that the others in both newsrooms will view these individuals as not part of either the print or broadcast newsroom. Thus, positive intergroup experiences between participants will more likely be limited to those individuals making up this small cadre.

## Common Ingroup Identity Model (CIIM)

The Common Ingroup Identity Model (CIIM) has been mainly advanced by Gaertner, Dovidio, and their colleagues (Gaertner & Dovidio, 2000; Gaertner et al, 1993; Gaertner et al., 1999, 2000). The CIIM operates exactly opposite the DCM. As its name implies, the CIIM seeks to create a common ingroup among previously competing factions. This will allow the transition from "us" and "them" to a more inclusive "we." This more inclusive identity not only reduces bias among the individuals present, but also has the potential to extend the benefits previously afforded only to ingroup members to others who had previously been part of an outgroup.

CIIM is rooted in some of the earliest work on intergroup conflict, namely, Sherif and colleagues' work (Sherif et al., 1961) with the Robbers Cave Experiments. In those studies, boys from around Oklahoma were brought to a campground and randomly assigned to one of two groups. The boys in each group quickly formed identities, assigned tasks and agreed on group norms. When the two groups engaged each other in competitive camp activities, bias was exacerbated and manifested through outgroup derogation and ingroup promotion. When the groups were brought together to engage in essential tasks that neither group could accomplish on its own, previous identities were abandoned and the boys worked cooperatively.

In examining this disbanding of group identities, Gaertner and Dovidio (2000) argue that the types of activities these boys participated in can be viewed as "superordinate," thus drawing "us" and "them" into a more collective identity. Superordinate tasks are one of the ways in which the CIIM proposes galvanizing the more inclusive identity. A brief look at the cooperation between the *Arizona Republic* and KPNX-TV offers an example of the benefits of applying this theoretical model. The convergence effort focused on "wins" for both sides (Callinan, 2001). The wins included larger projects, including some civic-journalism collaborations, that neither the newspaper nor the television station could do well on its own. Thus, in placing the groups in a cooperative environment with a superordinate task, both groups found benefits in collaborating. Duke Maas, a convergence leader in Tampa Bay, used superordinate goals when he explained that by working together, print, broadcast and online staff would tell better stories and operate more efficiently (Stevens, 2003). In a convergence symposium, Mindy McAdams, a professor at the University of Florida, also noted the importance of moving converged news operations toward a superordinate identity:

> The people—meaning print and broadcast reporters, photographers, editors and producers—Should not say "my station, your newspaper["]; they should say ["]our news operation, our TV news, our paper our website." (Glaser, p. 4)

Other approaches include the use of a common fate (positive or negative), making a secondary identity salient that is more inclusive, and mutually valued contact situations. A revision of the CIIM (Gaertner et al., 2000) argues that in cases where previous identities cannot be abandoned outright, the model can rely on a "dual identity" approach wherein individuals can still view themselves as members of both distinct units. For example, parents and children can view themselves as distinctly different groups while still acknowledging their collective identity as a family. In that regard, convergent media operations could reassign individuals to areas of coverage that place them in direct contact with others who cover similar stories for different media. By focusing on an area of specialization, journalists could refocus their energy away from their medium-based identities (print versus broadcast) and more toward promoting their collective efforts in their area of expertise (crime reporters versus education reporters). In this scenario, a "win" is then defined in the traditional format of a scoop, but a scoop for the team. In addition, positive esteem can be attained for the group in regard to things that are still medium-based, such as if the story leads the newscast or lands on the front page of the newspaper. In this case, individual journalists can receive positive distinctiveness within their convergence team. At the same time, the goals of the individual journalists are not in conflict with the larger convergence team goal. Realigning the identity of comparison among journalists would make competition a valuable asset among the journalists within a newsroom, which is akin to the way many current news operations work.

This reliance on subgroup identities allows the CIIM to sidestep much of the criticism levied against it, namely, that the attempt to homogenize groups in this fashion would meet with strong resistance by group members (Brewer, 2000). However, in making this adjustment, the model moves away from its original promise of a larger "we" and moves closer to the model proposed by Hewstone and Brown (1986): the Mutual Intergroup Differentiation Model (MIDM).

## Mutual Intergroup Differentiation Model (MIDM)

In contrast to the models previously outlined, the MIDM does not seek to break down boundaries between groups, but rather to accentuate the positive aspects of each group (Hewstone, 1996; Hewstone & Brown, 1986). The key to this model is to sponsor contact that does not threaten the identity of each group, but rather allows each group to do what it does best in the effort to reach a common goal. This process prevents each group from losing its distinctiveness while promoting interaction. In other words, each group can focus on its strength rather than stretching beyond its limits.

This model has been the most successful in showing generalization beyond those immediately involved in the intergroup dynamic (Brown, Vivian, & Hewstone, 1999). Furthermore, the model efficiently maximizes the different strengths that each group brings to the table. Group members do not need to subjugate their current identity to attain some higher-order goal. Rather, they can do what they do best in order to reach a common goal. Each member group can contribute equally and from a position of strength.

Early thoughts on convergence leaned heavily toward promoting the "backpack" or "platypus" model, in which reporters would be able to produce content for multiple media simultaneously (Stevens, 2002; Wenger, 2005). While some opportunities for this type of approach remain, such as the war in Iraq (Harmon, 2003), many individuals have cited problems and concerns with it. These concerns range from the watering down of the journalism (Sweet, 2001) to potential burnout among overworked journalists (Corrigan, 2002). Some of the current models of convergence under discussion involve specific individuals who cross over to promote stories or act as experts for the other's news product (Dailey, Demo, & Spillman, 2005).

The culture of both print and broadcast journalism is one of specialization. Newspapers have reporters who cover beats, or specialized reporting areas.

Assignment editors are present for city, state, national and international news. The copy desk is responsible for grammar, style and fact checking. Photographers are experts in capturing images who work in conjunction with the text editors and reporters, but also operate independently. Broadcast has anchors, reporters, videographers, producers and more, each of whom have specific tasks at which they are expected to excel. Specialization is an entrenched value, and, thus, it makes little sense to pry individuals away from their areas of expertise in order to do something that someone else can do better.

To that extent, the MIDM appears to hold the greatest promise for ameliorating cultural issues within the newsroom. This model will allow print and broadcast journalists to maintain a medium-specific identity as well as focus on their area of expertise. Besides bias reduction, an additional benefit of this model is that it promotes the idea of a deeper specialization rather than a surface-level understanding of multiple media. The level of depth that journalists can maintain while reporting in a convergent news operation is a concern among academics and professionals alike (Wenger, 2005).

Furthermore, the benefits extended to individual journalists from the other's medium are likely to be transferred to the whole group. Dovidio, Gaertner, and Validzic (1998) found that situations where individuals demonstrated high levels of competence and interacted on different facets of a task that were important to the group provided the most positive results in bias reduction. Thus, as individuals change within each newsroom due to hiring, and retirements and so on, there is a greater likelihood that the positive attitudes grown by these collaborative efforts will extend to new hires.

Brown and Wade's (1987) experimental examination of the MIDM also found positive results for journalistic endeavors. Students were placed into teams and required to complete a two-page magazine project. Individuals in one condition were expected to participate in each phase of the activity (writing the story, producing images, and laying out the spread) while the individuals in the other condition were assigned to work on one specific aspect of the project. Participants who operated in the specific-task condition were more positive regarding the project and their contact with other group members than those who were given no distinct role. The key to this approach in regard to convergence is not to create a convergence team, but rather to reassemble teams that relate to social identities currently maintained by the individuals. By accentuating those social identities, the movement away from medium-based identities will be easier and individuals will be better able to defend their sense of self.

The one negative that has been associated with MIDM is the possibility of increasing intergroup anxiety (Greenland & Brown, 1999). Individuals tend to express a level of anxiety when they will be interacting with individuals from another group. This anxiety can be due to a variety of factors, including a fear of being stereotyped or discriminated against because of group membership. Greenland and Brown (2000) note that when there has been a history of discrimination, this fear is most acute. Given the stereotyping and outgroup denigration found in much of the literature that explores the print-broadcast divide, this fear might be well founded in convergent newsrooms. (Killebrew's [2001] example of print reporters who wondered if their broadcast colleagues took a class in "Hairspray 101" would seem to illustrate that point.) Thus, the preconditions upon which bias reduction is met become exceptionally important, especially those that pertain to the quality of contact and equality among groups.

## Discussion and Conclusion

Singer (2004a) observed that journalists in converged news media operations are willing to adapt to the new media landscape when shown that "we" and "they" have congruent journalistic values. This is the first step toward a better convergence product as well as a better convergence environment. For the most part, convergence practitioners have been moving in a generally positive direction in regard to addressing important cultural issues within their newsrooms. That being said, this chapter provides a theoretical overlay that allows them to better manage future movement toward convergence and reduce the number of "false starts" they might experience.

Intergroup identity theories incorporate the positives and negatives that individual convergent operations have encountered into a larger framework. Drawing from specific models of intergroup relations, some possible suggestions can be offered to reduce bias and improve relations among journalists in a converged environment.

Newsrooms that hope to use convergence as a minor function of their news operations would be wise to look at the DCM for guidance. Creating a convergence beat to explore certain stories or report on certain topics can best be done by decategorizing the few individuals who are involved and allowing them to form more personalized relationships across medium-based boundaries. Managers can stress task-specific qualities in the convergence participants ("Jimmy is here because you need someone who can analyze a fund-raising database") and encourage the participants to find connections through personal interaction, such as discussion of their hobbies or family life. A full buy-in by both news agencies is not necessary for this kind of approach, so this might be the best model for these types of convergence efforts.

Conversely, if two news agencies are to merge their news operations and fully embrace convergence, the MIDM might be the better approach. Many of the news operations that sought a full conversion to convergence met with hostility and resistance. Not only does the intergroup literature outlined here explain why that might be the case, the MIDM outlines some potential points of intervention for convergence leaders to consider.

The MIDM debunks the idea that convergence means everyone needs to do everything. Rarely in a traditional journalism setting does a single journalist do everything to create the product that the audience receives, so it makes little sense to do it in a converged environment. The model also limits criticism that convergence will lead to weaker journalism. The MIDM will have individuals playing to their strengths while advancing a common goal of information gathering and storytelling. The best print reporters will still be writing the best possible newspaper story and the best videographer will still be shooting top-notch video. The only difference is that both of these individuals will be working with each other in a synergistic environment and the audience will be all the better because of it.

Finally, the model will help move individuals away from their medium-based identities while still allowing them to work cooperatively with journalists from other backgrounds. The MIDM helps harness the natural competition that journalists have honed over decades of fighting with one another for the "scoop." Convergence architects can facilitate this shift from foe to friend by emphasizing work groups based on topics (akin to the newspaper's beat system) instead of allowing the journalists to divide along medium-based lines. The "beat-based" approach to convergent journalism has yet to be formally tested while using the MIDM's underlying assumptions and guides. However, given the theoretical underpinnings of the MIDM and its success in various other areas, it seems the model would present a workable plan for convergence on a grand scale.

As stated earlier, there is no magic bullet for every scenario or every problem. However, by providing a theoretical basis for the outcomes obtained by these media groups, it is possible for interested parties to seek a deeper understanding of intergroup relations and find answers within the literature. Being able to replicate successful efforts with different people and various groups is the key to making convergence work. Using the theoretical models outlined here, convergence practitioners can approach each new endeavor systematically and produce successful results more often. Regardless of which theoretical model a convergence outlet embraces, many of the key elements remain the same. Journalists need to see each other in ways that do not perpetuate stereotypes and engage in behavior that breaks down medium-based biases. Convergence practitioners who need help in facilitating these and other bias-reduction measures can use the literature on intergroup relations as a "cheat sheet" of sorts.

It is also important to note that by rooting this discussion in group-based behavior, it is possible to transfer this outline to other groups within a newsroom, be it a converged or traditional newsroom. Conflicts arise between anchors and reporters, copy editors and writers, business staffs and editorial staffs and more. By examining the preconditions for a successful intergroup interaction, managers in any area can take steps toward improving harmony within their own group while working with other groups.

Convergence will likely continue changing directions as the media, the audiences and the advertisers seek ways to find a more parsimonious approach to news delivery. Regardless of the final outcome, understanding how people view themselves, each other and the product will likely be important in improving buy-in as well as the product.

## References


Allport, G. (1954). *The nature of prejudice*. Cambridge, MA: Addison-Wesley.

America, A. (2002, June). Cornfield convergence. *Presstime,* p. 22.

Ashburn-Nardo, A., Voils, C., & Monteith, M. (2001). Implicit associations as the seeds of intergroup bias: How easily do they take root? *Journal of Personality and Social Psychology, 81*(5), 789–799.

Bettencourt, B., Brewer, M., Croak, M., & Miller, N. (1992). Cooperation and the reduction of intergroup bias: The role of reward structure and social orientation. *Journal of Experimental Social Psychology, 28*, 301–319.

Billig, M., & Tajfel, H. (1973). Social categorization and similarity in intergroup behavior. *European Journal of Social Psychology, 3*, 27–52.

Breed, W. (1955). Social control in the newsroom: A functional analysis. *Social Forces, 33*, 326–335.

Brewer, M. (2000). Superordinate goals versus superordinate identity. In D. Capozza & R. Brown (Eds.), *Social identity processes* (pp. 117–132). Sage: Thousand Oaks, CA.

Brewer, M., & Miller, N. (1984). Beyond the contact hypothesis: Theoretical perspectives on desegregation. In N. Miller & M. Brewer (Eds.), *Groups in contact: The psychology of desegregation* (pp. 281–302). Orlando, FL: Academic Press.

Brown, R., Vivian, J., & Hewstone, M. (1999). Changing attitudes through intergroup contact: The effects of group membership salience. *European Journal of Social Psychology, 29*, 741–764.

Brown, R., & Wade, G. (1987). Superordinate goals and intergroup behaviour: The effect of role ambiguity and status on intergroup attitudes and task performance. *European Journal of Social Psychology, 17*, 131–142.

Bulla, D. (2002, August). *Media convergence: Industry practices and implications for education*. Paper presented at the Association for Educators in Journalism and Mass Communication conference, Miami, FL.

Callinan, T. (2001). TV reports next step in Phoenix convergence strategy. *Gannett News Watch*. Retrieved from http://www.gannett.com/go/newswatch/2001/august/nw0824–1.htm

Corrigan, D. (2002, October). Convergence—Overworked reporters with less news. *St. Louis Journalism Review, 32,* 20–23.

Dailey, L., Demo, L., & Spillman, M. (2005). The convergence continuum: A model for studying collaboration between media newsrooms. *Atlantic Journal of Communication, 13*(3), 150–168.

Dovidio, J., Gaertner, S., & Validzic, A. (1998). Intergroup bias: Status, differentiation and a common in-group identity. *Journal of Personality and Social Psychology, 75*(1), 109–120.

Downie, L., & Kaiser, R. (2002, May). Media synergy, if not now, when? *Presstime,* pp. 54–55.

Dyer, K. (2002, March). Faith in convergence. *Presstime,* p. 47.

Ensari, N., & Miller, N. (2002). The out-group must not be so bad after all: The effects of disclosure, typicality and salience on intergroup bias. *Journal of Personality and Social Psychology, 83*(2), 313–329.

Erickson, R., Baranek, P., & Chan, J. (1987). *Visualizing deviance: A study of a news organization*. Toronto: University of Toronto Press.

Fee, F. (2002, August). *New(s) players and new(s) values: A test of convergence in the newsroom*. Paper presented at the Association for Educators in Journalism and Mass Communication conference, Miami, FL.

Filak, V. (2004). Cultural convergence: Intergroup bias among journalists and its impact on convergence. *Atlantic Journal of Communication, 12*(4), 216–232.

Filak, V. (2006). The impact of instructional methods on medium-based biases and convergence approval. *Journalism and Mass Communication Educator, 61*(1), 48–64.

Filak, V., & Thorson, E. (2002, July). *Print and broadcast journalistic bias: An intergroup-conflict perspective on convergence*. Paper presented at the International Communication Association conference, Seoul, Korea.

Gabettas, C. (2000). Three's company. Retrieved August 29, 2005, from http://www.rtnda.org/technology/convergence.shtml

Gaertner, S., & Dovidio, J. (2000). *Reducing intergroup bias: The Common Ingroup Identity Model*. Philadelphia, PA: Psychology Press/Taylor Francis Group.

Gaertner, S., Dovidio, J., Anastasio, P., Bachman, B., & Rust, M. (1993). The Common Ingroup Identity Model: Recategorization and the reduction of intergroup bias. In W. Strobe & M. Hewstone (Eds,), *European Review of Social Psychology* (pp. 1–26). New York: John Wiley.

Gaertner, S., Dovidio, J., Banker, B., Houlette, M., Johnson, K., & McGlynn, E. (2000). Reducing intergroup conflict: From superordinate goals to decategorization, recategorization and mutual differentiation. *Group Dynamics, 4*(1), 98–114.

Gaertner, S., Dovidio, J., Rust, M., Nier, J., Banker, B., Ward, C., Mottola, G., & Houlette, M. (1999). Reducing intergroup bias: Elements of intergroup cooperation. *Journal of Personality and Social Psychology, 76*(3), 388–402.

Giles, R. H. (1991). *Newsroom management: A guide to theory and practice*. Detroit: Media Management Books.

Glaser, M. (2004). Lack of unions makes Florida the convergence state. *Online Journalism Review*. Retrieved August 29, 2005, from http://ojr.org/ojr/glaser/1081317274.php

Gonzalez, R., & Brown, R. (2003). Generalization of positive attitudes as a function of subgroup and superordinate group identifications in intergroup contact. *European Journal of Social Psychology, 33*, 195–214.

Greenland, K., & Brown, R. (1999). Categorization and intergroup anxiety in contact between British and Japanese nationals. *European Journal of Social Psychology, 60*, 509–517.

Greenland, K., & Brown, R. (2000). Categorization and intergroup anxiety in intergroup contact. In D. Capozza & R. Brown (Eds.), *Social identity processes* (pp. 167–183). Sage: Thousand Oaks, CA

Hammond, S., Petersen, D., & Thomsen, S. (2000). Print, broadcast and on-line convergence in the newsroom. *Journalism and Mass Communication Educator, 55*, 16–26.

Harmon, A. (2003, March 24). Improved technology turns journalists into a quick strike force. *The New York Times,* p. C1, col. 3.

Hewstone, M. (1996). Contact and categorization: Social psychological interventions to change intergroup relations. In C. Macrae, C. Stangor, and M. Hewstone (Eds.), *Stereotypes and stereotyping* (pp. 323–368). New York: Guilford Press.

Hewstone, M., & Brown, R. (1986). Contact is not enough: An intergroup perspective on the "Contact Hypothesis." In M. Hewstone & R. Brown (Eds.), *Contact and conflict in intergroup encounters* (pp. 1–41). Oxford, UK: Basil Blackwell.

Hornsey, M., & Hogg, M. (2000). Subgroup relations: A comparison of Mutual Intergroup Differentiation and Common Ingroup Identity models of prejudice reduction. *Personality and Social Psychology Bulletin, 26*(2), 242–256.

Killebrew, K. (2001, August). *Managing in converged environment: Threading camels through newly minted needles*. Paper presented at the Association of Educators in Journalism and Mass Communication, Washington, DC.

Kramer, R., & Brewer, M. (1984). Effects of group identity on resource utilization in a simulated commons dilemma. *Journal of Personality and Social Psychology, 46*, 1044–1057.

Lowrey, W., Daniels, G., & Becker, L. (2005). Predictors of convergence curriculum in journalism and mass communication programs. *Journalism and Mass Communication Educator, 60*, 32–46.

Miller, N., Brewer, M., & Edwards, K. (1985). Cooperative interaction in desegregated settings: A laboratory analogue. *Journal of Social Issues, 41*(3), 63–79.

Murphy, J. (2002). Hard news for hard times. *Mediaweek, 12*(14), 21–27.

Nesbitt, M. (2001). *Newspaper culture*. Readership institute, Media management center at Northwestern University. Available at www.readership.org

Peterson, D., & Hammond, S. (1999). *Converged media: Identity and commitment within a newsroom*. Paper presented at the American Association of Behavioral and Social Sciences conference, Las Vegas, Nevada.

Pettigrew, T. (1979). The ultimate attribution error: Extending Allport's cognitive analysis of prejudice. *Personality and Social Psychology Bulletin, 55*, 461–476.

Quinn, S. (2005). What is convergence and how will it affect my life? In S. Quinn & V. Filak (Eds.), *Convergent journalism: An introduction* (pp. 3–18). Boston: Focal Press.

Reese, S., & Ballinger, J. (2001). The roots of a sociology of the news: Remembering Mr. Gates and social control in the newsroom. *Journalism Quarterly, 78*(4), 641–658.

Robins, W. (2000). King of convergence. *Editor and Publisher, 133*(42), 12–17.

Rothbart, M. (2001). Category dynamics and outgroup stereotypes. In R. Brown & S. Gaertner (Eds.), *Blackwell handbook of social psychology: Intergroup relations* (pp. 45–64). Malden, MA: Blackwell.

Runett, R. (2004, February). Multimedia on the move: Cross-media partners are key. *Presstime,* p. 10.

Schank, R., & Abelson, R. (1977). *Scripts: Plans, goals and understanding*. Mahwah, NJ: Lawrence Erlbaum.

Schlesinger, P. (1978). *Putting "reality" together*. Sage: Beverly Hills.

Shearer, E. (2003). *Convergence*. Retrieved August 29, 2005, from http://www.asne.org/index.cfm?id=4941

Sherif, M., Harvey, O., White, B., Hood, W., & Sherif, C. (1961). *Intergroup conflict and cooperation: The Robbers Cave experiment*. Norman, OK: University of Oklahoma Book Exchange.

Sigelman, L. (1973). Reporting the news: An organizational analysis. *American Journal of Sociology, 79*(1), 132–151.

Silcock, B. W., & Keith, S. (2006). Translating the Tower of Babel: Issues of definition, language and culture in converged newsrooms. *Journalism Studies, 7*(4), 610–627.

Singer, J. (2004a). More than ink-stained wretches: The resocialization of print journalists in converged newsrooms. *Journalism and Mass Communication Quarterly, 81*(4), 838–856.

Singer, J. (2004b). Strange bedfellows? The diffusion of convergence in four news organizations. *Journalism Studies, 5*(1), 3–18.

Stevens, J. (2002). Backpack journalism is here to stay. *Online Journalism Review*. Retrieved September 2, 2004, from http://www.ojr.org/ojr/workplace/1017771575.php

Stevens, J. (2003). Moving online into the newsroom. *Online Journalism Review*. Retrieved August 29, 2005, from http://www.ojr.org/ojr/workplace/1069284495.php

Sweet, K. (2001). *Media convergence: The effect on newsroom roles*. Unpublished thesis, University of Nebraska-Lincoln.

Tajfel, H. (1969). Cognitive issues of prejudice. *Journal of Social Issues, 25*, 79–97.

Tajfel, H. (1978). *Differentiation between social groups: Studies in the social psychology of intergroup relations*. London: Academic Press.

Tajfel, H., & Turner, J. (1986). The Social Identity Theory of intergroup behavior. In S. Worchel & G. Austin, G. (Eds.), *Psychology of intergroup relations* (pp. 7–24). Chicago: Nelson-Hall.

Terry, D., Carey, C., & Callan, V. (2001) Employee adjustment to an organizational merger: an intergroup perspective. *Personality and Social Psychology Bulletin, 27*(3), 267–280.

Tompkins, A. (2001, February 28). *Convergence needs a leg to stand on*. Available at http://www.poynter.org/centerpiece/022801tompkins.htm

Tompkins, A., & Colon, A. (2000). Tampa's media trio. *Broadcasting and Cable, 130*(15), 46.

Wenger, D. (2005, September). The road to convergence and back again. *Quill,* p. 40.

Westin, A. (2001). You've got to "be carefully taught." *Nieman Reports, 55*(1), 63–65.

CHAPTER 8

# Developing Media Managers for Convergence

## A Study of Management Theory and Practice for Managers of Converged Newsrooms

Holly A. Fisher

Much of the discussion of media convergence has centered on its varying definitions, the technology and even the best way to train reporters across media platforms. Yet one key component has been left out of much of the convergence conversation—the role of media managers in shaping convergent news operations and guiding and managing convergence newsrooms. Once a news organization opts to take the convergence path to news, managers are the ones who lead the newsrooms through the process of change and development.

The idea of management and convergence is just beginning to take shape. Few have addressed the way in which media managers impact convergence. Yet it is critical for managers to retrain themselves and adjust their management styles to foster convergence and help reporters through this process of change.

This chapter looks at convergence from a management perspective, combining primary interviews with managers in converged newsrooms with management research and theory. To best analyze how management and convergence come together, this chapter begins by offering background and perspective as well as the prevailing media management theories. This chapter also addresses the structure and makeup of a converged newsroom, including the physical structure, the management structure, developing newsroom roles and titles applicable to the converged news outlet. The media manager's job and role in a converged newsroom is explained and a series of directives for media managers embarking on a convergence operation is outlined.

## Convergence in Action

Applying the varying definitions of convergence explored in Chapter 1, several dozen news outlets are opting to give convergence a try. As of early 2008, the Media Center, the American Press Institute's (2008) Convergence Tracker, lists 107 convergence relationships.

Yet few newsrooms are far along in the convergence process. Media General's News Center in Tampa, The Tribune Co.'s operation in Chicago and the Belo Corp.'s Dallas efforts are most often cited as solid examples of convergence in the United States. Even these pillars of the convergence community will admit they have a long way to go before reaching convergence utopia.

At the height of Belo's convergence work, Chris Kelley, former editor of Belo Interactive-Dallas, says he is pleased with the progress Belo has made but it is not perfect. "We have TV reporters filing short stories. We have newspaper reporters phoning in audio reports that we record and place online. Our newspaper photographers are shooting video. I even have a 6 and 10 p.m. news anchor writing a weekly online column. I'm pleased, but we're not where we need to be" (personal communication).

Yet in early 2007, Belo was not realizing the benefits it had hoped for in this partnership with television. "It was a failed experiment," said Tom Ackerman, one of the Belo TV reporters whose job was a casualty of the television-newspaper partnership. "There were great aspirations, but it was never followed through. It was a mismatch: The (Belo) bureau served 17 television markets, and the newspaper people only worked for the *Dallas Morning News*. It was not really seen as an asset (for newspaper journalists) to be put on the air anywhere but Dallas" (Ahrens, 2007, p. D1).

As evidenced by Belo's experiment with television, convergence is an evolving process, one that is having a big impact on traditional newsrooms. While agreement over what convergence is and how it should be implemented is still up for debate, most journalists and scholars agree convergence has staying power. Like the invention of the printing press, the computer, and digital television, convergence is taking up permanent residence in the journalism industry. This means media managers must determine the best methods and management skill sets that address the myriad issues of setting up a converged newsroom and ushering journalists through the convergence change.

## Perspectives on Newsroom Management

Media managers, for the most part, can draw on traditional management models when it comes to overseeing and leading a newsroom. Administrative management theory, created by mining and steel executive Henry Fayol, focuses on what he called "managerial activities," which remain common activities of managers today (Pringle & Starr, 1999, p. 4). Those activities are planning, organizing, commanding, coordinating, and controlling—all tasks newsroom managers can use for launching convergent operations in their newsrooms (Pringle & Starr, 1999, p. 4).

One factor to consider in implementing changes in a newsroom is the "Hawthorne Effect," the result of a 1920s and 1930s experiment at Western Electric's Hawthorne plant in Illinois (Pringle & Starr, 1999, p. 7). The lesson from the "Hawthorne Effect" is that if managers pay attention to the needs of their employees, productivity will most likely increase (Pringle & Starr, 1999, p. 7).

The Hawthorne Effect plays out in newsrooms where reporters value their work and want to feel appreciated. Even when workload increases without a pay increase, reporters will continue to pursue their love of journalism, provided managers are giving them attention and appreciation.

Related to the Hawthorne Effect is Frederick Herzberg's idea that employee attitudes and behaviors are, in part, influenced by motivators similar to self-esteem and self-actualization (Pringle & Starr, 1999, p. 8). According to this theory, "the critical task, therefore, lies in satisfying employees' needs for self-actualization by giving them more responsibility, providing opportunities for advancement and recognizing their achievement" (Pringle & Starr, 1999, p. 8).

These classic management theories do apply to newsrooms, yet media managers also have to be sensitive to the unique attitudes and working styles of journalists. Media managers can use these classic models, while supplementing them with skill sets and management techniques specific to media organizations.

Newsrooms are challenging to manage because they combine two distinct types of organizations (Redmond & Trager, 2004, p. 56).

> They are assembly lines with strict deadlines and involved processes that must be done in order, with speed, and with repetition.... But they are also work environments where employees have a lot of creative control of their work; are highly educated and use their mental abilities to do their work, which is full of judgment calls; and tend to see their work as a higher calling. (Redmond & Trager, 2004, p. 56)

That combination can be quite a challenge for media managers, and adding the uncertainties and newness of convergence only serves to foster the delicate way in which media managers have to handle their reporters, photographers, anchors, and graphics staff.

Reporters by nature tend to work as individuals. "News gathering is a cottage industry that relies on individual relationships with sources," explains Doug Fisher, an instructor at the University of South Carolina, and a former Associated Press editor (personal interview). But the production process relies on teamwork, he notes. Convergence is news production, but it has to start with the raw information, Fisher says. This duality is a challenge for news managers because they have to manage at two levels—it takes two sets of skills, he adds. They have to be skilled in managing creative talent as well as skilled in the production side or logistics management.

Because media organizations do rely on a level of teamwork in the production of the newscast, Web site, newspaper, or magazine, media managers should work on fostering a bond among staff members. This relationship should facilitate the convergence process, which depends heavily on teamwork and communication across media platforms.

Transforming a newsroom into a convergent operation is likely to create challenges in terms of training, organizational culture, and management styles. Staff members will have to adapt to new jobs, new production cycles, and a new way of thinking about the dissemination of news.

"To have effective bonding of people in such a dynamic environment, you have to get members of the organization to relate to one another, to build trust and confidence, and to have the ability to reach group decisions while they work independently and make crucial decisions on their own" (Redmond & Trager, 2004, p. 59). Implementing convergence is a drastic change for media organizations and their journalists. Convergence flies in the face of traditional journalism; it goes against the basic tenets of the profession most reporters have been operating under for decades. Convergence essentially eliminates a focus on "print" or "broadcast." It breaks down the walls of competition and puts print, broadcast, and online reporters on the same team. Cultures and backgrounds are brought together—sometimes under the same roof as in the case of the Tampa News Center. Staff members must learn the habits, techniques, and jargon of organizations with whom they once competed fiercely.

These changes are not always welcomed, and many reporters will balk at the idea of cross-platform delivery of news and information. This is where media managers and senior executives have a significant role in making the transition to convergence a smooth process. Just as journalists balked at the computer, many preferring to cling to a clattering typewriter, reporters of the 21st century may find themselves clinging to the titles of "print journalist" or "broadcast reporter." These titles will very likely end up in the storerooms and museums along with the typewriter. Not everyone embraces change, and it is up to media managers to develop a positive attitude about change within converging news operations.

According to LeNoir, R. (1987), managers have four ways to deal with substantial changes in an organization:

**1. Structuralists** see the organization as operating within the context of greater forces than itself, which require it to adjust to those forces. For example, media organizations are implementing convergence and bolstering their multimedia offerings as a result of economic forces. As online advertising continues to outpace print advertising (19.3 percent increase in second quarter 2007) and print readership declines, media companies will turn to convergence as a way to keep profits in the double digits (Associated Press, 2008). While media companies may be accused of bowing to economic pressures and shareholder interests, it is realistic to expect companies to recognize outside forces and adjust.

**2.** The **strategic choice** view perceives management as having considerable impact on organizational effectiveness through the choice of niches toward which the organization is directed. The environment is picked apart to find the best potential place for the organization to focus its energies. Media managers have to respect the changing media landscape and find where it fits best in the environment. Thanks to the Internet, online video and audio, blogs, and user-generated content, media managers must make a strategic choice in order to stay competitive in the media environment. "A consensus appears to be emerging in the industry: To achieve new growth, newspaper companies need to innovate and diversify. They need to create broadening 'portfolios' of products and services that get key jobs done for a wider range of customers" (American Press Institute, 2006).

**3.** In **collective action**, a change agent is perceived as dramatically altering the operating environment. The organization and environment must mutually redefine one another according to the context presented by the change agent, resulting in a collective mutually interdependent organization. In this case, convergence is the change agent and the organization and media environment must redefine themselves, creating an interdependent organization. Technology is not going away, so media managers have to figure out how to coexist and take advantage of new technology.

**4.** The **natural selection** concept is just Darwinism revisited, with an organization either being appropriate to its environment and surviving, or not appropriate and dying. Media managers who do not adapt to changes and who do not embrace the concept of convergence will find themselves left behind. By accepting the changes to the media environment and adapting to new ideas, media managers and their newsroom will ensure their future existence.

Media managers should assess their organization's corporate culture and attitudes to determine which change agent would work best (Redmond & Trager, 2004, p. 62). Journalists' occupational identity is tied to their ability to innovate or create (Wicks, et al., 2004, p. 65). Journalists are intrigued by new challenges, a promising prospect, as they will be faced with many changes as they become multimedia journalists. Motivating journalists and fostering their desire for creativity becomes an important part of a media manager's job (Wicks et al., 2004, p. 65).

When dealing with creative journalists, a participative management style is a good way to lead employees through a period of change and to demonstrate—from the top down—that convergence is important. Participatory leadership focuses on the collective, sharing power from the highest executive all the way down the ranks (Redmond & Trager, 2004, p. 183).

This management style fosters the concept of convergence by encouraging employees to "buy-in" to this changing form of journalism. Effective leaders can accomplish their vision—in this case, convergence—by creating an atmosphere in which organizational members are guided partners (Redmond & Trager, 2004, p. 183). Journalists are independent and work best when left to their own creative timeline. While they work under the constraints of deadlines and time pressures, they resist micromanagement techniques. Thus, research supports the idea that journalists prefer a participative approach to management and organizational leadership (Redmond & Trager, 2004, p. 183).

Chris Kelley, former editor of Belo Interactive in Dallas, agrees collaborative-style leadership is "the only leadership style that works in a converged newsroom" (e-mail interview). "The days of autocratic order-yelling city editors are over. Effective editors know not only how to craft strong, robust narratives for the newspaper, but also how to package them online with sound, video, and interactivity and to work closely with TV colleagues to create compelling TV packages."

## The Converged Newsroom

### Physical Structure

Once a news organization decides to take a step in the convergence direction, a host of additional decisions must be made. Those issues range from how to organize the staff to how to physically structure the newsroom.

In terms of how to design a converged newsroom, many news organizations have created multimedia or convergence desks to oversee the convergence process. It is unlikely a print reporter will automatically think of a story in terms of the best video or audio; a staff of editors trained to think about convergence and the best platform for the story will make this process smoother.

The *Orlando Sentinel* has a multiple-media desk for news coordination (Quinn, 2002). The desk is the hub for the organization's editorial decision-makers. "It allowed the individuals in a complex operation to work together efficiently. Editors for print, online, radio, and cable TV operations worked alongside production staff. When a major story broke, the aim was to cover it in as many formats as possible and practical" (Quinn, 2002).

The Ifra Newsplex at the University of South Carolina was built with an eye toward physically fostering convergence. Desks are positioned in a circular format to encourage discussion and the sharing of ideas. Founding Newsplex Director Kerry Northrup oversaw the design and construction of Newsplex. He says traditional newsrooms are usually divided into work areas that match the steps of newsflow and production—writing, editing, and design (Quinn, 2002).

Using the Newsplex model, traditional newsrooms can make adjustments to their physical structure to enhance convergence:

1. Appoint an editor to coordinate and coordinate the newsroom's convergence efforts (Northrup, 2002). This convergence or multimedia editor will be able to focus his or her full attention on implementing convergence and bringing multimedia into the newsroom.
2. Establish a central cross-media or multimedia news desk where key newsroom managers are co-located to best coordinate news efforts (Northrup, 2002).

3. Integrate similar staff functions from different media (Northrup, 2002). Weave reporters' responsibilities together with that of other reporters as a way to foster collaboration and team work (Northrup, 2002).

Similar to the *Orlando Sentinel,* another Florida newspaper has become the pillar of convergence. As discussed in Chapter 11 in this volume, The News Center in Tampa houses *The Tampa Tribune,* Tampa Bay Online, and WFLA-TV. All these editorial teams work in the same building and their convergent efforts are made better with a centralized news desk.

## Management Structure: Seven Best Practices

But simply moving some desks around or eliminating cubicle walls will not turn a newsroom into a hub of convergence. Although a necessary piece of the puzzle, organizational structure and management techniques are critical in the development of a successful converged newsroom. Gracie Lawson-Borders' research on the Tribune Co., Media General, and the Belo Corp. led her to develop seven "observations of convergence" (Lawson-Borders, 2003). These elements form a sort of "best practices" guide for media outlets embarking on convergence. They are: **communication**, **commitment**, **cooperation**, **compensation**, **culture**, **competition,** and **customer**.

Because convergence involves so many different elements, **communication** must flow throughout all facets of the newsroom. The print, broadcast, and online components have to bring their different cultures together. Most converged newsrooms have budget—or story planning—meetings that involve all elements. At the News Center in Tampa, managers and editors attend two budget meetings each day to discuss content across platforms. "A scorecard is marked as the discussion focuses on daily coverage to identify where convergence occurred across platforms, diversity in content, and placement of stories" (Lawson-Borders, 2003, p. 94). *The Chicago Tribune* has implemented a "content sharing" policy that outlines its rules for convergence. At *The Dallas Morning News'* afternoon page one meeting, editors gather with the interactive staff as pages of dallasnews.com are projected onto the wall. Editors exchange comments and ideas about convergence applications (Lawson-Borders, 2003).

Lawson-Borders' seven observations of convergence fall largely on managers to implement and support. A **commitment** to convergence has to resonate through all levels of the organization—externally and internally from corporate leaders to frontline workers and into the economics and technology of the organization.

**Cooperation** among managers and editors will translate into cooperation among photographers, reporters, videographers, and online staff members. Some outlets will have multimedia liaisons attending story meetings and discussions; others will put cameras on the newsroom floor so print reporters can do television spots.

Also important for managers to address is **compensation**. "Media managers must consider how to acknowledge and compensate for the additional skills and expertise expected of the changing role of staffers" (Lawson-Borders, 2003, p. 95).

Journalists are being asked to do more and to learn new skills, yet management has been slow to monetarily reward workers' efforts to learn and practice convergence. Convergence should not be seen as simply a way to save money by having one reporter handle multiple tasks. Asking reporters to produce one story for two or three media platforms can lead to more work, longer hours, and higher levels of burnout and frustration, particularly if reporters are not adequately compensated. Conflict over pay has been a part of the convergence effort in Tampa. When some *Tribune* reporters first appeared on WFLA, they received extra pay. But eventually this cross-platform reporting became part of the daily operation and the extra pay ceased (Silcock & Keith, Chapter 13, this volume). "The additional training and new responsibilities that are expected of journalists must result in pushing past merit pay into issues of enhanced compensation. Employees that are trained, allowed to grow, paid well for their services, and yes, happy with their work, will perform better in a converged environment" (Lawson-Borders, 2003, p. 95).

**Cultural changes**, **competition,** and having a **customer** focus are other aspects of Lawson-Borders' convergence best practices. Print, broadcast, and online media have their own cultures, jargon, and news practices; convergence brings all that together. Blending those cultural dynamics is key to convergence success.

In particular, television and print journalists find themselves working to bridge a significant cultural chasm. Definitions of news, competition, work styles, routines, and staffing issues are all cultural differences newsroom managers will have to address before convergence can flow smoothly (Silcock & Keith, Chapter 13, this volume).

Similar to cultural changes is **competition**. Journalists who have long competed against each other for the big story, the better story, and the first story are now working together. Newspaper reporters are scooping themselves on the Web or on the partner TV news station. This can be a bitter pill for many journalists to swallow, and managers need to be able to foster an attitude of cooperation over competition. Thanks to the Internet and 24/7 news flows, reporters also face competition from national media outlets and from the busy lives of readers and viewers.

Finally, convergence means putting the focus on the **customer**. No longer do reporters own their story; the owner is the audience. "Something we learned early on is that we have to get away from the notion of a particular reporter or platform owning a story," says former *Tampa Tribune* publisher Gil Thelen. "We have to say the community owns the story. What is the most effective way to return that story for the community's use?" (personal interview, October 21, 2004).

## Developing Newsroom Roles

Another aspect of structuring a converged newsroom is determining who does what, and that can result in entirely new roles. Kerry Northrup, founding director of Ifra Newsplex, created four roles that can work in a converged newsroom. These roles are regularly taught in training conducted at Newsplex.

**1.** *Newsflow Editor*—At the center of helping newsrooms manage convergence and tell stories across multiple platforms is the newsflow editor. The newsflow editor looks at the newsroom from a different perspective and decides how all the pieces fit together. "The key factor that differentiates the newsflow [editor] from these traditional roles is that the newsflow editor focuses on the story, not on the specific delivery platform" (Covington, 2004). The newsflow editor has his or her attention on each media platform, ensuring that none is an afterthought and the best ones are used to tell a story most effectively (Covington, 2004).

**2.** *Storybuilder*—This role takes assigning editors from the newspaper side and mixes them with producers from the broadcast side to create a new, converged manager. The storybuilder needs a copy editor's eye for detail and a producer's understanding of flow and pacing combined with the assigning editor's knack for seeing the different paths of a story and matching resources to those paths. Rather than managing many stories for just one medium, this new storybuilder manages just a few multimedia stories (either in print or broadcast form), filing those stories directly or making the various pieces available to other media-specific news desks (Fisher, 2004).

**3.** *News Resourcer*—A news resourcer is a super librarian who thinks like a journalist. This role "combines the best of journalism—writing, editing and news judgment—with the best of librarianship and information management—super searching, technology, training and content/knowledge organization and infrastructure" (Locicero, 2004). The news resourcer specializes in information, providing background, depth, and context to stories in any platform. News organizations are overflowing with information in all forms—stories, photographs, graphics, video, and audio, as well as searchable databases, the Internet, and wire services. Information flows into newsrooms in the form of e-mails, faxes, press releases, media kits, and personal interviews. Someone has to not only manage all this information, but make it a useful component of the news and information that is disseminated to the public (LoCicero, 2004).

**4.** *Multiskilled Journalist*—Sometimes referred to as "backpack journalists" or the "one-man-band," multiskilled journalists are adept at collecting interviews, photographs, audio, and video footage for dissemination in print, broadcast and online formats. Rather than thinking of themselves as "print" or "television" reporters, multiskilled journalists look at all elements of a story from the written article in a newspaper to the photo essay on the Web. Then these journalists craft a story in the appropriate formats for the best delivery method. "The primary concept of the multiskilled journalist is not the mastery of a particular set of skills but simply the mindset that the information being gathered will be distributed through a variety of media, with recognition of the individual elements that must be captured in order to bring the story to the consumer" (Grant, 2004).

These emerging roles are not employee-specific. One newsroom might have several newsflow editors and many multiskilled journalists. Likewise, one employee could handle multiple roles. Exactly how these roles develop and grow in converged newsrooms will be a source for further study.

Of these new roles, the newsflow editor might be considered the most "management-oriented" job. Yet managers will have to be familiar with each of these roles and how they work. Managers will need to understand hiring, compensation, and personnel management functions as well as governance issues in making these four roles work within the newsroom—either with other more traditional roles or as an entirely new newsroom structure.

## The Media Manager's Job

Journalists can receive endless amounts of training in photography, videography, and how to tell a story across multiple platforms, but the final measure of convergence success lies with the media managers. A critical factor in making convergence work is hiring managers who understand convergence, believe in the concept, and are willing to work with reporters to build a convergent relationship. Managers who are new to convergence should receive training in convergence and the appropriate way to manage a converged newsroom.

Former *Tampa Tribune* publisher Gil Thelen admits his early convergence efforts did not focus on the management aspect as much as they should have. "We devoted a lot of time to the camera training for print journalists and writing for print for broadcast and writing for online. I would say that we should have devoted more time to training the very frontline managers in how it is that they would need to manage this new environment" (personal interview, October 21, 2004).

For newsrooms embarking on a convergence operation, Thelen strongly suggests putting resources into middle management. "Focus on what your managers are going to need to do" (personal interview, October 21, 2004).

In addition to simply understanding the concepts behind convergence, managers must address personnel and resource management issues as well as determining the best methods for handling cultural differences between print and broadcast newsrooms.

As Silcock and Keith (2006) report in their study of 13 journalists working in converged operations, bridging the cultural gap between print and broadcast can be the greatest challenge to successful convergence. Although both are forms of journalism, print and broadcast newsrooms have entirely different newsroom cultures. Language and newsroom jargon can cause issues. For example, to print journalists, the term "budget" means a list of stories reporters are working on for the next day's newspaper, but to broadcasters, "budget" is a financial term.

Simple differences in training and presentation of news must be overcome.

> Print media reporter(s) . . . have been trained in the art and science of information gathering, the inverted pyramid (though sometimes denigrated today) and have placed emphasis on depth content while remaining literary where

> possible. Broadcasters have been trained to emphasize the visual aspects of their stories, to shorthand the details and bring the story's essence to the fore quickly. (Killebrew, 2001)

Print and broadcast journalists have ingrained in their respective cultures a sense of competition. Broadcast journalists, in particular, face more competition as most cities have more than one TV station; on the other hand, newspaper reporters face slightly less competition thanks to the demise of the two-newspaper town. So bringing print and broadcast journalists together under the same roof cannot help but cause some problems. It is up to media managers to create a collaborative environment. Media managers have to understand the newsroom environments and each side's news-making decisions—what is news and how to cover particular stories (Keith & Silcock, Chapter 13, this volume).

Another serious issue media managers must understand when implementing convergence is personnel and resource management, particularly compensation and staffing issues. Newspapers will have more employees than a TV station, and convergence operations run the risk of relying more heavily on print reporters for the bulk of the stories. Print reporters are often seen providing information and commentary on the sister TV station, yet few television reporters will be writing an investigative piece for the daily newspaper.

An early version of Keith and Silcock's study includes observations from Tom Wolzien, who provides investors with analysis of media companies. He notes that staffing problems can arise when convergence is seen as a money-saving venture because one reporter can do multiple tasks—in essence becoming that glorified multiskilled journalist as explained earlier in this paper. Wolzien notes:

> Where the problem exists is in deadlines and in the amount of time and productivity that you expect the reporter to have. One person is not going to be able to handle two media, two platforms in the same amount of time that two people going different directions can handle the two. Any thought that one person should be able to do both, is crazy. (Keith & Silcock, 2002, p. 21)

These staffing and personnel resources issues can lead to serious problems in dealing with reporters' compensation. Asking reporters to write one story for publication in two or three platforms yet paying them exactly what they made to produce one story for one platform can lead to burnout as well as resentment and frustration—feelings media managers should strive to avoid.

When news organizations decide to enter into a convergence partnership, media managers must be upfront in discussions about staffing, resources, and compensation. Those issues should be resolved before convergence is implemented; standards should be put into place. In Kerry Northrup's "Newsplex Convergence Guides," he notes how convergence should be integrated into personnel factors, such as hiring, job descriptions, performance evaluations, and career incentives, including salaries. "Media companies should decide up front what their remuneration policies are for cross-media performance to forestall deadlock on this issue" (Northrup, 2004, p. 105).

## Recommendations for Media Managers

Most people have a difficult time dealing with change. In bringing convergence into a newsroom, managers are asking their staffs to completely rethink everything they know about their profession. They are being asked to cooperate with competitors; they are being asked to embrace a medium that is foreign to them; and they are being asked to do more—and do it faster.

Helping reporters through this process of change becomes critical for media managers. "At the human level media managers must confront existing cultures, traditions, and conventions, while overcoming a frantic climate of uncertainty" (Killebrew, 2002a, p. 39). This means communication is key, and media managers must understand and accept convergence so they can share information. Employees will appreciate knowing what media managers are planning for the newsroom and what changes are in store. Whether it is an informal lunch meeting or a more formal staff meeting, media managers should communicate plans with the staff—even if those plans are tentative. This gives employees an opportunity to offer feedback and suggestions that could greatly improve the way convergence efforts will be implemented. "This downward flow of communication is important, but it must be accompanied by management's willingness to listen to and understand employees. Accordingly, it is necessary to provide mechanisms for an upward flow of communication from employees to supervisors, department heads, and the GM. Departmental or staff meetings, suggestion boxes, and an open-door policy by management permit such a flow" (Pringle & Starr, 1999, p. 17).

Team building and participative management techniques seem to work best in newsrooms where media managers are dealing with creative employees (Redmond & Trager, 2004). Media managers who are engaged in and understand the creative process can better relate to employees and are better equipped to guide them through this process of change (Killebrew, 2002a).

Media managers should consider the following steps as they begin to launch a convergent news operation:

*Step 1. Training of media managers.* The initial reaction may be to train the employees: to put a print reporter in front of a television camera or to ask the anchor of the evening news to write a column for the Web site. But convergence must take a "top-down" approach, with media managers receiving adequate training on how stories and resources will flow across media. Managers should be fully equipped with all the knowledge and information about how convergence will work, so they can better communicate with their employees. If media managers understand and embrace convergence, they will have a much easier time communicating that understanding.

"Managers who are ill-prepared to cope with the stress of change in the move to a converged news environment will increase the level of stress among employees who are being asked to work in a new cross-platform environment" (Killebrew, 2002a, p. 43–44).

*Step 2. Planning*. An undertaking as complex as convergence requires detailed planning and organization. Convergence is not something newsrooms should dive into without adequate preparation. The participating media outlets should craft a "well-designed plan of action to foster understanding among all employees and managers" (Killebrew, 2002a, p. 45). This plan will address the various media platforms to be used, discussing positive and negative implications as well as the best methods for implementation (Killebrew, 2002a).

*Step 3. Communication*. Media managers must spend as much time as possible on communication; it is rare for a manager to be accused of communicating too much. It is critical that employees understand each step of the convergence process and that media managers explain the steps to implementation and the expectations of employees. Managers must also be good listeners to find out what barriers exist that may impede employees:

> The degree to which an organization embraces openness to change is likely an outcome of the interrelationships between values, structure and climate. In instances where organizations have been open in the past and where organizational members feel there is a continuation of accurate and reliable information and actions, it is likely there will be a greater degree of acceptance to change. (Killebrew, 2002a, p.43)

Media managers should clearly communicate to employees who will actively participate in convergence what their workload will be, whether they will receive additional compensation, and how their daily schedule could change.

But communication should not stop once the initial steps of convergence have been explained. A newly converged newsroom will be a living, breathing, and changing organism—employees must be kept abreast of what is happening in their evolving workplace. It is best for media managers to be upfront and honest with their employees; this decreases the likelihood that incorrect information will be passed through the employee rumor mill.

*Step 4. Training of employees*. All reporters should be equipped with a basic skills set. Interviewing, writing, research capabilities, and an ability to turn a complex subject into a story that readers, viewers or listeners can digest are critical skills for journalists, regardless of media platform. Despite those commonalities, some newsroom employees will have a greater aptitude and interest in cross-platform training. Those employees should be identified and given the opportunity to learn and participate in convergent operations (Killebrew, 2002a).

All employees should receive information and training about convergence and how it will work and ultimately impact the flow of news and the structure of the newsroom. But not all print reporters will be clamoring to appear on television. Media mangers should target those with an interest in convergence; they will be the most receptive to change and new ideas. Forcing all reporters to break out of their comfort zones immediately will lead only to feelings of resentment and aggravation. As those who express initial resistant to convergence begin to see how this new element of journalism works and how it expands the level of creativity among their colleagues, they will be more likely to want to participate themselves.

*Step 5. Commitment of resources.* To successfully create a converged newsroom, media managers will have to commit the necessary resources—people, technology, and training. As discussed earlier, reporters may feel overwhelmed by added responsibilities of multimedia reporting. Media managers should assess the staffing situation and determine if additional reporters are needed. Also, media managers should consider whether to hire a multimedia editor who can oversee the transition to a convergent newsroom and focus solely on multimedia staffing needs and assignments. Technology is key to the development of a multimedia newsroom. Reporters should be equipped with the tools they will need to complete multimedia assignments. If you expect reporters to create audio and video packages for the Web, be sure you give them the equipment they will need. This also means media managers need to set aside time for training. Fancy tools will not do the reporters any good if they do not know how to effectively use them or are spending too much time trying to learn a video editing software program. Invest in training, so reporters are equipped with the knowledge and skills they need to turn your newsroom into a convergent operation. Providing training also demonstrates the news outlet's commitment to convergence and to the reporters tasked with doing the job.

*Step 6. Implementation and introduction of new newsroom cultures.* Once everyone is on the same page, media managers can begin to implement convergence. Beyond just telling a story using different media platforms, the journalists and media managers will need to work with their partners from other newsrooms to understand their work flow and organizational culture.

Print, broadcast, and online newsrooms will need to work carefully on communication and understanding. Each newsroom will have its own schedule, deadlines, and jargon that will need to be explained. Again, communication becomes a key element in smoothly implementing convergence and in getting different media platforms to work together.

*Step 7. Continued training and communication.* Media managers should continue to train employees. Employees who are new to the media outlet will need training in convergence, and employees who were resistant to convergence at first may change their minds and desire more training in other media platforms, which media managers should encourage.

As issues, problems, and opportunities arise in the new convergent operation, those should be communicated to the staff. Questions should be answered and employees should have a solid understanding of changing workloads, expectations, and deadline schedules. Employees will appreciate and respect a culture of openness and sense of teamwork in the organization.

## Conclusion

Convergence is not going away, and managers, particularly new, up-and-coming managers, need the training and understanding that will help them move into leadership positions within a converged newsroom. Employees will look to management for understanding and guidance through this period of change. If the managers have not been properly trained and have not developed the appropriate skills sets, how will the employees know what to do? Management cannot expect more from its employees than it is willing to give.

Communication will be critical as the newsroom transitions to a convergent operation. Media managers will have to understand all facets of the newsroom, serving as a vehicle for communication as stories are adapted for the best medium. No longer will managers be thinking about a story from a purely print or broadcast perspective. Communication is needed in the entire process—from assigning stories to determining the most effective media platform.

Professionals working in convergence and academics studying and teaching convergence must shift some of their focus from definitions, technology, and training journalists in different media platforms to the management issues discussed in this chapter. While definitions and technology are important to the development and success of convergence, equal attention must be given to media managers and their role in convergence. Media managers must be trained and equipped with the skill sets needed to communicate, implement and practice convergence successfully.

## References

Ahrens, F. (2007, January 11). Newspaper-TV marriage shows signs of strain. *The Washington Post*. D1.

American Press Institute (2006, September). Newspaper next: The Transformation Project. *Blueprint for Transformation*, 37. Retrieved March 15, 2008 from http://www.newspapernext.org/2006/03/newspaper_next_blueprint_for_t_1.htm.

American Press Institute (2008). *Convergence tracker*. Retrieved March 2, 2008, from http://www.mediacenter.org/convergencetracker

Associated Press (2008, February 25). Internet ad revenue exceeds $21B in 2007. Retrieved March 15, 2008 from http://ap.google.com/article/ALeqM5hccYd6ZuXTns2RWXUgh6br4n1UoQD8V1GGC00

Covington, R. (2004, June 2). Newsflow editor focuses on journalism, not delivery method. *The Convergence Newsletter, 1*(10) Retrieved March 2, 2008, from http://www.jour.sc.edu/news/convergence/issue11.html

Fisher, D. (2004, April 6). 'Storybuilder' embodies new roles in evolving newsrooms. *The Convergence Newsletter, 1*(9). Retrieved March 2, 2008, from http://www.jour.sc.edu/news/convergence/issue9.html.

Grant, A. (2004, July 7). Multiskilled journalists are prepared to tell stories in many forms. *The Convergence Newsletter, 2*(1) Retrieved March 2, 2008, from http://www.jour.sc.edu/news/convergence/issue12.html.

Keith, S., & Silcock, B. (2002, November 15–16). *Translating the Tower of Babel: Issues of language and culture in converged newsrooms: A pilot study*. Paper presented to The Dynamics of Convergent Media, Columbia, SC.

Killebrew, K. (2002a). Culture creativity and convergence: Managing journalists in a changing information workplace. *The International Journal on Media Management, 5*(1), 39–46.

Killebrew, K. (2002b, November 15–16). *Distributive and content model issues in convergence: Defining aspects of "new media" in journalism's newest venture*. Paper presented to The Dynamics of Convergent Media, Columbia, SC.

Killebrew, K.C. (2001, August 5–8). *Managing in a converged environment*. Paper presented to Media Management and Economics Division, Association for Education in Journalism and Mass Communication Convention, Washington, D.C.

Lawson-Borders, G. (2003). Integrating new media and old media: Seven observations of convergence as a strategy for best practices in media organizations. *The International Journal on Media Management, 5*(2), 91–99.

LeNoir, R. (1987). An analysis of adaptation in a regulated industry: Responses of the domestic trunk airlines. Unpublished doctoral dissertation, University of Colorado.

LoCicero, G. (2004, May 5). News resourcer is key information chief. *The Convergence Newsletter, 1*(10). Retrieved March 2, 2008, from http://www.jour.sc.edu/news/convergence/issue10.html

Northrup, K. (2002, June). Preparing your newsroom for the digital world. *Newspapers and Technology*. Retrieved March 2, 2008 from http://www.newsandtech.com/issues/2002/06-02/ifra//06-02_dotink.htm

Northrup, K. (2004, September). The Newsplex convergence guides. *Newspaper Techniques,* pp. 102–105.

Pringle, P., & Starr, M (1999). *Electronic media management*. Boston: Focal Press.

Quinn, S. (2002). *Knowledge management in the digital newsroom*. Oxford,UK: Focal Press.

Redmond, J., & Trager, R. (2004). *Balancing on the wire: The art of managing media organizations*. Cincinnati: Atomic Dog.

Silcock, B. W., & Keith, S. (2006). Translating the Tower of Babel? Issues of definition, language, and culture in converged newsrooms. *Journalism Studies, 7*(4), 610–627.

Wicks, J., Sylvie, G., Hollifield, C., Lacy S., & Sohn, A. (2004). *Media management: A casebook approach*. Mahwah, NJ: Lawrence Erlbaum.

CHAPTER 9

# A Feminist Perspective on Convergence

August E. Grant
Jennifer H. Meadows
Elizabeth Jordan Storm

Most studies of convergent newsrooms (e.g., Dupagne and Garrison, Chapter 11, this volume; Keith and Silcock, Chapter 13, this volume; Quinn, 2004; Singer, 2004) indicate that these newsrooms adapt a hierarchical newsroom structure that is similar to those in traditional newsrooms. In hierarchical newsrooms, content flow is controlled and characterized by competition and power structures. This structure, however, is just one of many possible structures for converged newsrooms. Rather than simply adopting the organizational structure that has been used in the past, managers implementing convergence in a newsroom have an opportunity to reconsider the structure of the newsroom, examining alternatives that may facilitate the flow of news and information from the news organization to the audience, and vice versa.

Alternative newsroom structures may be created phenomenologically, or they may be derived from theoretical conceptions of organizational structure. Many theories of management, organizational behavior, media structure, and media management may be applied to this question. This chapter provides an application of feminist theory to suggest alternate conceptions of converged newsrooms. In the process, the chapter demonstrates how theory may be applied to address the structure of newsrooms.

The application of feminist theory to the structure of a converged newsroom presents an interesting set of options regarding the structure and flow of information within these newsrooms. Although the present media's culture of contribution (Willis & Bowman, 2003) presents an opening for multidirectional conversations and cooperation within newsrooms, those with hierarchical structures potentially silence opportunities for the introduction of additional input, diversity of voices and questioning of content. This structural barrier proves particularly problematic for community journalism outlets and other participatory newsroom models that welcome citizen content through online media.

This chapter applies a variety of perspectives rooted in feminist theory to a review of converged newsrooms in order to identify optimal as well as flawed practices, resulting in the formulation of alternative conceptions of structuring newsrooms and facilitating newsflow, with an emphasis on collaboration and cooperation in nonhierarchical structures rather than on hierarchy and competition. The framework of feminist theory, which is divided into different perspectives, offers a network of individual and distinct applications from which one can gain a better understanding of what is happening around and within (hooks, 1984) in regards to difference, power, and the production of domination, with the goal of improving understanding and awareness in order to activate change. Although the popular belief is that feminist theory is primarily concerned with issues of gender, many feminist theorists today argue that the network of oppression and inequality is supported by interweaving issues of race, gender, class, disability, nationality, and age (Heywood & Drake, 1997).

The chapter begins with an introduction to feminist theory and an exploration of dimensions of feminist theory that are especially relevant to the structure of converged newsrooms. These theoretical dimensions are then applied to create an alternate conception of how a convergent newsroom may be structured. The implications of this theory for community and participatory journalism are then discussed.

## Introduction to Feminist Theory

It will be impossible to thoroughly review feminist theory in this chapter. This body of theory is constantly growing and evolving, with multiple branches and differing, even oppositional, perspectives. Even the term feminism can be difficult to define. bell hooks (2000) offers one simple definition of feminism: "Feminism is a movement to end sexism, sexist exploitation and oppression" (p. viii). Feminist theory focuses on issues of gender and oppression but also incorporates social change. This body of theory examines ways to break through barriers created by hierarchical structures and patriarchy to allow a multiplicity of voices and true participatory community.

The history of feminist theory is often discussed using waves. The first wave of feminist theory was the period of liberal feminist theory, which ended around the early 1960s. The first wave was primarily concerned with obtaining equal opportunity in the public arena. For example, the suffrage movement grew out of the first wave.

The second wave of feminist theory is generally agreed to have begun with Betty Freidan's *The Feminine Mystique* in 1963. The second wave looked at equality, access, and equal opportunity beyond the public arena, but also at the unspoken inequalities. In the second wave feminists also began to critique the white, middle-class, and heterosexual positions of feminism. Radical, lesbian, and multicultural feminist theory looked at issues of race, age, class, sexual preference, and disability. Second-wave feminists also began to go beyond maximizing opportunity under the existing patriarchal system and began to consider alternative structures that encouraged cooperation, collaboration, and openness.

The third wave of feminist theory emerged when feminists began to critique the second wave as not being inclusive of the issues with which they were concerned. That said, second-wave feminism certainly influenced the third wave, but these latter feminists do not feel addressed by it (Johnson, 2006). Koyama (2001) suggested that third-wave feminism be defined as feminism outside of the second wave and that it is concerned with power imbalances between women, not just between men and women. Another useful definition explains that this movement is concerned with the equity and gender feminism of the second wave but also wants to change the structures within which people work (Heywood & Drake, 1997). This "branch" of feminism is difficult to define and is a contested term. What seems to unite these feminists is a concern for multiple ways of addressing problems of inequality, political activism, global and ecological concerns, and racial and sexual inclusivity (Rowe Karlyn, 2006).

Thus, feminist theory looks at a variety of issues including but not limited to bodies, race, class, gender, language, pleasure, power, and sexual division of labor (Kolmar & Bartkowski, 2005). Power and its exercise is an important variable to examine in relation to feminism. As Allen (1999) writes, "Feminists are interested in studying power... because we have an interest in understanding, criticizing, challenging, subverting and ultimately inverting the multiple axis of stratification affecting women in contemporary western societies" (p. 2). Allen suggests that there are three types of power: power over, power to, and power with. Power can be a relationship between domination and subordination. Power can also be a resource. Finally, power can be empowerment or transformation (Allen, 1999).

Feminists study the way power is used by men to dominate women and the way women use power to dominate other women based on age, class, racism, and so on. This notion of power is addressed in feminist communication theorists' critiques of the sender/receiver model, with the center transmitting to the receiving masses. The center has the power over the masses. Feminist communication theory argues for the consideration of other models of communication. For example, the inherent interactivity of online communication breaks up this traditional sender/receiver model, opening the study of online discourse to alternate models and defining new roles, including overlapping roles.

Feminist theory is concerned with the boundaries created in patriarchal/hierarchical structures. For example, in traditional newsrooms, news is separated into boxes: political, lifestyle, health, sports, and so forth, or even grassroots and professional, and user-generated and traditional. Application of feminist theory suggests that these boundaries may be fluid, allowing for multiple voices and experiences. The boundaries inherent in a hierarchical structure block these voices and experiences and, perhaps more importantly, stifle difference and conflicting interpretations. For example, feminist pedagogy supports classrooms that focus on collaborative learning, learning with multiple strategies, decision making through negotiation and consensus and working individually and in groups (Rowe Karlyn, 2006).

## Liberal Feminist Theory

Arguing for equality of opportunity and rights for women and men, liberal feminist theory grew out of Locke's and Rousseau's concepts of liberalism, which were first explored in the 18th and 19th century. Liberal feminist theory essentially argues against socially constructed gendered roles of difference between the sexes, particularly in the public arena (Tong, 1989).

For over 200 years, liberal feminists have challenged discrimination and fought for equal rights between men and women, fighting against barriers to women's education and economic opportunities (Saulnier, 1996) by encouraging women to replace competitive strategies with cooperative initiatives (Tong, 1989).

Liberal feminists argue that women and men, as rational creatures, are essentially the same, despite their outward appearances (Jagger, 1983), and therefore that hierarchical structures that separate men and women based on their physical properties should be replaced with a fair and equal meritocracy (Eisenstein, 1981). As Saulnier (1960) states, when women enter the workforce, they are often viewed as caregivers rather than equal workers, and pushed into caregiving roles and low-paying jobs as well as paid less when holding the same positions as men. In addition, careers that women entered were often "downgraded" in terms of pay and status. Broadcast journalism, for example, has been called a "pink ghetto" because as young women entered the field, job status and pay seemed to decline and/or stagnate. This reality resulted in the liberal feminist cry to action, "Equal pay for equal work" (Saulnier, 1996, p. 13).

Liberal feminist theory is widely criticized for its narrow focus on gender issues. Feminists today are concerned with the intersectionality of forms of oppression, and feel that liberal feminist theory is too simplistic, that it is concerned only with women achieving equality with men in the ruling class without challenging or changing the cultural basis of group oppression (hooks, 1984). Jaggar (1983) also criticizes liberal feminists' failure to address adequately the split between public and private issues by ignoring the diversity of oppression. She states, "Real human beings are not abstract individuals but people of a determinate race, sex or age, who have lived different histories, who participate in different systems of social relations, and who have different capacities and different needs" (Jaggar, 1983, p. 46–47).

While an application of liberal feminist theory to convergent newsrooms is severely limited, it is important as it introduces the need for management structures that are based on egalitarianism and qualification rather than privilege, as well as cooperative interactions among all employees, rather than cooperative strategies among management.

Convergent newsroom structures are fluid beasts, nonrigid and flexible in manner, whereas nonconvergent newsroom practices are primarily controlled and maintained through exclusionary practices. Thus, the liberal feminist fight against socially constructed norms in the workplace, particularly in hiring practices and network structuring, remains relevant in our discussion of convergent newsrooms.

## Radical Feminist Theory

Screaming "revolution" rather than "reformation," radical feminists disagree with liberal feminists' desire to achieve gender equality through the elimination of discriminatory policies and systems, such as the workplace. Radical feminists also disagree with the liberal feminist strategy of incremental change in the fight against women's oppression (Echols, 1989). Instead, radical feminism, which was modeled after the civil rights movement (Saulnier, 1996), fights for women's liberation, arguing the fundamental cause of women's oppression is the current sex/gender system (Tong, 1989). Saulnier's framework of the five fundamental components of radical feminist theory, as discussed in the following paragraph, highlight the theory's dedication to activism and social change (Saulnier, 1996). The underlying philosophy of radical feminism may be compared to one of the tenets of journalism: to challenge the status quo, providing a check on those in power–a role journalist Finley Peter Dunne described as to "comfort the afflicted and afflict the comfortable."

Saulnier's (1996) first component of radical feminist theory is that the personal is political. Radical feminists insist that separating public issues from private issues masks the reality of a male power system that operates in both spaces. By engaging in consciousness raising, radical feminists attempt to form bonds of sisterhood and identify similar sociopolitical struggles. Saulnier's second component of radical feminism is that all women are all oppressed, as society is patriarchal. Radical feminists argue that society is a cultural universal wherein every institution reinforces the social order and patriarchal structures privilege men through complex manipulations of identity, social interactions and structures of power. Saulnier's third component of radical feminist theory is that women are controlled and damaged psychologically by the internalization of oppressive patriarchal messages. She states that women's personalities and sexuality have been constructed to serve men's purposes, particularly in regard to women's ability to reproduce. In response to oppression based on women's biological differences, some radical feminists argue for women's abandonment of patriarchal reproduction and movement toward technological reproduction. Others argue that technological reproduction would still be controlled by men, as men control technology. Saulnier's fourth component of radical feminism is that radical feminists view women and men as fundamentally different, particularly in regard to how they conceptualize power. While men seek to dominate and control power, women are more interested in sharing and nurturing power. Saulnier's final component of radical feminism is that society must be completely altered, as sexism is a social system that is pervasive in every facet of life and that hierarchies must be eliminated in women and men's public and private lives.

Critics argue that the most notable flaws within the framework of radical feminist theory, as well as liberal feminism, is its failure to identify the diversity of women's oppression. Additionally, its anti-male stance repels many women who do not define their interests solely by sex and gender lines (hooks, 1984). Perhaps the most important lesson from radical feminist theory for understanding newsroom management and structure is the manner in which patriarchal structures have come to dominate newsrooms, as they have other areas of society. The important question to address is whether a patriarchal structure serves to oppress those within the organization, limiting both the individual and the organization.

## Marxist and Socialist Feminist Theory

Unlike liberal feminist theory, Marxist and socialist feminist theory rejects individualism and positivism (Rosser, 2005), arguing that women's oppression is not the result of individuals' interactions but rather larger political, social and economic structures within which individuals live (Tong, 1989). Marxist and socialist feminist theory is essentially focused on issues of capitalism and patriarchy, arguing against oppressive systems of production (Saulnier, 1996). Building on the argument that capitalism and patriarchy reinforce divisions of labor by gender, class and race, Marxist and socialist feminist theory suggest a focus on the erasure of production divisions in newsrooms.

Numerous studies have documented the need for more diversity in newsrooms, as well as newsrooms' small yet concerted effort to move from being bastions of white male power to newsrooms that are more reflective of the communities they serve. Still, diversity in ownership, workforce, and news production is not the norm today (Barton, 2002; Dates, 2003; Hardin, 2006).

Convergent newsrooms offer opportunities to challenge inequalities of diversity in the organizational structure and workplace of newsrooms, as well as opportunities to potentially introduce a more diverse and reflective voice to the newsroom production process and ultimately the newsroom product through technological advances such as the Internet. Through interactive features, conversations and collaboration between newsrooms and the public are becoming a reality (Storm, 2006; Sundar, Kalyanaraman, & Brown, 2003). This collaborative relationship is evidenced by newsroom Web sites and blogs, which enable the public to communicate directly with newsrooms and potentially introduce a diversity of voices into the production of news.

## Multicultural Feminist Theory

Recognizing that oppression worldwide requires joint efforts to combat, multicultural and global feminists work across and despite national borders, through an application of a larger international framework that recognizes the unique understanding of racism, classism and sexism held by women in different nations (Saulnier, 1996). As a theoretical conception that explores diversity, multicultural feminism maintains that women and men are not all created or constructed equally. Rather, every person experiences oppression differently depending on her race, class, and sex, sexual preference, age, religion, education, occupation, marital status, and health condition (Tong, 1989). Unlike early feminists who argued that the fight against sexism must take precedence over all other forms of oppression, multiculturalist feminists find oppression to be an interlocking network that cannot be separated or understood as a single variable (Tong, 1989). According to bell hooks (1984), people do not share a common lot of oppression, but rather endure a diversity of experience that contributes to oppression through the absence of choices.

While critics of multiculturalism argue that solidarity is needed among vast groups of minorities, Tong states people should not have to look, act, speak and think alike, but rather cultivate mutual toleration, respect and knowledge of each other (Tong, 1989).

This perspective provides insight into both the structure of the newsroom and the conception of the public's relationship to the newsroom. Promoting collaboration, access, and horizontal mobility, convergent newsrooms resist homogeneity by promoting workforces with varied intellectual and technical backgrounds and by producing content that not only speaks to the community, but also reaches it (Rosser, 2005). This perspective provides a theoretical explanation for an observation discussed elsewhere in this volume,that the role of reporter cannot and should not evolve into becoming a "one-man band." Rather, there should be plenty of opportunity for individuals with diverse skills and expertise.

The issue of access across communities is more problematic. Commonly referred to as the "digital divide" (Compaine, 2001), members of the public do not all have ready access to the information resources that enable participation in the process of contributing to the news production process. Since some differences in access are related to ethnicity and socioeconomic status, converged journalists should be encouraged to make extra efforts to seek out contributions and participation from all members of the community, regardless of race, gender, or access to technology.

## Postmodern Theory

As an epistemology that questions essentialism, postmodern feminist theory examines how the power inherent in language acts upon social reality (Saulnier, 1996). Postmodern feminists seek to avoid all "reinstations of phallogocentric thought" and invite women to become the "kind of feminist(s)" they want to be (Tong, 1989, p. 193).

By analyzing the construction of the concepts of women, gender, and sex and often deploying deconstructionist tactics, postmodern feminists are able to redefine and separate these concepts from their roots of oppression. Striving to include many points of view, postmodern feminist theorists are able to think "nonbinary, nonoppositional thoughts" and possibly, through their speaking and writing, "help overcome binary opposition, [and] phallocentrism" (Tong, 1989, p. 210).

The structure of news organizations and definitions of news have been characterized as rigid and unconforming. In the process of restructuring a newsroom as a convergent newsroom, postmodern feminist theory can be applied to the deconstructionist processes of reinvention and conversation. By resisting divisions of power structure and access, a restructured newsroom has the potential to discard the shackles of homogeneity (Rosser, 2005; Rothfield, 1990), resulting in a newsroom structure and product that reflect the situatedness of every member of the newsroom's workforce and readership base.

## Applying Feminist Theory to Converged Newsrooms

Each of the dimensions of feminist theory discussed earlier suggests specific implications and considerations for the converged newsrooms. Looking beyond the contradictions that are inherent across such a multidimensional field, there are a few central themes that can be applied to guide physical design, organizational structure, information flow, and other characteristics of a converged newsroom.

(At this point, it should be acknowledged that many feminists would view the above "prescription" as a patriarchal exercise. These feminists might appropriately argue that these elements must be allowed to emerge in an organic fashion, with the impetus coming from empowerment of the members of the news staff and the public. However, given organizational inertia toward imposition of structure from outside, combined with time constraints related to deadline pressure, perhaps a prescriptive solution is required in order to begin the change process.)

One of the difficulties of maintaining a hierarchical organizational structure in converged newsroom is the sheer number of inputs and outputs in the newsroom. Where a hierarchical structure is quite functional and controls the flow of information from many sources to be delivered through a single source, the fact that a converged newsroom is dealing with multiple output media complicates the organizational structure. One choice is to create multiple hierarchies, with separate flows and decision makers for each output medium. But the theoretical discussion that began this chapter suggests that the complexity of the information flows into a converged newsroom might be better served by a more organic and egalitarian structure, one in which all members of the organization play a more active role in the myriad decision processes regarding the flow of news to individual media.

There is a practical reason for adopting a more collaborative structure in a converged newsroom as well: the time variable. When a news story is processed in a converged newsroom, there is always a decision to be made regarding which medium should receive the story at what time. Immediate updates might go on the Web site or be distributed through cell phone alerts, with more detailed information then provided in a television newscast or a newspaper story. By empowering individuals in the newsroom to make decisions, a collaborative structure allows information to be processed more quickly, and then more quickly delivered to consumers.

In addressing the structure of newsrooms, it is important to recognize that the power structure goes far beyond the confines of the newsroom. Editors and publishers may be at the top of this hierarchy, but it is not reporters and photographers who are at the bottom. Rather it is the members of the public who consume the content.

If, rather than seeing the public as passive consumers of news content, they are seen as active participants in creating the agendas, identifying the issues, and even providing content, an alternate conception of the newsroom emerges. Indeed, this principle is at the heart of community journalism (Rosen, 1999).

According to Kovach and Rosenstiel (2001), the concepts of journalism and community cannot be separated, as journalism initiatives help establish and maintain communities by reporting local concerns and interests. Focused on local coverage, community journalism initiatives define and reflect the perspectives of their fellow community members better (Byerly, 1961; Husselbee & Adams, 1996) through their involvement with the communities they serve (Lauterer, 2000).

To define community journalism, an understanding of community must first be formed. In 1983 Stamm and Fortini-Cambell defined community according to three domains: community as a place, community as a social structure, and community as a social process, all of which facilitate connections or ties people form. Community journalism occurs when journalists are involved in a community domain, usually through community organizations and activities (Coble-Krings, 2005).

One fundamental challenge that may thus be affecting community journalism practices is the structure of the newsroom itself. Given a traditional, hierarchical newsroom structure, members of the public are less likely to be truly empowered to create, contribute, and have a voice in the news organizations.

An example of this challenge can be found in the study of *Bluffton Today,* a local newspaper considered a pioneer in community journalism efforts. Analysis of this news organization using a combination of in-depth interviews and content analysis revealed that this newspaper's community journalism efforts played a more important role in setting agendas than in empowering community members to share and participate in the news gathering and reporting process (Storm, 2006).

In this newsroom, reporters were found to be using the blogs, forums, and other participatory elements of the Web site as sources for their own stories. In essence, the contributions and efforts of these community members within the hierarchical newsroom structure provide a false empowerment of the community members, but play a relatively minor role in contributing to the information flow and product of the newsroom.

Although an examination of the range of community journalism efforts is beyond the scope of this chapter, we argue that adding the community journalism level to the existing newsroom hierarchy is less effective than creating an alternative organizational structure that is designed to more fully empower members of the community as equal partners in news processes.

In most cases there is no active decision process regarding the structure of a converged newsroom. Rather, the structure is simply an adaptation of the previously existing structure. Presuming that theory—whether feminist theory or another useful paradigm—may be applied to suggest alternative organizational structure, physical design or information flows in a converged newsroom, one question that must be addressed is *when* the structure of the newsroom should be changed.

Management studies indicate that changing behavior in an organization is a difficult process. The process is more complex than just changing the behavior: The existing behavior must be "unfrozen"; the new behavior must be established; and the new behavior must be "frozen" (Lewin, 1974). The process of unfreezing and refreezing behavior is facilitated by other changes, both physical and behavioral. The best time to implement a change from a traditional hierarchical structure to a more collaborative structure is therefore when the convergent practices are introduced into the newsroom.

## Next Steps

Early in this chapter, we suggested that feminist theory could be applied to the structure of converged newsrooms. There are many motivations for this application, including empowerment of the newsroom staff, empowerment and reconceptualization of the public, and (perhaps) economic efficiencies.

The process is not limited, however, to converged newsrooms. The Internet has enabled the emergence of new journalistic forms and organizations that also have the potential to apply these same principles. In a nascent study of online newsrooms, Stern (2007) talked to women working in editorial leadership positions in online newsrooms. Interesting, these women discussed the ability to set up alternative newsroom structures that were more progressive and cooperative. They found the online newsroom had less attention to hierarchy than traditional newsrooms. They also described themselves as mentors and role models rather than supervisors and editors.

## Conclusions

It should be noted that this chapter is not arguing for a change in the structure of all newsrooms; rather purpose of this chapter is to help illustrate the manner in which theory may be applied to address fundamental issues related to convergent journalism. There are certainly other theoretical conceptions that could be equally informative. The lesson is that the application of communication and social theory to convergent journalism provides an opportunity to examine and rethink practices and procedures and converged newsrooms.

The implications of this discussion are important. Through a theoretical application of the goals of feminist theory, namely the elimination of patriarchal/hierarchical power distributions, inequalities of value, and oppression within convergent newsrooms, this chapter identified original conceptions of newsroom structures and newsflow that would protect and promote a diversity of thought and voice within newsrooms, as well as multidirectional conversations and collaboration between staff members and citizens within convergent newsrooms.

## References

Allen, A. (1999). *The power of feminist theory*. Boulder, CO: Westview Press.

Barton, G. (2002). Is diversity making a difference? *Quill, 90*(2), 16–21.

Byerly, K. R. (1961). *Community journalism*. Philadelphia: Chilton.

Coble-Krings, L. (2005, September). *Weekly dilemmas: A study of ethics and community journalism in small towns*. Symposium conducted at the 199th Annual convention of the National Newspaper Association, Milwaukee, Wisconsin.

Compaine, B. M. (Ed.). (2001). *The digital divide: Facing a crisis or creating a myth?* Cambridge, MA: MIT Press.

Dates, J. L. (2003). Diversity advancements require change. *Quill, 91*(6), 8–9.

Echols, A. (1989). *Daring to be bad: Radical feminism in America 1967–1975*. Minneapolis, MN: University of Minnesota Press.

Eisenstein, Z. (1981). *The radical future of liberal feminism*. Boston: Northwestern University Press.

Hardin, M. (2006). Fewer women, Minorities work in Sports departments. *Newspaper Research Journal, 27*(2), 38–51.

Heywood, L., & Drake, J. (1997). Introduction. In L. Heywood & J. Drake (Eds.), *Third wave agenda* (pp. 1–20) Minneapolis: University of Minnesota Press.

hooks, b. (1984). *Feminist theory: From margin to center*. Boston: South End Press.

hooks, b.(2000). *Feminism is for everybody: Passionate politics*. Cambridge, MA: South End Press.

Husselbee, L. P. & Adams, A. (1996). Seeking instrumental value: Publication of disturbing images in small communities. *Newspaper Research Journal*, 17, 39–52.

Jaggar, A. (1983). *Feminist politics and human nature*. Totowa, NJ: Rowman & Allanheld.

Johnson, M. (2006). Introduction: Ladies love your box: The rhetoric of pleasure and danger in feminist television studies. In M. Johnson (Ed.), *Third wave feminism and television* (pp. 1–27. New York: Tauris.

Kolmar, W., & Bartkowski, F. (Ed.). (2005). *Feminist theory: A reader*. Boston: McGraw-Hill Higher Education.

Kovach, B. & Rosenstiel, T. (2001). *The elements of journalism*. New York: The Crown Publishing Group.

Koyama, E. (2001). *Third wave feminisms*. Retrieved February 27, 2008, from http://eminism.org/interchange/2001/20010622-wmstl.html

Lauterer, J. (2000). *Community journalism. A personal approach (2nd ed.)*. Ames, IA: Iowa State University Press.

Lewin, K. (1974). Frontiers in group dynamics: Concept, method, and reality in social science: Social equilibria and social change. *Human Relations, 1*(1), 5–41.

Quinn, S. (2004). Better journalism or better profits? A key convergence issue in an age of concentrated ownership. *Pacific Journalism Review, 10*(2), 111–129.

Rosen, J. (1999). *What are journalists for?* New Haven, CT: Yale University Press.

Rosser, S. V. (2005). Through the lenses of feminist theory: Focus on women and information technology. *Frontiers: A journal of women's studies*. 26(1), 1–23.

Rothfield, P. (1990). Feminism, subjectivity, and sexual difference. In S. Gunew (Ed.), *Feminist knowledge: Critique and construct* (pp. 121–146). New York: Routledge.

Rowe Karlyn, K. (2006). Feminism in the classroom: Teaching towards the third wave. In J. Hollows & R. Moseley (Eds.), *Feminism in Popular Culture* (pp. 57–75). New York: Berg.

Saulnier, C. F. (1996). *Feminist theories and social work: Approaches and applications*. Binghamton, NY: Haworth Press.

Singer, J. (2004). Strange bedfellows? The diffusion of convergence in four news organizations. *Journalism Studies, 5*(1), 3–18.

Storm, E. J. (2006, October). *Converging the conversation: The introduction of Web-generated citizen content into newsrooms*. Paper presented to "Convergence and Society: Ethics, Religion, and New Media," Columbia, SC.

Stern, S. (2007). Increased legitimacy, fewer women? Analyzing editorial leadership and gender in online journalism. In P. Creedon & J. Cramer, (Eds.), *Women in mass communication* (3rd ed., pp. 133–146. Thousand Oaks, CA: Sage.

Sundar, S., Kalyanaraman, S., & Brown, J. (2003). Explicating Web site interactivity: Impression formation effects in political campaign sites. *Communciation Research, 30*(1), 30–59.

Tong, R. P. (1989). *Feminist thought: A more comprehensive introduction*. San Francisco: Westview Press.

Willis, C., & Bowman, S. (2003). *We media: How audiences are shaping the future of news and information*. Retrieved June 15, 2006, from http://www.hypergene.net/wemedia/weblog.php

CHAPTER 10

# On Linkages and Levels

## Using Theory to Assess the Effect of Converged Structures on News Products

George L. Daniels

The number of case studies and surveys of convergence activity in news organizations has increased steadily since 2002. A good many of the studies that have been presented or published in academic journals seem to focus on describing what various news operations are or are not doing in the area of convergence. In the face of these studies, some of those focusing on convergence and training others to work in converged environments have suggested convergence journalism leverages the strengths of each medium to tell a more complete story than any one medium could tell on its own (Tompkins, 2005). This conception of the effect of convergence contrasts with those who say the "content provider" of convergence can only ricochet mindlessly, trying to meet the demands of myriad media forms (Corrigan, 2004).

Many question whether good journalism or the bottom line is the reason that organizations converge (Quinn, 2004, 2005). This chapter reports on a research project designed to answer that question. For those in search of better journalism, two theories or conceptual models are particularly helpful in understanding the effect of convergence on the resulting news product. In studying news production process, McManus (1992) juxtaposed a purely economic theory of news production, where news is a commodity like any other raw material, with a purely journalistic theory of news production. He suggested actual commercial news production represents a compromise between the two. Meanwhile, Tuggle, Carr, and Huffmann's (2007) seven levels of convergence permit us to look at converged news content in terms of intensity and impact.

Using cases from one of the most talked-about models for newsroom convergence, this chapter reports an analysis of work product of converged partners as compared to their competitors' product. The competition extends beyond just the television competitors to the newspaper and online competitors. Market-driven journalism theory and the levels of convergence are used together as a framework for comparing the news products. This theoretical and conceptual discussion adds to a growing body of academic research on converged operations, much of which has appeared in mainstream journalism and mass communication journals.

## Theories and Frameworks

Conceptually, convergence is no different from the optimal mixture of organizational breadth and depth that comes from synergy, where there is coordination of parts of a company so that the whole actually turns out to be worth more than the sum of its parts acting alone (Turow, 1992). Increased fragmentation of media channels has threatened the ways that media firms have gotten resources and used them efficiently, making synergy a viable tool for organization (Turow, 1992). Conceived as synonymous with the media management strategy of synergy, convergence may be conceived as merely a tool for increasing one's bottom line. Indeed, critics of convergence have lumped it in with consolidation, which until recently, happened at a greater pace as companies including AOL, Time Warner, Viacom, NBC, and Vivendi sought the advantages that came with conglomeration.

However, another way to conceptualize convergence is as a means of doing better journalism. The synergistic advantage is better news and information for audiences. If one defines theory as a set of interrelated statements asserting how and why two or more variable concepts are related among a class of objects (McLeod & Pan, 2005), McManus' (1994) market-driven journalism theory is most appropriate for a discussion of convergence. As discussed above, the theory juxtaposed a purely economic theory of news production where news is a commodity like any other raw material with a purely journalistic theory of news production.

The *market theory of news production* says the probability of an event becoming news is inversely proportional to the cost of uncovering it and the cost of reporting it. Furthermore, the more the expected breadth of appeal of a story to audiences that advertisers will pay to reach, the more likely the event becomes news. On the other extreme is *pure journalistic theory of production,* which suggests that the probability of an event or issue becoming news is directly proportional to the expected consequence of the story and size of the audience for whom the news item is important. Looking at news production in three stages–discovery, the selection of events for coverage, and reporting of the story–McManus (1994) suggested that actual commercial news production represents a compromise between the two. As other scholars have noted, McManus's (1994) work formalizes the relationship between business consideration and local television news content. In particular, it accounted for economic considerations such as profit maximization (Lacy & Bernstein, 1992).

News operations operating under purely market theory treat news like any other commodity used to generate profits. On the other hand, those operating under pure journalism theory treat news as a means through which an organization serves the public interest. Money is no object in journalism theory, while minimizing expense is a major objective in market theory. Underpinning the market theory is a market structure that economists have termed oligopolistic competition, where there are a few major competitors and high barriers to entry.

McManus (1994) suggested that one of the solutions to market-driven journalism was the notion of change in the public demand for market journalism through consumer education. Along those lines, he offered a survey for consumers to rate newscast "nutrition." While not giving a fully operational or conceptual definition of newscast nutrition, McManus does state that the primary purpose of news is to explain how one's environment is working so that a person can make good decisions, particularly civic decisions. Thus, a more nutritious newscast would aid viewers in fulfilling this purpose. A less nutritious program would be less helpful or not helpful at all in fulfilling this purpose. High-quality news is generally associated with nutrition, while lack of quality is associated with malnutrition.

Rather than viewing it as just synergy for the purpose of increasing the bottom line, market-driven journalism theory allows one to push the discussion about convergence into somewhat uncharted territory, where the focus is on news products and less on practices. While Huang et al.'s (2004) research focused on the quality of the *Tampa Tribune's* product, such an analysis did not account for the competitive environment in which news products are produced. To contextualize the convergence in the larger discussion of media processes (Grant, 2005), the competition in which some converged products are produced is a relevant area for discussion, especially for the television partner in a convergence relationship. Conceptual models such as Dailey, Demo, and Spillman's (2005) convergence continuum account for competition between convergence partners, but not between those converged news operations and those nonconverged operations. Market-driven journalism theoretical perspective also looks at the degree to which news is treated as a commodity.

Not to be confused with a theory per se, the conceptual model developed by the authors of the *Broadcast News Handbook* (Tuggle et al., 2007) to describe the convergence operation at Media-General's News Center is one that can be used to identify and compare degrees of intensity of convergence in a fashion somewhat similar to the convergence continuum. The difference between Dailey et al.'s (2005) convergence continuum and Carr's (2007) Seven Levels of Daily Convergence is that the latter speaks to the day-to-day within an established converged operation. The former is more relationship-based, while the latter is operation-based. As a conceptual framework for assessing day-to-day convergence activity, the Seven Levels of Daily Convergence is appropriate for evaluating and assessing news content, the goal of this study.

According to Carr (2007), the most common and least obvious form of daily convergence occurs at *Level One: Daily Tips and Information*. This level is characterized by a lot of talking between staff from each of the converged partners. This "cross-platform sharing" is the process whereby the converged news operation would "gang up on the competition and deliver what they think is better service to the end users" (Carr, 2007, p. 260).

When the converged partners move from talking to sharing, they have moved to *Level Two: Resources*. One platform covers from another in cases where both are covering the same story. According to Carr, this is quite likely to happen in the area of photography, where still photographers carry small digital video cameras.

The sharing of resources on spot news is what Carr calls *Level Three: Spot News*. In his estimation, this is where converged news coverage "really shines," as the TV station faces the task of going on the air immediately with live coverage and the newspaper and Web staffs help the converged partners "flood the field with crews." Mentioned particularly at this level is the ability for the newspaper partner to bring the resources of the newspaper archive and research desk into the converged coverage strategy for spot and breaking news.

Sometimes the breaking story may not be known to other media in the market. *Level Four: Enterprise Reporting* involves what Carr (2007) calls a breaking enterprise story, which he called "the most powerful journalism—enterprise reporting" (p. 262), whereby an exclusive story eventually may be told on all three platforms. This level of convergence may also occur in what is called "planned co-publication," where editors from converged partners time the release of the story on each platform in a coordinated fashion (Carr, 2007).

When converged partners have a standing commitment to air particular content on all three platforms, they have reached what Carr (2007) calls *Level Five: Franchises*. This most often occurs when reporters on certain beats make regular contributions to other media within the convergence relationship. Examples of this might be the consumer or health reporters doing Web, television, and newspaper reports.

A step above the franchises that regularly appear are those big stories for which the converged partners may plan. Instead of just planned co-publication (Level Four convergence), the convergence has moved to Carr's *Level Six: Events and Special Coverage,* when the converged partners "showcase joint coverage" of major events. Reporters crossing platforms is not a necessary condition for Level Six convergence. Instead, promotion is the major outcome of this level of convergence.

Finally, there is convergence at *Level Seven: Public Service,* where the converged partners bond with the community by explaining their values in a way the public can understand and appreciate. Examples of this type of convergence would be jointly sponsored town hall meetings or community forums. In this level of convergence, the public sees and may appreciate and/or criticize the partners for their news coverage.

## Research Questions

At least one scholar asking the fundamental question of the reason behind the push toward convergence noted that ultimately convergence is about better journalism (Quinn, 2005). The key criterion appeared to be the editorial and social values of the news managers. On the other hand, some have suggested that convergence among media firms is simply a strategy of value creation (Rolland, 2002). Researchers studying the environment of television news, in particular, have found that local newsrooms tend to air a higher percentage of stories with low assembly costs than high assembly costs (Lacy & Bernstein, 1992) and that they operate under a competitive ethos that has become both a practice and value system (Ehrlich, 1995). So, what does that mean for a television operation such as WFLA-TV, in the highly competitive 14th largest media market in the nation? If convergence is ultimately about better journalism while at the same time making one's firm more valuable, how are these objectives achieved in the everyday work of the media outlet? By linking the perspectives of McManus's (1994) *Market-Driven Journalism* and Carr's (2007) *Seven Levels of Daily Convergence,* to actual news content, one can begin to assess the effect of convergence on news content.

In particular, this study sought to answer the following research questions:

**R1:** Was there an obvious difference between a story generated by the converged news operation and a stand-alone news operation?

**R2:** What is the nature of the collaboration between the two different media to produce a news story for each respective outlet?

**R3:** Does the converged news product move closer to pure journalism than the story from each of the stand-alone operation(s)?

**R4:** Which level of convergence is most often seen on a night-to-night basis in the late local newscast?

**R5:** How well does the converged news product exemplify the characteristics of synergy as outlined in the management literature?

## Methodology

A set of tape-recorded newscasts and newspapers as well as the LEXIS-NEXIS Academic Universe database were used to build 13 convergence case studies in which the news product could be analyzed using *market-driven journalism theory* and the *Seven Levels of Convergence*. The cases were analyzed qualitatively, applying these two perspectives.

Because it has been analyzed in multiple research projects (Flanagan & Hardenbergh, 2003; Huang et al., 2004; Killebrew, 2005; Lawson-Borders, 2002; Singer, 2004) and was the case from which Carr's (2007) *Seven Levels of Daily Convergence* was derived, Media-General's News Center was the focus of this first study of converged news content across multiple media. This included not only News Center-generated content from the converged partners—WFLA-TV, *The Tampa Tribune* and Tampa Bay Online (TBO.com), but also the content of its competitors. The Tampa-St. Petersburg media market is part of an increasingly rare breed of two-newspaper towns. Both the *Tribune* and its more widely circulated competitor, the *St. Petersburg Times,* publish editions for Tampa and St. Petersburg. WFLA-TV is one of seven network-affiliated stations. For this study, primetime or late newscasts on WFTS (ABC affiliate), WTVT (FOX affiliate), WTTA (WB affiliate), WTSP (CBS affiliate) were analyzed. In addition to the newspapers and television stations, the converged TBO.com has an arch–Web competitor in the portal Web site, tampabay.com, which is owned by the *St. Petersburg Times* and until Fall of 2005 also included Gannett-owned WTSP in a three-way partnership. WTVT.com, WTSP.com, WFTS.com, and WTTA38.com also provide local news Web content online and compete directly with TBO.com

In the interest of assessing news content in highly competitive news environment, the television partner, WFLA, along with its competitors, was the starting point for building 13 "everyday" convergence cases. A convenience sample of newscasts included primetime and late news programs videotaped on March 2004 (nonsweeps) and February 2005 (sweeps). Another week of webcasts provided by WFLA, WTSP, and WFTS were monitored on the World Wide Web only during the first week of October 2005 (nonsweeps). Sweeps periods are traditionally times when stations put their "best foot" forward in an attempt to attract the highest audiences and ratings, which transfer into greater advertising revenue. Previous analysis of converged newscast content was only done during the month of August 2001 (Flanagan & Hardenbergh, 2003). It included both 6 p.m. and 11 p.m. newscasts. Because it immediately follows primetime, the late newscasts are often considered the "signature" program of the day. It also follows the primetime newscasts of competitors who do a 9 p.m. or 10 p.m. newscast. A third reason to focus on the late newscasts is the fact that content from the newspaper partner is more likely to be ready for inclusion and/or promoted in this particular news program.

Each newscast on WFLA was screened for on-air mention of the stories produced with newspaper partner, *The Tampa Tribune*. A listing of those stories in the seven WFLA newscasts screened was made, and newscasts on WFTS, WTVT, WTTA, and WTSP were also screened for whether or not they aired the same stories. Transcripts were generated from all television stories that appeared on the list as having been produced with the *Tribune*. Additionally, newspaper stories matching each of the WFLA stories were either clipped from the actual newspaper or located via the LEXIS-NEXIS Academic Universe database. The same search was conducted on the designated dates for *The St. Petersburg Times*. The search included the same time periods for stories in the *Tribune* where a WFLA-TV reporter was involved as a coauthor. Finally, TBO.com content was located for each of the stories mentioned either in the WFLA screening or *Tribune* search. Along with the TBO.com content, searches were conducted on October 2, 2005, for competing Web content.[1]

As an additional source of data, to assess the degree to which everyday convergence was occurring, the study analyzed more recent rundowns taken from webcasts posted on WFLA/TBO.com, WTSP, and WFTS[2] during the week of October 5, 2005. See Table 10.1 for a complete list of record dates. Where a convergence case was identified, the television report was transcribed and online content located on the Web page.

## Convergence Cases

### Case 1 *Coronet Report*

This story focused on a news report by the Florida Department of Health that found cancer rates were no higher near the Coronet Industries plant than anywhere else in the state or country. The report was the latest in an ongoing battle between officials

**Table 10.1** ▰ Convergence Monitoring Dates

| Date | Monitoring Method |
|---|---|
| Wednesday, March 4, 2004 | Over-air |
| Thursday, March 5, 2004 | Over-air |
| Friday, March 6, 2004 | Over-air |
| Sunday, February 13, 2005 | Over-air |
| Monday, February 14, 2005 | Over-air |
| Tuesday, February 15, 2005 | Over-air |
| Wednesday, February 16, 2005 | Over-air |
| Monday, October 3, 2005 | Web |
| Tuesday, October 4, 2005 | Web |
| Wednesday, October 5, 2005 | Web |
| Thursday, October 6, 2005 | Web |

1. It should be noted that because the Web searches were conducted after the fact, some Web sites did not provide search engines where archived stories could be located.

2. Neither WTVT nor WTTA provide webcast content. The news content on their stations' Web sites consisted only of short versions of Web stories.

at the Coronet phosphates processing plant and nearby residents, who have long held that their sicknesses were the result of pollutants from the plant.

This story was the subject of multiple WFLA NewsChannel 8 On Your Side Investigative Reports, which were completed in cooperation with *The Tampa Tribune* and published in a "Special Report" on Coronet Industries on TBO.com The special report included not only all the stories in this ongoing debate, but links to government reports associated with the story. All of WFLA's competitors reported this story, mostly in news package format. Ironically, all except WFLA and WTTA led their late evening newscasts with the story.

Even though WFLA had aired special reports on the Coronet Industries investigation in cooperation with *The Tampa Tribune,* its coverage of this latest development included only comments from the Hillsborough Health Department and a Coronet statement. The fewer sides on a controversy exemplify more market journalism than pure journalism. At the same time, the reporter maintained a neutral stance, and the sources did give factual information. The story mentioned background, but did not handle it as well as several of the other television station reports. The strong point of this converged coverage was the Special Report on TBO.com. This case exemplifies *Level Three: Spot News.*

## Case 2 *Dollar Daughter*

This story of alleged abuse became national news. In this particular case, the daughter of John and Linda Dollar, a couple accused of torturing and starving their seven adopted children, wants to gain custody of the children in the case.

Lawyers for Shanda Rae Shelton talked with WFLA-TV on the day before a story about her court hearing was to appear in *The Tampa Tribune*. The WFLA-TV reporter received a rare double byline for her work in gathering the interview in the case. The same lawyers also spoke with WTTA, which also ran the story in their primetime newscast. Both WFTS and WTSP ran brief stories on Shanda Shelton's upcoming hearing, but did not include interviews with Shelton's lawyers.

The John and Linda Dollar case was clearly a crime story, making it market journalism. By obtaining an interview early, WFLA gave *The Tampa Tribune* an opportunity to advance the story in a way that *The St. Petersburg Times* did not. However, the only sources provided both in *The Tampa Tribune* and the WFLA stories were the lawyers. The stories did not include comments from the prosecution side of the story. This is an example of convergence at *Level Four: Enterprise Reporting*.

## Case 3 *Clearwater Plane Crash*

This story focused on the deaths of two people killed after a single-engine Beechcraft Debonair crashed into a home shortly after take-off from the nearby Clearwater Airpark. While no one on the ground was hurt, it was the latest in what appeared to be several crashes involving aircraft using the Airpark.

This was a story reported by WFLA-8 on its early evening newscast and developed further as the lead story at 11 p.m. As a tag (story after the main television news report), WFLA used the resources of its coverage partner, *The Tampa Tribune,* to look for incidents involving aircrafts at the Clearwater Airpark. The timeline of previous incidents was carried as an infographic in the next day's edition of the *Tribune*.

Not surprisingly, all the stations carried this story, some in a more in-depth way than WFLA. While all of WFLA's competitors (WTVT, WTTA, WTSP, and WFTS) had reporters live on the scene of the crash, WTSP and WFTS had two reporters on the story. WTTA produced a separate news package on neighbors' concerns about the incidents related to being close to the Airpark.

Using its coverage partner, the *Tribune,* WFLA-TV provided some background on the breaking news, which was an accident, a market journalism story. The WFLA-TV report had multiple sources. It is unusual to see various sides of stories in a tragedy like a fatal plane crash. However, the stations that provided a second reporter were able to provide more sources who could give facts in their comments to reporters. Additionally, compared to its competitors, WFLA-TV's story with *The Tampa Tribune* lacked, in terms of description, a discussion of the problem of living near an airfield. This is an example of *Level Three: Spot News* convergence.

### Case 4 *Mobile Home Eviction*

This was an advancer on a planned shutdown by the City of Tampa of Wuiza Bug, a mobile home park that had multiple code violations. The shutdown was expected to happen the following day. More than 10 residents were still at the park, many of them elderly and/or military veterans.

This story was reported by *The Tampa Tribune* in its Monday editions and its coverage partner, WFLA-Channel 8, on its Sunday night newscast. WFTS also reported the story in much the same way as WFLA, including sound bites from some of the same residents who were featured in the WFLA story.

Being evicted from one's home can be a traumatic story. The story deals with an important social issue—the conditions of housing for elderly and military veterans. This topic falls into the pure journalism area. By taking the story and developing a television package with comments from those being evicted, WFLA extended the story using the visuals and sound of broadcast media. At the same time, no city officials or the landlord (a key figure in the story on WFTS) were included. This story is an example of *Level Four: Enterprise Reporting convergence*.

### Case 5 *NHL Season Ends*

A "big story" of the day for the Tampa Bay area, involved the defending Stanley Cup champion team, the Tampa Bay Lightning. National Hockey League officials announced there would be no hockey season due to an impasse in negotiations between owners and players. WFLA in its reports included *The Tampa Tribune* sportswriter, who covers the Lightning as his beat. Erik Erlendson had stories in *The Tampa Tribune,* which were also carried on TBO.com's Web reports.

All of WFLA's competitors carried this particular story with multi-element coverage (more than one story). *The St. Petersburg Times,* which owns the St. Pete Times Forum, where the Lightning play, also had extensive coverage. However, there was no evidence that *The St. Petersburg Times* staff worked directly with its broadcast partner, WTSP.

As with other sports stories analyzed, this one included reporters from the converged partner in the broadcast story. This extended the expertise that would have only come from the print side. However, the coverage itself was superior to the other television stations, which featured mostly the same Lightning players and NHL and team officials, only because it provided some analysis and content beyond the emotion of the story. This business or economic story is classified as more pure journalism, which WFLA improved with its convergence. It is *Level Three: Spot News convergence.*

## Case 6 *Port of Tampa*

As a move to strengthen the security at the area's ports, new security systems were being installed as part of a $16 million new operations center at the 5,000 acre Port of Tampa. This story was apparently broken by *The Tampa Tribune* in a feature on its Business Page. WFLA ran a reader on the news of the new security center, citing the report by its coverage partner. Neither of WFLA's competitors, nor *The St. Petersburg Times,* had this story.

The actions of local government fall into the category of pure journalism. The newspaper story uses multiple sources that provide more background and understanding about this new security measure. The story addresses the degree to which local officials are guarding against terrorism, a public safety issue. The WFLA story only reports the headline and lead of the newspaper story without any video or other more compelling way to explore the public affairs issue of local security against terrorism. This story did NOT take advantage of the television medium to improve the journalism. Although this was a promotion for the newspaper, journalistically, this story falls into the category of *Level One: Daily Tips and Information* convergence.

## Case 7 *Coronet Injunction*

This was a follow-up to an ongoing story about the alleged harmful effects of living near Coronet Industries (see Case 1), a phosphate processing plant. A circuit court judge dissolved an injunction that kept Coronet from removing soil and water from the plant. Lawyers for residents suing the company had obtained the order after alleging the company was trying to dispose of evidence. This story appears to have been broken by *The Tampa Tribune,* which originally conducted an investigative report that included a multimedia report on TBO.com and special WFLA NewsChannel 8 On Your Side Investigative Report. Only WFLA and the *Tribune* reported this story.

The ongoing criminal investigation places this story in the category of market journalism. The story itself may have been investigative and involved the health and welfare of citizens. But the story in *The Tampa Tribune* simply updates the legal battles between victims and the company. WFLA's report provided even less insight on what the story means or what the different sides are saying about the judge's ruling. It did provide background. However, overall, this amounts to just *Level One: Tips and Information convergence*, because it is based on a tip from the *Tribune* to WFLA.

## Case 8 *Lefave*

This story involves an update on the story of Debra Lefave, the Bay area woman charged with having sex with a 14-year-old student. A judge has ruled that Lefave's mental health records must stay sealed. This was a story on which TBO.com prepared a special report, which included a number of supplemental materials in the case. From audio of recorded phone calls in the case to backgrounders on Lefave, her husband and her attorney, the TBO.com multimedia report could go with the WFLA story.

WTSP did a "10 News Extra" which was also listed on its Web site. The Extra appears to be a "sweeps piece" prepared and promoted during the February sweeps, ironically on the same night that the new developments were reported on WFLA and WTVT, which aired a VOSOT (video with a sound bite) from the attorney.

The convergence element of this story comes from the Web site TBO.com. The Special Report on Debra Lefave provides important background and context in a multimedia manner. The story itself is little more than a crime story, making it market journalism. However, the converged content provides factual evidence via additional sources, including access to materials used by lawyers involved in the case (i.e., audio of the phone calls). This is important background for the reader or user to understand the larger issues. It is a resource page each converged partner could point to at any time when a new development occurred in the case. This story is an example of *Level 6: Events and Special Coverage* convergence on TBO.com.

## Case 9 *Judge Holder*

This story focused on new developments in the case against a Hillsborough County Circuit Judge who was to be tried on charges he plagiarized parts of a military research paper. Both *The Tampa Tribune* and TBO.com were involved in initially reporting results of the *Tribune*'s investigation of documents from the Judicial Qualifications Commission, which the *Tribune* had reviewed by three independent experts. The *Tribune*'s investigation showed the questionable paper was not a forgery. The decision to try Holder was only reported on WFLA, but *The St. Petersburg Times* did report on the case later.

Despite the fact this was an earlier investigation of a judge by WFLA's newspaper partner, the update on this story contained little balance. There was no comment from the judge or his attorney. No sources were cited in the television story except the *Tribune*. As a crime story, this fell into the category of market journalism. It is merely *Level One: Daily Tips and Information* (to WFLA).

## Case 10 *Are We Ready?*

In the wake of the floods and tragedy in New Orleans, *The Tampa Tribune* conducted an investigation using Census records to see how many people would need to be evacuated should a deadly storm surge threaten to flood Tampa. It appears the *Tribune* was the lead partner on this story, which included an interactive look on TBO.com at what floods would do to downtown Tampa. WFLA-TV used footage from a local summit held earlier in the month along with interviews with key emergency management officials to produce its version of this story. While not covered on the same day, a story on lack of emergency preparedness was published in *The St. Petersburg Times* before the *Tribune* ran their package.

Exploring the impact of a natural disaster has become more commonplace recently due to the tragic events surrounding Hurricane Katrina. Newspapers and television stations around the country have examined this topic. However, the *Level Four: Enterprise Reporting, planned copublication*, of this story by TBO.com, *The Tampa Tribune,* and WFLA was an illustrative example of benefits of convergence for doing better journalism. The enterprise nature of the actions of government and reporting resources allocated to do the analysis reflect pure journalism. The three organizations worked together to take advantage of the Web's interactive capacity to show visually what a flooded city might look like and a television reporter's unique way of walking and talking and showing highlights of government officials at work.

## Case 11 *Baghdad Bomb Victim*

This story featured an interview that was a follow-up with a Tampa Bay area woman who was injured in a rocket attack in 2003 in Baghdad. WTSP worked with *The St. Petersburg Times* to develop a feature on a key figure in this local story on an international crisis related to the Iraq War. There was no indication that any of WTSP's competitors or *The Tampa Tribune* covered this story.

The fact that this story involves an injury places it in the category of market journalism; however, the story was *not* about the injury itself, it was about the recovery from the injury. Seeing the bombing victim, who was a local person, and hearing her story audibly added a great deal to the coverage that otherwise would have appeared only in the newspaper. The television story was the element *added* to a newspaper package of stories that included multiple sources, such as the surgeons involved in the explosion victim's recovery and her relatives. The secondary story in the package helped give background and context. This was an exampled of planned copublication, which is *Level Four: Enterprise Reporting*.

## Case 12 *Ober Plot*

While prosecutors were not saying much about an alleged plot to kill Hillsborough County State Attorney Mark Ober, local police confirmed that Joe Wynperle, 31, was involved. *Tribune* reporter Candance Samolinski apparently broke this story when it appeared in the newspaper on February 17, 2005. WFLA reported the basic details in a "reader" during the 11 p.m. news on February 16. There is no indication that either *The St. Petersburg Times* or any of the other broadcast media reported this story on February 16 or 17.

Although WFLA-TV received this story from its coverage partner at *The Tampa Tribune,* the story does not reflect any additional reporting or expansion of the story for the broadcast medium. The only source that is cited is an arrest report, which was in the *Tribune* story. No comments were provided by the accused or his lawyer. As a crime story, it falls into the category of market journalism. The lack of balance, depth, and background place it the routine crime story category, which is more market journalism than pure journalism. Convergence does not appear to have added anything to this story, which is classified as a *Level One: Daily Tips and Information*.

## Case 13 *Devil Rays Shake-up*

New Yorker Stuart Sternberg became majority owner of the Tampa Bay Devil Rays in what some observers called the biggest day in history of the franchise since it started in the 1990s. It appears TBO.com and *The Tampa Tribune* were ahead of their competitors on reporting the story the night before the major announcement. Not only did TBO.com have a story on its flash page just after midnight on the day of the announcement, but it included an interview with one of the *Tribune* sportswriters who cover the Devil Rays.

Not to be outdone on the convergence efforts of WFLA, which used a *Tribune* sportswriter in their coverage, WTSP called on the *St. Petersburg Times* sportswriter who covers the Devil Rays to comment in their story. WFTS also had a story, which was available online. It is unclear what exactly WTVT and WTTA did in covering this story. WTVT had a brief story on its Web site the day of the news conference.

A sports franchise is a business and the actions of this business and economic effects fall into the category of pure journalism. By breaking the story online the night before the major announcement and later providing audio of an interview with a *Tampa Tribune* sportswriter, TBO.com set the tone for this coverage that was followed the next day when all the other media covered the official announcement. The team reporting on WFLA-TV that included a *Tribune* sportswriter as an analyst was imitated on the competing converged partnership between the WTSP and *St. Petersburg Times*. In both cases, the stories for the television stations had added depth. The newspaper beat reporters were able to provide context. Pure journalism includes reporters who maintain a neutral stand. This was not the case in reporting by some of WFLA's competitors, which focused on the troubled outgoing Rays owner, Naimoli, to the point of being critical in their report. By breaking the story the day before the announcement on TBO.com, the converged relationship was an example of *Level Four: Enterprise Reporting*.

# FINDINGS

As noted earlier, only 13 cases were generated from the 11 days of recordings that were made (see Table 10.1). Of the 13 cases, five exemplified *Level 4: Enterprise Reporting* and four were examples of the simplest form of convergence, *Level 1: Daily Tips and Information* (Table 10.2). Three cases showed convergence journalism in the reporting of *Level 3: Spot or Breaking News*. Noticeably absent from this particular sample were examples of *Level 5: Franchise Reporting* or the highest form of convergence, *Level 7: Public Service*.

## Obvious Differences

The first research question (R1) asks whether it was easy to see the difference between stories that were generated from converged news operations and

**Table 10.2** Convergence Case Summary

| CASE | STORY | LEVEL OF CONVERGENCE | MARKET JOURNALISM THEORY |
|---|---|---|---|
| 1 | Coronet Report | Level 3: Spot News | Semi-pure journalism |
| 2 | Dollar Daughter | Level 4: Enterprise Reporting | Semi-market journalism |
| 3 | Clearwater Plane Crash | Level 3: Spot News | Semi-pure journalism |
| 4 | Mobile Home Eviction | Level 4: Enterprise Reporting | Semi-pure journalism |
| 5 | NHL Season Ends | Level 3: Spot News | Semi-pure journalism |
| 6 | Port of Tampa | Level 1: Daily Tips and Information | Mid-range journalism |
| 7 | Coronet Injunction | Level 1: Daily Tips and Information | Semi-market journalism |
| 8 | Lefave | Level 6: Events and Special Coverage | Semi-pure journalism |
| 9 | Judge Holder | Level 1: Daily Tips and Information | Semi-market journalism |
| 10 | Are We Ready? | Level 4: Enterprise Reporting | Pure Journalism |
| 11 | Baghdad Bomb Victim | Level 4: Enterprise Reporting | Semi-pure journalism |
| 12 | Ober Plot | Level 1: Daily Tips and Information | Market journalism |
| 13 | Devil Rays Shake-up | Level 4: Enterprise Reporting | Semi-pure journalism |

stand-alone news operations. Based on the qualitative examination of the transcripts and broadcasts, one would have to say that "it depends" on the type of story. With ongoing investigations initiated by one of the partners in a convergence partnership, the converged operation was noticeably aggressive in reporting the latest details on multiple platforms. On the other hand, with breaking news, converged operations were not necessarily more equipped to respond than the nonconverged or stand-alone news operations. When those stories involved sports, the difference between converged and nonconverged was more pronounced. The converged broadcast operation had the benefit of the sports reporting expertise of both the broadcast sports team plus the more specialized beat reporting on a specific sports team from the newspaper. But on a nonsports breaking story, the converged operations were more even with their nonconverged competitors.

## Collaboration

It appears that collaboration (R2, the second research question) was less of a factor in the converged cases identified here, and conferring and consulting was more prevalent. Except in cases where *Level 4: Enterprise Reporting* copublication took place, the converged partners seemed to do their own thing after conferring with one another about the stories each was doing. By consulting ahead of time, some original content was generated by a second partner. On the other hand, consulting after one partner generated a story resulted in a simple republication or broadcast of the content gathered by the other partner. Where collaboration did take place, the result was a coordinated effort to use the same sources and work on the same type of story with a planned date for presenting the respective journalism to the viewers or readers.

## Type of Journalism

The 13 convergence cases demonstrate that convergence does not necessarily mean a turn away from market-driven journalism. In response to the third research question (R3), the tenets of market journalism theory play out in how the stories in various media are reported. As McManus (1994) reported in presenting the market theory versus journalism theory, most news operations fall somewhere in between the two extremes. Of the 13 convergence cases, 11 were either in the mid-range or semi-range of either market or pure journalism. Only two stories fell on either extreme. The most common type of journalism occurring in these cases was semi-pure journalism, which meant that a story showed the possibilities when the appropriate resources are applied and the journalist is careful to include such things as context, background, and several sources.

## Convergence Levels

Perhaps the best news from this study for journalism is the number of cases of enterprise reporting, either through copublication or the efforts of one converged partner to break news. In reference to the fourth research question (R4), of all the levels of convergence, *Level 4: Enterprise Reporting* was the most frequently occurring level. This finding suggests when that convergence is a part of the equation, there are opportunities to extend an exclusive story (breaking or planned) a lot further if one has the benefits of a convergence partner. Sometimes the exclusive may come from the print side, while other times it may be broken first on the Web or on the air in a television broadcast.

## Synergy

Among these 13 convergence cases, the whole was greater than the sum of the parts in the two cases of Level 4 convergence where enterprise local reporting was augmented by the use of a second partner and/or third partner in a strategically planned date of release. The focus of the fifth research question (R5), synergy, was more difficult to come by without the intense reporting that could be done on *all* platforms. If one platform does all the reporting (as was the case in some of these stories), one part is definitely greater than another part. That is not synergy in the way management scholars have previously conceived its use in the operation of local media.

# Discussion

Instead of asking those who are converged what they are doing or why they are doing it, this study focused on the product of those converged efforts. From this somewhat unique vantage, one can bring together theory and examine the elements of practice. McManus's market-driven journalism was a most appropriate theory to apply to the converged news operations' practice of producing news. Not only did it allow for the evaluation of those products as to their journalistic value, it also provided a way to cast convergence in a new level of discussion that integrates with those theorists like McManus who are interested in an overall improvement in journalism.

Developed by one of the managers at a converged operation that was part of this study, the *Seven Levels of Daily Convergence* were helpful in determining the intensity of converged collaboration that was or was not occurring. A mere mention of a convergence partner on-air or in the newspaper does not equate with an improved level of journalism. This means the higher, more sophisticated levels of convergence or best journalism is even more rare than the 13 instances identified in 11 days of programming monitored for this study.

Like most studies, this study has several weaknesses or limitations. Chief among them is the number of days that were monitored. While Flanagan and Hardenbergh (2003) monitored WFLA and WTSP content for a 30-day period and Huang et al. (2004) analyzed 90 issues of *The Tampa Tribune* from three 30-day periods, this study only included 11 recording dates across three or five television stations and two newspapers. The logistical challenges of obtaining recordings would need to be overcome before a larger sample could be examined. Given the difficulty in obtaining multiple recordings in a single market, future funded research might obtain recordings from a video monitoring service.

The design of this comparative, qualitative study could easily be adapted to other markets where converged and nonconverged news operations produce news products. Efforts to examine multiple markets of similar or dissimilar sizes might also improve the results' applicability.

## References

Carr, F. (2007). The brave new world of multimedia convergence. In C.A. Tuggle, F. Carr, & S. Huffman (Eds.), *Broadcast news handbook: Writing, reporting and producing in a converging media world* (3rd ed., pp. 257–272). New York: McGraw-Hill.

Corrigan, D. (2004). Convergence works for media owner, but not news consumer. *St.Louis Journalism Review, 34* (271), 14–15.

Dailey, L., Demo, L., & Spillman, M. (2005). The convergence continuum: A model for studying collaboration between media newsrooms. *Atlantic Journal of Communication,13*(3), 150–168.

Ehrlich, M.C. (1995). The competitive ethos in television newswork. *Critical Studies in Mass Communication, 12*, 196–212.

Flanagan, M.C., & Hardenbergh, M. (2003). Central Florida's convergence triangle: A qualitative analysis of two major converged local television news operations. *Feedback, 44*(1), 23–44.

Grant, A. (2005, May 4). Convergence research needs larger focus on theory. *The Convegence Newsletter, 2*(10). Retrieved October 1, 2005, from http://www.jour.sc.edu/news/convergence/issue21.html

Huang, E., Rademakers, L., Fayermiwo, M.A., & Dunlap, L. (2004). *Uncovering the quality of converged journalism.* Paper presented at the AEJMC Annual Convention, Toronto, OH.

Killebrew, K.C. (2005). *Managing media convergence: Pathways to journalistic cooperation.* Ames, IA: Blackwell.

Lacy, S., & Bernstein, J.M. (1992). The impact of competition and market size on assembly cost of local television news. *Mass Communication Review, 19*(1–2), 41–48.

Lawson-Borders, G. (2002). Integrating new media and old media: Seven observations of convergence as a strategy for best practices in media organizations. *The International Journal of Media Management, 3*(1), 91–99.

McLeod, J.M., & Pan, Z. (2005). Concept explication and theory construction. In S. Dunwoody, L.B. Becker, D.M. McLeod, & G. M. Kosicki (Eds.), *The evolution of key mass communication concept*. (pp. 13–76). Creskill, NJ: Hampton Press.

McManus, J.H. (1992). What kind of commodity is news? *Communication Research, 19*(6), 787–805.

McManus, J.H. (1994). *Market-driven journalism: Let the citizen beware?* Thousand Oaks, CA: Sage.

Quinn, S. (2004). Better journalism or better profits?: A key convergence issue in an age of concentrated ownership. *Pacific Journalism Review, 10*(2), 111–129.

Quinn, S. (2005). Convergence's fundamental question. *Journalism Studies, 6*(1), 29–38.

Rolland, A. (2002). Convergence as strategy for value creation. *The International Journal of Media Management, 5*(1), 14–24.

Singer, J.B. (2004). Strange bedfellows? The diffusion of convergence in four news organizations, *Journalism Studies, 5*(1), 3–18.

Tompkins, A. (2005, January 25). Combining forces for a converged investigation. *Poynter Online*. Retrieved June 15, 2005, from http://www.poynter.org/content/content_view.asp?id=77528&sid=8

Tuggle, C. A., Carr, F., & Huffman, S. (2007). *Broadcast news handbook: Writing, reporting and producing in a converging media world* (3rd ed.,pp. 257–272). New York: McGraw-Hill.

Turow, J. (1992). The organizational underpinnings of contemporary media conglomerates. *Communication Research, 19*(6), 682–704.

# CHAPTER 11

# The Meaning and Influence of Convergence

## A Qualitative Case Study of Newsroom Work at the Tampa News Center

Michel Dupagne
Bruce Garrison

## Introduction

When Media General's Tampa News Center opened its doors with fanfare in March 2000, it stimulated considerable interest and trepidation among media professionals, but also a sense of uncertainty. On the one hand, some in management felt that this type of cross-platform consolidation was inevitable in the highly competitive news business and could herald a new model for newsrooms across the country (e.g., Steinberg & Sorkin, 2003; Thelen, 2002). Others in the profession, however, were more skeptical and expressed concern about the impact of this convergence experiment on the culture

This article originally appeared in Journalism Studies, Vol. 7, No 2, 2006, ISSN 1461–670X print/1469–9699 online/06/020237–19—2006 Taylor & Francis DOI: 10.1080/14616700500533569

The authors are grateful to Edward Pfister, former Dean of the School of Communication at the University of Miami, for funding this project. They also thank Yvonne He and Ambar Hernandez for their assistance with data collection and transcription.

of the newsroom and employment opportunities (e.g., Sanders, 2003; Strupp, 2000). Three years after the launch of this unique initiative, it is time to revisit how some employees at the Tampa News Center construct the meaning of media convergence and how they perceive it has affected their work environment and skills development. The examination of these issues is not only important for the evolution of the news industry, but also for the future of academic programs that increasingly believe in the merits of media convergence education (e.g., Duhé & Tanner, 2003).

There are, of course, other convergent newsrooms in the country (see Killebrew, 2005; Singer, 2003), but we chose to study the News Center in Tampa, Florida, for several reasons. Not only is the News Center one of 21 newspaper-television combinations grandfathered by the Federal Communications Commission (FCC, 2003), but it has upped the convergence ante by moving its three news units under the same roof. In March 2000, *The Tampa Tribune*, the NBC-affiliated WFLA-TV, and the Tampa Bay Online (TBO.com) service began operation in a brand new $40 million, 120,000 square foot building (Strupp, 2000). Furthermore, the News Center's owner, Media General, announced that it intends to replicate the News Center's convergence strategies with other current and future properties. A week before the FCC (2003) lifted the ban on newspaper-television cross-ownership,[1] Media General's chairman J. Stewart Bryan spelled out his plans: "Any of the places where we have a newspaper, we'd like to have a TV station . . . Any of the places we have a TV station, we'd like to have a newspaper" (Steinberg & Sorkin, 2003, p. C6). Thus, the News Center was an ideal setting to investigate qualitatively the meaning of media convergence, changes in newsroom practices and culture, and optimal job functions in a convergent newsroom that could impact on the journalism curriculum.

This descriptive case study will examine three research questions that are particularly relevant to journalism researchers and educators: First, how do employees at the News Center define media convergence? Second, what changes have journalists experienced on their jobs and in the newsroom since the creation of the News Center? Third, what skills do news staff members need to function optimally in the convergent environment of the News Center? We begin by placing the study within a broad convergence framework. Next, we will review the historical, organizational, and strategic aspects of the Tampa News Center. We will then supplement

1. In June 2004, the U.S. Court of Appeals for the Third Circuit affirmed in part and remanded in part the FCC's revised ownership rules (*Prometheus Radio Project v. F.C.C.*, 2004). It ruled that the new cross-ownership rules, including the newspaper-broadcast combination rule, were not unconstitutional per se, but also held that cross-media limits were insufficiently justified. Therefore, the court instructed the Commission to rewrite the rules. In January 2005, the U.S. Department of Justice declined to seek review of the case before the U.S. Supreme Court. But broadcast groups, such as Fox, NBC Universal, andViacom, petitioned the High Court for review of the appellate decision. In June 2005, the U.S. Supreme Court denied these petitions. The old cross-ownership rules remain in effect while the FCC reevaluates and rewrites the rules based on the Third Circuit's ruling.

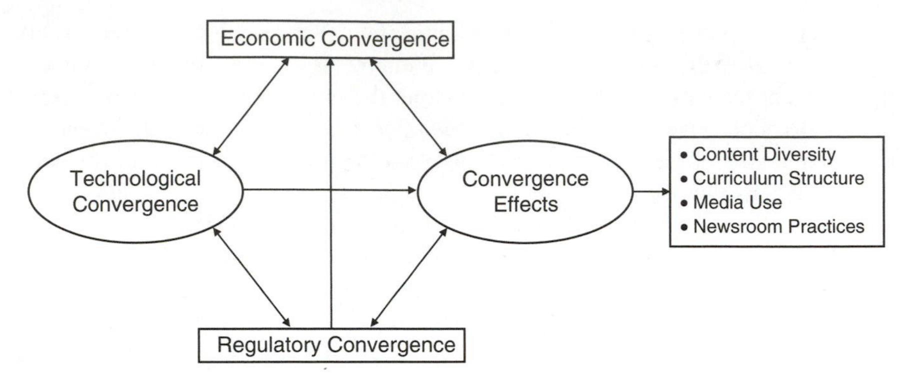


**Figure 11.1** Model of Media Convergence

this document analysis with a series of in-depth interviews with news directors, editors, producers, reporters, and technical personnel from the News Center.

## A Model of Media Convergence

In proposing this general model of media convergence (see Figure 11.1), we seek to situate our research questions within a larger frame of reference and provide an integrated look at different conceptions of media convergence. Although our study focuses specifically on the newsroom environment, it is important to realize that media convergence represents more than a common technical platform, a business strategy, or a regulatory action (see Dennis, 2003). There is no single definition for convergence. Instead, this construct is multidimensional and has different conceptions and contexts. Below we define and discuss three main types of media convergence: technical, economic, and regulatory.

## Technical Convergence

More than 20 years ago, de Sola Pool (1983, p. 24) coined the term "convergence of modes" and offered an early conceptualization of media convergence as a process "blurring the lines between media." In his view, the traditional divisions between media industries, such as the press, broadcasting, and telephone networks, were slowly collapsing due to the growing use and influence of digital electronics (see also Mueller, 1999). Therefore, the term "technical convergence" has come to signify the "coming together of all forms of mediated communications in an electronic, digital form, driven by computers" (Pavlik, 1996, p. 132; see also Blackman, 1998;

Vallath, 2000). Technological convergence is not without its critics, though. For instance, Noll (2003) argued that because television sets increasingly incorporate digital components, this does not automatically mean that television and computers are merging into a single home appliance.

## Economic Convergence

As the European Commission's (1997) *Green Paper* plainly stated, convergence goes beyond technology: "It is about services and new ways of doing business of interacting with society" (p. ii). Economic convergence, also known as market or industrial convergence, can take place at the user or institutional level. User-oriented economic convergence may or may not involve significant vertical integration activity because it targets products and services, not acquisitions. For instance, some cable operators can offer cable and broadcast programming, video-on-demand, voice-over Internet Protocol, and Internet access in a bundled fashion on the same wired platform.

At the institutional level, economic convergence emphasizes *multiple* but integrated platforms. John Haile, a former *Orlando Sentinel* editor and a media consultant, dubbed this form of convergence "complete convergence" and defined it as "a single business operating with multiple platforms: common management, ads sold across multiple media, and a shared news operation" (Aaron, Robinson, & Smothers, 2002, p. 18). As Killebrew (2005) points out, the cross-promotional value of sharing and presenting information across platforms could bring about greater advertising revenues. But even such convergence advocates as Robert Decherd, chairman and chief executive of Belo Corporation (a leading media group that owns a newspaper-television combination in Dallas), questioned the economic expectations of this type of integration. "There is a belief on the part of some people in the financial community that there are tremendous efficiencies and revenue opportunities associated with any cross-ownership.... That is just not correct" (Steinberg & Sorkin, 2003, p. C6).

Thus far, institutional economic convergence, which is supposed to produce cross-media mergers, has not fared well in the marketplace, often because of philosophical and management differences. The failures of the AOL Time Warner and AT&T Broadband mergers are two frequently mentioned examples. Mueller (1999) contends that most of the mergers and acquisitions after the Telecommunications Act of 1996 have fallen within the horizontal integration category–which involves the acquisition of companies or units that operate at the same level of production as the acquiring company (see also Dennis, 2003). Chan-Olmsted (1998) reported that despite a 200 percent increase in the number of transactions in the radio, television, cable, and telephone business between 1991 and 1996, the primary merger and acquisition strategy remained intra-industry.

## Regulatory Convergence

There is an important distinction between full and partial regulatory convergence. Full regulatory convergence "refers to the confluence of previously separate industry based laws and regulations into a single legal and regulatory framework" (Garcia-Murillo & MacInnes, 2003, p. 58). On the other hand, partial regulatory convergence is more limited in scope and involves deregulatory actions that allow media organizations to engage into economic or technological convergence. For instance, the Telecommunications Act of 1996 and the FCC's revised ownership rules have the potential to generate cross-media services for consumers and create interindustry alliances among media companies (Labaton, 2003; Steinberg & Sorkin, 2003).

de Sola Pool (1983) was primarily concerned with the question of full regulatory convergence and its First Amendment implications: which of the three main regulatory models–print, broadcast, and common carrier–would govern these media once they have converged (See also Blackman, 1998)? Contrary to de Sola Pool's expectations, full regulatory convergence has never materialized in the United States. None of the regulatory models has prevailed over others in the public policy sphere. Instead, these industries continue to be regulated differently without a unifying structure even when they offer similar services. For instance, McGregor (1994) noted that "three competing wireline distribution industries are regulated under much different regimes: local exchange carriers as common carriers, cable systems under a complicated three-way jurisdictional scheme, and SMATV systems as largely unregulated" (p. 139).

## Convergence Effects

Some studies have investigated whether media convergence yields direct effects on media use, content diversity, newsroom practices, and curriculum structure (see Figure 11.1). In recent years, a growing number of convergence-type products and services, such as TV/PC combos, home media centers, voice-over Internet Protocol, multifunction cell phones, and streaming technology, have been introduced on the market. But the evidence of convergence effects on individuals' media use has been mixed: a single device or Web-based application offering multimedia attributes does not necessarily replace or displace traditional media usage. For instance, while 22 percent of Americans reported using Internet radio or video in December 2004, the same survey also indicated that 82 percent of the respondents plan to continue to listen to over-the-air radio in the future as much as they do now (Rose & Lenski, 2005). In the Lin (2004) study, respondents who expressed greater interest in streaming were also more likely to consider newspaper and radio content substitution (but not television content substitution), to watch less television, and to spend less time on magazines (but not on newspapers). Kayany & Yelsma (2000) concluded that "time displacement effects are not uniform but different by medium and by communication activity" (p. 224).

At the corporate level, a few studies have assessed the impact of newspaper-television cross-ownership on local news coverage bias, quality, and output. For instance, Pritchard (2002) analyzed whether 10 cross-owned newspaper-television combinations displayed a consistent slant in their news coverage in favor of a particular presidential candidate during the 2000 election. In five combinations, including Tampa, the overall slant of newspaper coverage was significantly different from that of television coverage. While most of the coverage of Media General's WFLA-TV was coded neutral, coverage of *The Tampa Tribune* was deemed pro-Bush. Another FCC report found that affiliated television stations that own daily newspapers aired more local news and public affairs programs and received more awards than those without such newspaper ownership (Spavins et al., 2002; see also Project for Excellence in Journalism, 2003).

Little research has examined how convergent news operations, such as those at the News Center in Tampa, have affected newsroom practices, roles, and culture. Zavoina & Reichert (2000) reported that photo editors working for the printed version of a daily newspaper handled photographic content differently from their Web director counterparts working for the online version of the same newspaper. Killebrew (2003) contends that "Reporters, editors and the supervisors charged with making convergence or 'new media' journalism a reality are finding a great deal of dissonance in the workplace today" (p. 43). Although he does not present specific empirical evidence, he hypothesizes that convergence will disrupt news operations due to differences in cultures and decision-making procedures. This present qualitative study will focus on media convergence at that corporate level.

At the educational level, a growing number of journalism schools have modified their curricula to adapt to trends in media convergence (see South & Nicholson, 2002). These schools believe that students must be able to present news across different platforms–newspaper, television, and the Web. Duhé & Tanner (2003) found that 72 percent of their surveyed school administrators "believe convergence is the future of mass communications" (p. 8). Another recent study of journalism and mass communication school administrators revealed that nearly 85 percent reported that their curriculum emphasizes either cross-media learning or both cross-media and specialization learning (Lowrey, Daniels, & Becker, 2005). But if a convergence curriculum is to be successful, it is important to determine what new practices, if any, are being implemented in convergent newsrooms. Lowrey et al. (2005) found that perception of industry hiring was the most important predictor of faculty interest in pursuing a convergence curriculum.

## The Tampa News Center

As part of the case study, this section briefly reviews the historical, organizational, and strategic accounts published in the trade press about the News Center (e.g., Colón, 2000; Downie & Kaiser, 2002; Fitzgerald, 2001; Gabettas, 2000; Garrison,

2000; Sanders, 2003; Strupp, 2000). Much has been written about the News Center in 2000, but we will make a special effort in this section to synthesize these industry insights according to logical themes. Not only do such documents provide a necessary context to describe the unit of analysis, but they also play a vital role in corroborating data sources from the subsequent in-depth interviews (see Yin, 2003).

Historically, Media General's News Center has lent itself to becoming a convergent newsroom. In 1975, the FCC (1975) banned common ownership of a broadcast station and a daily newspaper in the same market, but grandfathered Media General's *Tampa Tribune/Tampa Times*/WFLA-TV combination in Tampa (Media General, 2003a). In 1982, the operations of *The Tampa Times* were consolidated into those of *The Tampa Tribune*. In 1994, *The Tampa Tribune* launched Tampa Bay Online (TBO.com) as its newspaper's Web site. At that time, the *Tribune*/TBO.com and WFLA-TV were located in separate downtown buildings. In March 2000, Media General inaugurated the $40 million News Center to house its three Tampa news properties under the same roof (Strupp, 2000). Organizationally, the 120,000 square foot, four-story News Center was designed to foster interaction and coordination between the staff members of TBO.com, WFLA-TV, and The Tampa Tribune (Garrison, 2000). The first floor (and, by extension, part of the second floor) houses two large WFLA production studios. The second floor provides space to both the WFLA and TBO.com newsrooms. The third floor is home to the *Tribune* newsroom and TBO.com executive offices. The fourth floor houses the WFLA executive offices. A central piece of the building is an atrium, which rises through the second and third floors. Lying in the middle of the atrium on the second floor is the so-called "superdesk" a circular multimedia assignment desk where editors of the three news organizations work side-by-side (Downie & Kaiser, 2002; Gabettas, 2000; Garrison, 2000). The atrium is often an area bubbling with activity where employees interact and even pass on videotapes.

Strategically, Media General believes "that the best way to ensure the production and delivery of strong local news is to allow companies like ours to practice good journalism across various media platforms" (Media General, 2003c, p. 3; see also Colón, 2000). For this corporation, "Convergence brings together the depth of newspaper coverage, the immediacy of television and the interactivity of the Web" (Media General, 2003c, p. 4; see also Gabettas, 2000). For instance, WFLA can rely on the *Tribune*'s archives and in-depth knowledge of beat reporters, while TBO.com can borrow content from both the newspaper and the television station. Thus, the primary mission of the News Center is to cultivate cooperation and resource sharing in its news-gathering operations (Thelen, 2002), thereby creating a form of "editorial gestalt" in which the convergent newsroom is greater than the sum of its individual parts. By February 2002, about three-fourths of the *Tribune* and WFLA photographers were equipped with both still and video cameras (Stevens, 2002).

While the News Center management team rarely mentions, or even downplays, the financial benefits that could result from this convergence experiment (see Beard, 2003), the potential economic windfall is not ignored. In May 2003, Bryan commented that "When quality improves, circulation and audience share increase, all of which create revenue growth" (Fitzgerald & Moses, 2003, p. 11). The Media General 2002 Annual Report (Media General, 2003c) painted a relatively rosy picture of the News Center's financial results in 2002: The daily circulation of The Tampa Tribune increased by 5.8 percent from December 2001 to December 2002 and NBC-affiliated WFLA solidified its No. 1 position in the Tampa television market (13th Designated Market Area [DMA]) with an audience share of 12 percent in November 2002. In 2004, despite losses attributed to heavy hurricane activities in Central Florida, the circulation of *The Tampa Tribune* grew by 0.9 percent for daily editions compared to 2003 and by 1.5 percent for Sunday editions (Media General, 2005). WFLA remained the top-rated television station in the market that year. In a press release, Media General (2003b) also reported that the number of TBO.com page views rose by 11 percent from 2001 to 2002. In 2004, the number of page views increased 10 times during peak storm and hurricane periods, and TBO.com became profitable for the first time in its short history (Media General, 2005). But it is not entirely clear whether the convergence itself is responsible for these positive results and whether it has led to significant revenue growth. According to Media General, the Tampa integration only produced an additional $6.4 million in revenue in 2001, and cross-media sales rose by a mere 2 percent during the same year (Beard, 2003). Gil Thelen, publisher and president of *The Tampa Tribune*, admitted that "commission-sensitive sales teams have had trouble finding ways to co-operate" (Beard, 2003, p. 28). Still, he believed that "the editorial benefits of convergence will translate into greater financial ones over time" (Beard, 2003, p. 28).

Not surprisingly, the Tampa venture has its share of critics who have voiced concerns that a convergent newsroom would damage the editorial independence of news operations, reduce the amount of original content, and augment employee workloads without proper compensation. Despite common ownership of the three news outlets, the News Center has repeatedly pledged to retain its journalistic independence (see Gabettas, 2000; Thelen, 2002). In practice, however, it is not always immune from subtle or not-so subtle editorial interference. For instance, *Tribune* television reporter Walt Belcher acknowledged being asked not to cover several WFLA-related stories (see Strupp, 2000). Critics also fear that the unification of the three news operations will limit the amount and type of local news stories (Gabettas, 2000; Strupp, 2000). *The Washington Post* executive editor Leonard Downie and associate editor Robert Kaiser (2002) have argued that much of the News Center's news sharing is cross-promotional in nature and consists of repurposed stories with little original content. Rival *St. Petersburg Times* managing editor Neil Brown has contended that a convergent newsroom could dilute content and adversely affect how the news is being covered. Thelen has acknowledged that "the Tampa News Center still housed three different and often conflicting news cultures" (Downie & Kaiser, 2002, p. 55). He wryly added: "Convergence is a contact sport that is shaped one staff collision at a time" (Gabettas, 2000, p. 28). Along with issues of journalistic integrity and quality, there has been concern about fair compensation and working conditions (Sanders, 2003; Strupp, 2000). Economies of scale in a convergent newsroom could easily invite management to demand more from fewer employees. Thus far, some of these fears have not materialized. In fact, total employment has remained stable between March 2000 and June 2005. In June 2005, staffing authorizations were 15 journalists at TBO.com, 100 at WFLA, and 300 at the *Tribune* (G. Thelen, personal communication, June 21, 2005).

## Methodology

Case study research refers to "an empirical inquiry that investigates a contemporary phenomenon within its real-life context" (Yin, 2003, p. 13). As such, it is best used to understand complex social and organizational issues. In this study, we focused on a single news-producing organization as an intrinsic case study–also our unit of analysis–to explore and to gain insight into news media convergence. Methodologically, case study research is inherently qualitative because it is bounded to understanding a specific case rather than seeking generalization beyond that case (see Stake, 2000). In addition, it often relies on multiple qualitative data sources, such as documentation, archival records, interviews, and direct observations to provide corroborating evidence on a phenomenon (Yin, 2003).

In this study, we used a combination of documents and in-depth interviews to address our three research questions. In-depth or intensive interviews have been called "one of the most powerful methods" in qualitative research because they allow investigators to "step into the mind of another person, to see and experience the world as they do themselves" (McCracken, 1988, p. 9). Participants are able to tell their stories and discuss their involvement in detail. In-depth interviews will enable us to better understand the meaning of media convergence, the changes in the newsroom culture, and the necessary job skills at the News Center.

## Selection of Respondents and Data Collection

Prior to conducting the in-depth interviews, we completed a pilot field study, also known as briefing interviews, to discover the organization's goals, identify the types of personnel who would be potential respondents, construct the open-ended questionnaire, and determine the interview schedule (see Lindlof & Taylor, 2002; Wengraf, 2001; Yin, 2003). This preliminary field visit took place on June 6, 2003. It included observation of facilities and attendance at several daily staff meetings of the three different News Center units. The authors also participated in a lengthy overview panel discussion held with six news, circulation, and advertising staff members who were presently involved in day-to-day convergence activities. The panel discussion was audiotaped for later review in preparation of the in-depth interviews.

Following the pilot field study, we first developed a short open-ended questionnaire for the in-depth interviews. These questions, which are available from the authors, dealt with the meaning of media convergence (e.g., "How would you define media convergence?"), changes in newsroom practices and culture (e.g., "What benefits have you experienced since the newspaper, television, and online operations were moved under the same roof?") and recommended job skills in a convergent newsroom and implications for the journalism curriculum (e.g., "What do students need to succeed in such a convergence environment?"). These questions were unstructured to allow probing flexibility and encourage a thick narrative description.

We then purposefully selected one dozen individuals with different specialized roles in the newsroom (see Wengraf, 2001). There is no firm rule on a recommended sample size, but McCracken (1988) indicates that recruiting eight respondents is generally sufficient for many in-depth interview projects. Our respondents were selected based on their contributions to the news process and their involvement in news convergence. To further diversify the sample, we also sought respondents in different types of news decision-making roles as well as those in news content development and news production roles. For balance of perspective, we chose respondents from each of the three news organizations in the News Center. The final 12 respondents included the News Archives and Research Center manager, the Tampa Bay Online news team manager, the Tampa Bay Online sports and weather team manager, *The Tampa Tribune* business and real estate reporter, *The Tampa Tribune* news photography team leader, *The Tampa Tribune* senior editor of photography and convergence technology director, *The Tampa Tribune* senior editor of multimedia, *The Tampa Tribune* senior features editor, the WFLA 5:30 newscast producer, the WFLA assignments manager, the WFLA motion graphics artist/designer/animator, and the WFLA news director. A more detailed profile is available from the authors.

We conducted the in-depth interviews on June 19 and June 20, 2003. All interviews but one took place on site in the WFLA and *Tampa Tribune* newsrooms, the News Research Center, or News Center conference rooms. Because one respondent, a key news manager, was on vacation at the time of the other interviews, that single interview was conducted by telephone, when he became available, on July 2, 2003. Each interview averaged 45 minutes and was audiotaped with the expressed permission of the respondent (see Berg, 2001).

## Data Analysis

All recorded interviews were transcribed verbatim by two supervised and experienced typists using a professional transcription recorder (see Berg, 2001; McCracken, 1988). Transcriptions were manually reviewed by the authors using procedures recommended by McCracken (1988) and Yin (2003). McCracken suggests five stages to analyze in-depth interview content, each representing a higher level of generality to identify news convergence themes and theses. These stages include observation of a useful utterance, development of expanded observations, examination of interconnection of observed comments, collective scrutiny of observations for patterns and themes, and review and analysis of the themes across all interviews for development of theses (McCracken, 1988).

## Findings

### Research Question 1: Meaning of Media Convergence

The interviewed News Center journalists viewed media convergence in terms of their experience over the past three years. The single recurring theme was resources. Regardless of whether the journalist was part of the newspaper, television station or online staffs, the dominant perception focused on the availability of combined and additional resources in terms of people, equipment, and ideas. For instance, the multimedia desk manager talked at length about the additional resources created through convergence:

> Over time, one of the things that I've seen was that the best use for convergence for us... has been the sharing of resources. We still have reporters writing stories for print, going on air for broadcast, writing and doing things in print and broadcast through the Web. But... the best way for us to converge is through resource-sharing. Making sure reporters talk to each other, how do they feel, making sure that they create these bonds that have us, as a News Center, to get the information out to our users, viewers, or readers.

The assignments manager for WFLA-TV, responsible for selecting reporters, photographers, and other television news resources used to cover stories on any given day, offered this example of shared resources:

> If we run out of photographers to respond to a story, they [the *Tribune*] may have a photographer at a story that we wouldn't have covered otherwise. Now we're at a position where *The Tampa Tribune* photographers are also shooting video for us, so a lot of times we're able to shoot video of assignments that we're not able to get just because we're out of people. The same goes for them.

Discussion of resources by respondents went well beyond people. The shared new building housing the News Center was mentioned often. There is less duplication of facilities with the new model. A *Tribune* business section reporter who specializes in real estate coverage spoke about the importance of shared building and facilities:

> I think that of the few things that say convergence, number one is the building, how it's set up. The facility says a lot about convergence. It's still a treat

> to bring people into this building for the first time and point out, "This is the only place in the country where you're going to see a TV station and a print newsroom in the same building."

Respondents also cited the shared daily news story budgets (i.e., the day's stories that are being developed for publication). With a database developed internally BudgetBank editors and other gatekeepers at all three platforms now know what is going to happen and what stories will be covered at any time. This information is taken to daily news meetings and discussed to maximize the use of reporters, photographers, and others in the field. Shared reporters and photographers have increased the breadth and, perhaps, depth of coverage of the three news organizations. Journalists discussed a perceived growth in sharing story ideas and story tips, a process facilitated by BudgetBank.

The WFLA-TV news director said that tracking of convergence "acts" by the News Center was also revealing, noting hundreds of such acts each month, many involving shared resources:

> We chart something like 300 different acts of convergence per month. Most of them we had some level of cooperation that probably wasn't even visible to the home viewer. Sometimes the levels of cooperation are much more explicit like in the Super Bowl; we worked with each other hand in hand. The TV reporters were frequently assisted by newspaper reporters in getting the coverage on the air. In the anniversary of the 9/11 attacks in September, we aired a series of nearly two dozen stories in concert.

The news research center has become an icon of resource sharing. Previously used and funded exclusively by the *Tribune*, the research center now serves journalists working for all platforms. Respondents also often mentioned increased availability of new equipment such as desktop and portable computers, software, telephones and telephone systems, and cameras. The News Center also shares a single cellular telephone service arrangement. The single provider situation, which results in considerable savings across all three platforms, provides similar services and equipment to all journalists to work on their assignments.

Convergence was rarely discussed in terms of distribution of content involving a single delivery platform or cross-promotional activity. When it was, comments were limited to shared or loaned journalists who prepared content for another platform. For example, newspaper reporters often make themselves available for television news "talkbacks," where they discuss a story they have covered and are often interviewed by television reporters. Some newspaper journalists function even more like television reporters and prepare reports for air.

Other themes in defining convergence included enhanced communication within the newsroom, a changing understanding of news competition and the market served, and the process of covering and reporting stories, especially breaking news. These thoughts, however, were secondary in almost all cases.

## Research Question 2: Changes in Newsroom Practices and Culture

Respondents often echoed themes uncovered in discussing their definitions of convergence when they were asked about how convergence has impacted on their work. The jobs and responsibilities have changed because convergence has brought additional new resources and duties. At the core, in many ways, however, their tasks are much the same as they have always been.

The multimedia desk manager saw the change involving jobs and roles as a gradual process:

> What we have determined over time is that convergence sort of changed for us. Initially, the convergence was going to be making sure that we complemented all coverage with another platform. You'll see a lot of broadcast reporters writing stories for print, a lot of print reporters going on air. In fact, TBO and Web producers are doing both, going on air and writing stories for the newspaper.

In some cases, respondents noted that their positions are now more demanding because there is more to do. Television reporters, for example, are given the opportunity to write for the newspaper and vice versa. Print and broadcast photographers are asked to shoot images for both platforms. These demanding situations have left some individuals without sufficient time to do everything, some respondents noted. At least one journalist wondered aloud if managers realized the amount of time and effort required writing across platforms. The manager of the news research center was well aware of the additional workload in her department:

> A lot more work . . . I would say in terms of me personally as a manager, it's juggling the balls. It's been a real culture shock for the library staff. . . . When I first got here, there were people who only took care of the photo archives, people who only took care of the text archives, and those people never did research. We have a really talented group of people in here that just weren't being utilized. So there were changes that I made, in terms of if you are going to enhance text for the digital archive, why not do the photos at the same time, the same person, that way you know how to work the photo database. . . . We're not going to get a lot more people, so let's diversify the job duties here and it's working so much better now.

Jobs have changed, of course, for those who work across the three platforms. One interesting observation by several respondents centered on increased collegiality and the willingness to work across platforms in covering both major news stories and routine news on a daily basis. While it might be expected by everyone in the News Center for major stories such as the Tampa Bay Buccaneers' Super Bowl championship in early 2003, jobs and responsibilities have also been affected by cross-platform approaches to daily news coverage. There is now a routine, many respondents observed, to meeting with, seeing, talking to, and interacting in other ways with journalists from platforms other than the one within which an individual works. More than one individual attributed this new sense of community to the new facility, its design, and the fact that all news platforms are located in a two-floor area opened up with few walls and the common atrium. This has led to a team approach and a declining sense of internal competition and conflict over approaches and resources.

The Tribune senior editor of multimedia explained, using shared reporting sources as an example:

> I think one of the things that initially is always difficult for reporters is to open their notebooks, open up their source list to another reporter. I think that was one of the hurdles that initially we had to get over because, as a reporter myself, we used to compete against reporters at WFLA. It was just the idea of sharing the information that I worked so hard to gather, to open it up and share it with another reporter.... But I think that our environment–we've been doing it for three years, which is not a long time, but the strides that we've made as far as sharing resources and information have come a long way. It's a second nature now.

Gatekeepers have certainly felt the impact of convergence. Respondents sensed an increase in the number of choices about coverage. These decisions involve the number of stories covered in a given day, the number of reporters and photographers involved, and even the basic range of stories covered. It is the view of several respondents that convergence has increased this form of work. Furthermore, at least one television news producer observed that her position has changed because there are more ideas to share and consider over the course of any news cycle.

The multimedia environment has also brought change to many newsroom jobs. Interviewed journalists said that they think more about multimedia at several levels. Multimedia approaches to coverage of a given news story may be first on this list. Multimedia, especially for the online platform, is at the heart of storytelling and the coverage a story receives. Journalists stated that they think differently about how to cover a story and how to report it, and this perception is common across all platforms.

The news team leader for TBO, a veteran local government reporter for the *Tribune* prior to taking his current position three years ago, made this point:

> It doesn't make me think differently about journalism. It makes me think differently about presentation and opportunity. I brought to TBO my value system and my journalism experience as a print reporter and I still adhere to that as an online journalist because I don't know any other way. What I look at is the opportunity to be more creative. I look at it as layering... [like] those old overhead projectors: they would take the first one and there's a map and then they do an overlay. Well, that's how you build a story online. It is one overlay after another as you fill in that puzzle.

In addition to the multimedia thinking about news coverage, some journalists felt that jobs have changed because they have been required to learn about the other platforms with which they work on a regular basis. One print journalist remarked about the need to learn television jargon just to be able to communicate with his WFLA colleagues:

> ...learning more about the other platforms. Learning what they are looking for, what their wants and desires are, essentially just understanding how they talk their language. You know, the language they use...So this is not just an understanding across...but also understanding their language.

Respondents also noted that convergence has required some jobs in some departments to be realigned. This is especially true in central departments serving all platforms, such as the news research center. New positions and departments that did not exist prior to the creation of the News Center are an obvious example of significant job changes. The multimedia desk, which serves all platforms, has been staffed with individuals brought from previous positions at one of the three platforms, clearly creating a substantially different work situation. This has led to a sense of evolution for new departments or jobs impacted by convergence. Respondents noted that positions have matured as needs have forced refinement or modification of original job descriptions. They also observed that job changes were most intense during the first year of the convergence of newsrooms and were less significant in the past two years. There was little, if any, discussion by respondents concerning gain or loss of jobs during the past three years.

## Research Question 3: Skills Needed in a Convergent Newsroom

Interviews pointed to five themes of interest to journalism and mass communication educators. First, respondents felt that students must be very good at one task or skill, but able to do others as well. Second, they emphasized writing and reporting as fundamental elements for any platform. Third, they identified the need for adaptability and collegiality across platforms. Fourth, News Center journalists highlighted a need for students to be good communicators regardless of platform. Finally, they stated that students must obtain experience working in a converged newsroom.

The news team manager for TBO elaborated on the need for new journalists working in a converged newsroom to be strong in at least one skill and able to handle others as well:

> I think that whatever concentration they want, become really good at it. And focus your attention there and become a valuable employee to the market, but at the same time understand your partners and what their wants and needs are and how they go about collecting information and presenting information. I know people that are very good online that are able to do video and broadcast, but they can't write worth a lick. And I've seen writers who are tremendous and either don't feel comfortable in front of TV or don't know anything about the Web.

While having a specialization is important, other respondents noted concern for basics, such as reporting and information gathering. Fundamentals, respondents agreed, remain important. For them, journalism is a business of information gathering and distribution. The weather/sports team leader at TBO was typical of individuals who emphasized the need to gather news:

> I think they should still be reporters, first and foremost. Newsgathering should be the most important thing; still be a reporter. It is fine to know all the bells and whistles in video and audio and Flash and all that. Those skills are useful in what you're doing, but I think being able to write a tight lead [is important].

Receptivity to new technologies and convergence is important for graduates, a *Tribune* business reporter stated:

> The biggest skill is you need to be open to it.... People need to be receptive to the environment. I don't think people who come and say, "I'm going to be a print reporter and that's all I'm going to do" are going to get very far.... That works two ways. On the first hand, a student starting out in journalism needs to realize that the image of what a journalist does has changed, but at the same time it's incumbent on the media organizations to let them know.

Beyond specialization in a skill and information gathering, respondents noted that these skills and tools will not go far without the ability to communicate at several different levels. Communication within the newsroom is necessary, but general ability to communicate a message to a print, broadcast, and online audience is essential as well. Good communication basics and abilities should remain key attributes for new graduates, the director of *Tribune* news photography argued:

> [Graduating students] need to be good communicators as well. They need to be able to communicate with a lot of different types of people. And if you are a journalist of any sort, you should have those skills as well coming in–being able to communicate with a lot of different types of people.

Respondents also emphasized the need for graduates to be adaptable across platforms and offer significant collegiality with newsroom coworkers across platforms. This includes understanding the work of others and willingness to contribute when possible. While this is partly personality driven and cannot easily be taught in a classroom, it remains important. Perhaps another important theme, experience working in a converged newsroom, can offer development in the area of collegiality and reduction of platform-related loyalties. Respondents repeatedly mentioned the value of internships and summer work in converged newsrooms before graduation.

# Discussion and Conclusions

## Summary

### Meaning of Media Convergence

Shared resources benefit all interviewed journalists in the Media General News Center in Tampa, but the real winner seems to be the television news operation.

It benefits from the depth of resources of the newspaper that did not exist when the operations were housed in separate locations and did not work together. The business or financial sides of the platforms have yet to completely merge under convergence, although there is more effort to share costs across platforms. There are still, for example, *Tribune* employees, not News Center employees, paid by the *Tribune* budget, not by a News Center budget. This problem of financial integration will likely be corrected as employee and other expenses for operating and maintaining the news research center, for example, are increasingly shared by all three units. There is also a desire to continue growth of sharing through a single, common content writing, editing, and production computer system, which is the primary function of BudgetBank. Movement in this direction is apparent and will continue during the next several years. In fact, as of June 2005, the News Center was beta testing CCI NewsGate, an integrated editorial management solution that would support the entire news process (G. Thelen, personal communication, June 21, 2005).

## Changes in Newsroom Practices and Culture

Job and role changes were expected given the scope and depth of the News Center operation. Most changes related to additional duties or responsibilities beyond those originally stipulated in a single-platform environment. The fact that journalists saw their core work as generally unchanged is somewhat surprising, however. This outcome may be explained by the fact that convergence has brought additional efficiency through shared resources that allow the same number of people to get more done in a given time period such as a news day. Not only has media convergence at the News Center fostered a greater sense of community among the different units, but it has also brought to the forefront the importance of versatility in news-gathering operations. Respondents are now thinking about their job duties at different levels. Most of them view these changes as an evolutionary trend rather than a complete overhaul of the existing newsroom culture.

## Recommended Job Skills

Convergence, as anyone in education knows, has implications for the next generation of journalists. The impact of convergence will cut at the heart of organizational charts, budgets, and even personnel. The challenge for faculty and school administrators will be to decide how to adapt their curricula to the new convergence expectations while taking into account the fluctuating economics of higher education. Given the continuous importance of the journalism education fundamentals, which was confirmed in this study (see also Brill, 1997; Zoch & Collins, 2003), adjusting existing curricula rather than designing new ones could be the preferred route of many administrators. Our respondents indicated that while future journalism graduates must become increasingly versatile and knowledgeable about multimedia, good communication, reporting, and writing skills remain the bedrock of the news profession. Based on these interviews, those students who plan to work in a convergent newsroom would be well inspired to take courses in cognate television and visual communication fields.

### Theoretical Implications

On a more theoretical level, how do these findings relate to the model of media convergence presented in Figure 11.1? The News Center could be described as the confluent product of regulatory, economic, and technical convergence. First, partial regulatory convergence in the form of FCC grandfathering allowed the existence of *The Tampa Tribune*-WFLA combination. Then, the creation of the News Center in 2000 led Media General to anticipate economic convergence derivatives, such as cross-media advertising sales. However, our examination of trade press articles and our in-depth interviews suggest that these expected economic advantages have not yet reached their full potential. Instead, the results reveal a gradual emphasis on organizational and technical convergence, whereby employees of the three units cooperate more frequently, share resources and equipment, and use uniform database systems.

## Future Research

The findings of this qualitative inquiry prove to be heuristic and suggest multiple avenues for subsequent quantitative research. Obviously, we cannot generalize the responses of 12 individuals to an entire news operation or other convergent newsrooms, nor is generalization an actual goal of in-depth interview methodology. But it is also clear that these qualitative insights lend themselves to a follow-up quantitative phase that would benefit from a larger sample. One must recognize, however, that a self-administered survey inside a large newsroom has its own methodological hurdles. Among other things, distribution of questionnaires generally requires permission from management to secure the highest possible response return. Prospective employee respondents, on the other hand, may view the survey instrument with some suspicion if the company endorses it and may not answer questions as truthfully as expected.

These challenges notwithstanding, it would be valuable to administer a survey to employees in one or more convergent newsrooms to determine the perceived importance of traditional and convergence job skills. Zoch and Collins (2003) recently found that respondents from media organizations involved in cross-platform initiatives favored traditional journalistic values (e.g., understanding of ethical issues, writing skills) over visual communication knowledge (e.g., graphic and Web design) for new reporters. Brill's (1997) survey of online journalists reported similar results. More research is needed to determine whether these findings represent an actual pattern among news managers and whether they hold true in converged newspaper-television operations.

Future research could also focus on the relationship between changing job responsibilities in convergent newsroom and job satisfaction. This present study has pointed out that some News Center journalists have voiced concern about the increasing workload that a convergent newsroom created. To what extent these additional duties could affect and further reduce the satisfaction level of journalists is an open question that warrants further study. In their latest survey, Weaver et al. (2003) found that only 33 percent of U.S. daily newspaper journalists reported to be "very satisfied" with their jobs in 2002, up by 22 percent from 1992, but still 32 percent down from 1971. The literature on job satisfaction in mass communication industries is vast, but no study has examined the impact of convergent news operations on job satisfaction. Daniels and Hollifield (2002) analyzed how organizational changes at CNN Headline News affected newsroom personnel and found that the staff responded negatively to these changes. They concluded that "[t]he degree to which change affects morale and drives talented professionals from news organizations and the industry should be a question of serious concern to media scholars and professionals" (Daniels & Hollifield, 2002, p. 676). These are some of the questions and challenges that await convergence researchers in the years ahead.

## References


Aaron, L., Robinson, J., & Smothers, S. (2002, May). Convergence, defined. *Presstime*, p. 18.

Beard, A. (2003, June 3). Tampa takes to integrated media. *Financial Times,* p. 28. Retrieved June 3, 2005, from LexisNexis Academic database,

Berg, B.L. (2001). *Qualitative research methods for the social sciences,* (4th ed.). Boston: Allyn & Bacon.

Blackman, C.R. (1998). Convergence between telecommunications and other media: How should regulation adapt? *Telecommunications Policy, 22,*163–170.

Brill, A.M. (1997). Way new journalism: How the pioneers are doing. *Electronic Journal of Communication* 7(2). Retrieved June 22, 2005, from http://www.cios.org/www/ejcmain.htm.

Chan-Olmsted, S.M. (1998). Mergers, acquisitions, and convergence: The strategic alliances of broadcasting, cable television, and telephone services. *Journal of Media Economics 11*(3), 33–46.

Colón, A. (2000 May/June). The multimedia newsroom. *Columbia Journalism Review,* pp. 24–27.

Daniels, G.L., & Hollifield, C.A. (2002). Times of turmoil: Short- and long-term effects of organizational change on newsroom employees. *Journalism and Mass Communication Quarterly 79,* 661–680.

Dennis, E.E. (2003). Prospects for a Big Idea–Is there a future for convergence? *International Journal on Media Management, 5*, 7–11.

de Sola Pool, I. (1983). *Technologies of freedom*. Cambridge, MA: Harvard University Press.

Downie, L., Jr. & Kaiser, R.G. (2002, May). Media synergy, If not now, when? *Presstime*, pp. 54–55.

Duhé, S., & Tanner, A. (2003, November). *Convergence education: A nationwide examination of ACEJMC accredited schools*. Paper presented to the meeting on Expanding Convergence: Media Use in a Changing Information Environment, Columbia, SC.

European Commission (1997). Green Paper on the convergence of the telecommunications, media and information sectors, and the implications or regulation. Towards an information society approach. *COM, 97*, 623. Brussels: European Commission.

Federal Communications Commission (1975). Amendment of Sections 73.34, 73.240, and 73.636 of the Commission's Rules Relating to Multiple Ownership of Standard, FM, and Television Broadcast Stations. Second Report and Order, 50 FCC 2d 1046.

Federal Communications Commission (2003). 2002 Biennial Regulatory Review: Review of the Commission's broadcast ownership rules and other rules adopted pursuant to Section 202 of the Telecommunications Act of 1996. Report and Order and Notice of Proposed Rulemaking, 18 FCC Rcd. 13620.

Fitzgerald, M. (2001, January 15). Media convergence faces tech barrier. *Editor and Publisher,* p. 32.

Fitzgerald, M., & Moses, L. (2003, June 23). At the crossroads. *Editor and Publisher,* pp. 17–18.

Gabettas, C. (2000, June). Three's company. *Communicator,* pp. 24–28.

Garcia-Murillo, M., & MacInnes, I. (2003). The impact of technological convergence on the regulation of ICT industries. *International Journal on Media Management*, *5*, 57–67.

Garrison, B. (2000). Convergence to the Web is no longer just the future. *College Media Review, 38*(3), 28–30.

Kayany, J.M., & Yelsma, P. (2000). Displacement effects of online media in the socio-technical contexts of households. *Journal of Broadcasting and Electronic Media, 44*, 215–229.

Killebrew, K.C. (2003). Culture, creativity and convergence: Managing journalists in a changing information workplace. *International Journal on Media Management, 5,* 39–46.

Killebrew, K.C. (2005). *Managing media convergence: Pathways to journalistic cooperation*. Ames, IA: Blackwell.

Labaton, S. (2003, June 20). Senators take steps to reinstate limits on media holdings. *The New York Times,* pp. A1, C7.

Lin, C.A. (2004). Webcasting adoption: Technology fluidity, user innovativeness, and media Substitution. *Journal of Broadcasting Electronic Media, 48*, 446–465.

Lindlof, T.R., & Taylor, B.C. (2002). *Qualitative communication research methods* (2nd ed.). Thousand Oaks, CA: Sage.

Lowrey, W., Daniels, G.L., & Becker, L.B. (2005). Predictors of convergence curricula in journalism and mass communication programs. *Journalism and Mass Communication Educator 60*(1), 32–46.

McCracken, G. (1988). *The long interview*. Newbury Park, CA: Sage.

McGregor, M.A. (1994). Toward a unifying regulatory structure for the delivery of broadband telecommunications services. *Journal of Broadcasting and Electronic Media, 38*, 125–143.

Media General (2003a). *Corporate history*. Retrieved June 22, 2005, from http://www.mediageneral.com/history/index.htm.

Media General (2003b, January 29). *Media General reports December revenues* [press release]. Retrieved June 22, 2005, from http://www.mediageneral.com/news/2003/jan29_03_revenues.htm

Media General (2003c). *2002 Annual report*. Richmond, VA: Media General. Retrieved June 22, 2005, from http://www.mediageneral.com/reports/annual/2002/mg2002ar.pdf.

Media General (2005). *2004 Annual report*. Richmond, VA: Media General. Retrieved June 22, 2005, from http://www.mediageneral.com/reports/annual/2004/mg2004ar.pdf

Mueller, M. (1999). Digital convergence and its consequences. *Javnost, The Public 6*(3), 11–27.

Noll, A.M. (2003). The myth of convergence. *International Journal on Media Management, 5*, 12–13.

Pavlik, J.V. (1996). *New media technology and the information superhighway*. Boston: Allyn & Bacon.

Pritchard, D. (2002). *Viewpoint diversity in cross-owned newspapers and television stations: A study of news coverage of the 2000 presidential campaign*. Media Ownership Working Group, Report 2. Washington, DC: Federal Communications Commission. Retrieved June 22, 2005, from http://hraunfoss.fcc.gov/edocs_public/attachmatch/DOC-226838A7.pdf.

Project for Excellence in Journalism (2003). *Does ownership matter in local television news: A five-year study of ownership and quality*. Retrieved June 22, 2005 from http://www.journalism.org/resources/research/reports/ownership/Ownership2.pdf

*Prometheus Radio Project v. F.C.C.* (2004). 373 F.3d 372 (3rd Cir. 2004).

Rose, B., & Lenski, J. (2005). *Internet and Multimedia 2005: The on-demand media consumer*. New York: Arbitron/Edison Media Research, Retrieved June 22, 2005, from http://www.arbitron.com/downloads/IM2005Study.pdf

Sanders, E. (2003, May 7). Journalism's future may start in Tampa. *Los Angeles Times,* p. A-1. Retrieved June 22, 2005, from NewsBank database.

Singer, J.B. (2003). Strange bedfellows? The diffusion of convergence in four news organizations. *Journalism Studies, 5*(1), 3–18.

South, J., & Nicholson, J. (2002, July/August). Cross-training: In an age of news convergence, schools move toward multimedia journalism. *Quill*, pp. 10–15.

Spavins, T.C., Denison, L., Roberts, S., & Frenette, J. (2002). The measurement of local television news and public affairs programs. Media Ownership Working Group, Report 7. Washington, DC: Federal Communications Commission. Retrieved June 22, 2005, from http://hraunfoss.fcc.gov/edocs_public/attachmatch/DOC-226838A12.pdf

Stake, R.E. (2000). Case studies. In N.K. Denzin & Y.S. Lincoln (Eds.), *Handbook of qualitative research* (2nd ed., pp. 435–454). Thousand Oaks, CA: Sage.

Steinberg, J., & Sorkin, A.R. (2003, May 26). Easier rules may not mean more newspaper-TV deals. *The New York Times,* pp. C1, C6.

Stevens, J. (2002, April). TBO.com: Then and now. *Online Journalism Review*. Retrieved June 22, 2005, from http://www.ojr.org/ojr/workplace/1017859157.php

Strupp, J. (2000, August 21). Three-point play. *Editor and Publisher,* pp. 18–23.

Thelen, G. (2002, July/August). Convergence is coming. *Quill*, p. 16.

Vallath, C. (2000). The technologies of convergence. In M. Hukill, R. Ono, & C. Vallath (Eds.), *Electronic communication convergence: Policy challenges in Asia* (pp. 33–47). New Delhi: Sage.

Weaver, D., Beam, R., Brownlee, B., Voakes, P., & Wilhoit, G.C. (2003). The face and mind of the American journalist. Retrieved June 22, 2005, from http://www.poynter.org/content/content_view.asp?id_/28235

Wengraf, T. (2001). *Qualitative research interviewing*. Thousand Oaks, CA: Sage.

Yin, R.K. (2003). *Case study research: Design and method* (3rd ed.). Thousand Oaks, CA: Sage.

Zavoina, S., & Reichert, T. (2000). Media convergence/management change: The evolving workflow for visual journalists. *Journal of Media Economics 13*(2), 143–151.

Zoch, L.M., & Collins, E.L. (2003, August). *Preparing for a career in the unknown: What convergent newsroom managers need and want*. Paper presented at the meeting of the Association for Education in Journalism and Mass Communication, Kansas City, MO.

CHAPTER 12

# News Convergence Arrangements in Smaller Media Markets

Tony DeMars

Discussions of converging media often focus on the larger markets and media groups, such as Disney's ownership of movie studios, local market TV stations, the ABC television network, ABC Radio Networks, and local market radio stations. These mergers of complementary production and distribution outlets tend to focus on entertainment distribution and maximizing competitiveness and profits. However, as Landler and Fabricant (2002) note, media conglomerates lost billions attempting to position themselves for convergent technologies like the Internet, seen as the next platform for profits and business operations. David Geffen, cofounder of the DreamWorks movie studio, commented, "Convergence may be the most expensive word in history" (Landler & Fabricant, 2002, p. 1).

Compared to this national media conglomerate image of convergence, news-based partnerships in smaller media markets may be as important to the audience as they are to station owners and stockholders. Redding (2002) notes that cost and profit issues, combined with the shrinking audience and changing habits of news consumers, led to about 60 noteworthy print-broadcast partnerships in local markets across the country. These were usually not cross-ownership cases, but rather business arrangements for newsgathering and promotion. Redding suggests that most of those experimenting with convergence in 2002 were doing so only casually and mainly for cross-promotion purposes. This chapter describes these local-market attempts at news convergence, rather than national media convergence, and summarizes complementary research projects conducted on small-market news convergence.

The model within which we can view convergence is elusive, but two factors stand out in leading news managers to engage in convergence activities. One factor is increased competition in local news markets, and the second factor stems from the changing dynamics of technology. Some researchers (Kaufhold, 1998; Singer, 2004) thus have looked at convergence from a diffusion of innovations perspective. Filak (2004) looked at the effect of intergroup bias among journalists on convergence. Silcock and Keith (2006) investigated issues of language and culture in converged newsrooms. There has been no common instrument to measure convergence efforts.

Dailey, Demo, and Spillman (2005) saw these incongruities as necessitating a model for studying convergence. While the industry continued developing convergence partnerships and educators continued creating convergence curricula, these same practitioners and scholars struggled with defining convergence and establishing either professional or educational parameters. Dailey and colleagues' "convergence continuum model" arose from this investigation.

Dailey, Demo, and Spillman defined five stages of activity among news organizations involved in convergence, moving left to right (from least- to most-involved), across the continuum: (1) cross-promotion, (2) cloning, (3) "coopetition," (4) content sharing, and (5) full convergence. Cross-promotion features basic acknowledgment of the convergence partner's content. With cloning, content is copied from one media to another. "Coopetition" is a term borrowed from Ray Noorda, founder of the networking software company Novell, to describe efforts where converged media entities both cooperate and compete with each other. With content sharing, partners share repackaged content and sometimes even budgets, and, with full convergence, the partners fully share in gathering and disseminating news, with a common goal of using each medium's unique strengths to best tell the story.

According to Tatro (2002), the most common news-related convergence model is a link, by agreement or through ownership, of a newspaper, a television station and a Web site. For example, the *Chicago Tribune* puts reporters' stories on the Web and asks reporters to appear on their co-owned WGN-TV and their local cable news channel CLTV. Although the operation in Chicago exists in part due to common ownership of the media outlets, more often than not the convergence partnerships are between media owned by different companies.

One particular, often recognized, example is the relationship between the *Star* newspaper and TV station WTHR in Indianapolis. The partnership started in 1996 and, without the benefit of co-ownership, developed to where *Star* reporters were making on-air television appearances, and WTHR staffers were submitting stories to the newspaper. Did this partnership provide better news content and quality? Jon Schwantes, Associate Editor for the cross-media efforts between the two companies, said his efforts to represent the opposite ends of the spectrum represented by print and broadcast did not "always [create] beautiful music" (Robins, 2001, p. 16).

## Teaching News Convergence

As news organizations have developed such convergence relationships, university journalism and mass communication departments have struggled to prepare students for possible changing career expectations. Birge (2004), for example, reviewed previous studies and conducted interviews as to what convergence really was; the conclusion was that journalism schools should not rush to teach convergence as part of their curricula. Birge suggests adding too much to the traditional areas of study in a journalism or broadcast journalism program can weaken the quality of classes. Adding work in existing classes to create some type of so-called super journalist is also questionable. Even industry professionals working in converged newsrooms suggest that the consequences outweigh the benefits in trying to create a newsperson who does it all, for every medium.

In contrast, Kraeplin and Criado (2005) build a case for convergence, equating convergence to a multidisciplinary approach to journalism education. In one of the challenges for a converged journalism curriculum, Kraeplin and Criado report that journalism educators placed a higher belief in the importance of convergence than did newspaper editors or television news directors. Kraeplin and Criado also report that, while a large percentage of the universities they surveyed were adjusting their curriculum in response to industry convergence, the changes were described as minor instead of being complete redesigns of the curriculum.

Castaneda, Murphy, and Hether (2005) documented the development of a converged journalism curriculum at the University of Southern California. They found widespread agreement that such traditional components of journalism as critical thinking, writing and fairness should remain a part of the curriculum. However, as has been the debate in journalism education for years, disagreement was more prevalent when educators attempted to judge the proper quantity of practical, technology-teaching–based courses. These researchers also note that the industry is doing little to train its workers in convergence; the bulk of responsibility seems to lie with journalism educators.

Mohl (2004) has pointed to the decline in local radio news as an example of changes in journalism. Changes in media ownership and the emphasis on the bottom line are interrelated with the rise in apathy among news consumers. The editorial by Mohl maintains that educators normally try to prepare students for the type of journalism they plan to enter, but that this is an increasingly daunting task considering the uncertainty of what journalism may be like in coming years.

The traits expected of a journalist are at the heart of what critics say is wrong with convergence. Some scholars have tried to determine how and why universities have adopted a convergence model (Lowrey, Daniels, & Becker, 2005), while others have assessed the success of university convergence programs (Castaneda et al., 2005). Virtually all have sought to identify the benefits of convergence to university education and to the public. Corrigan (2004) asserts that convergence benefits the media owner but not the consumer, and he suggests there is a loss of insightful journalism as a result of convergence efficiency. Corrigan concurred with *Boston Globe* columnist Ellen Goodman, who said "good journalism requires time to think about what is being written, while the new 'content provider' of convergence can only ricochet mindlessly trying to meet the demands of myriad media forms" (Corrigan, 2004, p. 14).

Thelen, Kaplan, and Bradley (2003) debated convergence from the different perspectives that have emerged from academia and industry. Executive Editor of *The Tampa Tribune*, a well-known converged newsroom, Thelen said the industry does not expect college graduates to be fully trained in a wide range of convergence-driven skills. Instead, journalism schools should continue to train students to be skilled in a particular area while being prepared to handle the demands of convergence.

Syracuse University professor Joel Kaplan holds that although convergence is here to stay, there are dangers. For one, Kaplan warns that the companies trying to increase their audiences may actually lose them by doing convergence badly. The so-called bells and whistles of fancy convergence operations cannot hide shallow, superficial reporting. To underscore this point, Kaplan suggests that one of the nation's so-called best examples of convergence in action, in Tampa, Florida, is far from a quality news product. He asserts that *The Tampa Tribune* was inferior to its *St. Petersburg Times* competition on the newspaper side and its WFLA TV station did not appear to be any better than its nonconverged competitors. So, while owner Media General may benefit from the sharing of newsgathering by reporters for the newspaper, TV station, and Tampa Bay Online Web site, the promise of better content for the consumer is dubious. According to Kaplan, "Bottom-line public companies like Media General and Gannett do not view convergence as a way of improving their product; they view it as a way to make more money" (Thelen et al., 2003, p. 518).

How journalism schools teach convergence is important because most graduates will begin their careers in smaller markets. According to Tanner and Duhé (2005), fewer than four out of 10 news directors said their stations have training programs in place. Of the ones with training programs, more than half described these programs as "broadcast writing classes"—again stressing the importance of writing. Tanner and Duhé suggest that although *media convergence* may be the "buzzword" in journalism programs and professional newsrooms, students training for a future career in broadcast journalism still need to specialize in television news.

A number of research texts encourage triangulation as a means of ensuring the validity of research findings (Nunnally & Bernstein, 1994, p. 37; Wimmer & Dominick, 2006, p. 50). In the case of small-market media convergence, results of two research studies—one qualitative and one quantitative—provide similar insights into this phenomenon from two different perspectives.

## Study One: A Survey of News Convergence

Based on the recent convergence developments noted earlier, research is needed to identify the views of broadcast and print news managers in the smaller markets. One such study examined levels of news convergence as well as opinions of relative success from news directors and editors at daily newspapers and local market TV stations. The survey was conducted in Texas, which by itself has 16 separate local television markets, ranging from small (Victoria, number 205) to large (Dallas-Fort Worth, number six), out of the 210 markets designated by Nielsen Media Research (n.d.).

Three assumptions guiding this research were (1) there would be more resistance to news convergence by print people more than broadcast people; (2) there would be more negative attitudes about the relative contribution of broadcast news compared to print;, and (3) converged operations would tend to be co-owned rather than being partnerships between two independent companies.

The data were collected in the fall of 2003 through an online survey with television news directors and newspaper editors in Texas. Through an Internet search, 48 television stations were identified as having locally produced newscasts in 16 local television markets, then 32 newspapers were identified as having circulation in these market areas. From this initial list, 48 television news directors and 32 newspaper editors were identified and contacted by e-mail. A total of 17 TV stations and nine newspapers responded, for a combined 32.5 percent response rate. These numbers reflect that there are typically multiple television stations in any given market, but only one newspaper. Regarding work experience, 70 percent of the news managers reported they had worked in the news business more than 20 years and almost 90 percent were over 35 years of age.

Although the initial contact was with a television news director or newspaper editor, of those actually responding, just over half (54 percent) identified themselves as television news directors and 20 percent said they were newspaper editors. Other titles given included newspaper publisher, radio news director, or TV assignments editor. Station or newspaper ownership was varied, but a variety of media companies were represented, including Belo Corporation; Hearst Corporation; Drewry Communications; Ramar; Community Newspaper Holdings, Inc.; Disney/ABC; Freedom Communications, Inc.; and NBC.

Almost half of the respondents (47 percent) reported being involved with convergence for more than three years. More than 15 percent said they had been doing some form of news convergence for one to two years. As expected, the three major types of media news outlets where content was shared were local TV stations, local newspapers and Internet sites. Whether the converged operation was print-broadcast versus broadcast-Internet or print-Internet was not measured in this study, although the questionnaire focused on evaluating local market TV news/newspaper partnerships.

## Evaluation of Convergence Success

Although the number of responses was too small for testing hypotheses, some general indications of attitudes toward media convergence emerged. The responses are provided here to generate thought and discussion for further research. They can also be considered in tandem with the case study that follows.

### Print Respondents

Seven out of nine newspapers reported they were working with other media to produce content. Of those seven, six indicated they believed their employees were positive toward "doing converged news work." These seven were also nearly unanimous ($n = 6$) in the belief that the quality of the coverage was improved and audience reactions to convergence were positive. The responses were mixed, however, regarding how well college programs were preparing students for a converged news environment. This observation applied to both print and broadcast college programs.

### Broadcast Respondents

A total of 10 out of 17 broadcast respondents said they were working with other news media, but only seven provided additional information about specific aspects of convergence. Six out of seven rated their success with convergence as positive or very positive. Broadcast respondents were also generally positive about viewer reaction (five "positive," two "no opinion") and how broadcasters feel about "doing converged news work" (five "positive," one "no opinion," one "negative"). The rating for the converged news product was less clear in that four were positive/very positive but three were noncommittal, indicating "no opinion."

### Combining Print-Broadcast Respondents

When combining the responses of the two groups, a clearer picture emerges for some items. Thirteen out of fourteen, for example, rated the success of their convergence efforts as positive (93%).

There is also a degree of consensus indicated for the notion that convergence journalism is valued by the audience for news. (See Table 12.1.)

**Table 12.1** ▰ Perceived Audience Reactions to Convergence

| Question: "How would you rate your viewers/readers reaction to convergence?" | Number | Percentage |
|---|---|---|
| Negative/very negative | 0 | 0 |
| No opinion | 3 | 21 |
| Positive/very positive | 11 | 79 |

**Table 12.2** ▰ Perceived Impact on Quality of News Coverage

| Rating: "What is your opinion of how the quality of news coverage has changed?" | Number | Percentage |
|---|---|---|
| Negative/very negative | 0 | 0 |
| No opinion | 4 | 29 |
| Positive/very positive | 10 | 71 |

**Table 12.3** ▰ Perception of College Training: Print Versus Broadcast

| Rating | How Well College Degree In *Print Journalism* Prepares Student For Convergence | How Well College Degree In *Broadcast Journalism* Prepares Student For Convergence |
|---|---|---|
| Negative/very negative | 4 | 6 |
| No opinion | 2 | 5 |
| Positive/very positive | 8 | 3 |

Likewise, most respondents believe news coverage is improved through the practice of convergence journalism. As in the previous table, at the very least, no one said it made things worse (see Table 12.2).

Comparing the opinions on how well print/newspaper journalism programs prepare students versus broadcast journalism programs, there was some indication that print programs do a better job than broadcast programs (see Table 12.3)

## Implications

The results overall suggest that many print and broadcast news managers will publicly state that convergence is important, audiences like it, and they are doing a good job of it. The view of how well educators are preparing students is not as clear (print programs seem to be rated a bit better). Print newsroom employees are rated marginally higher than their broadcast counterparts in adapting to a converged news environment.

**Table 12.4** ▰ Convergence and Advertising Sales Strategies

| Selling Advertising | Percentage |
|---|---|
| Usually separately on each outlet | 32 |
| More often as a cross-platform package | 11 |
| More often as primary sales on one with complimentary or discounted ads on other(s) | 11 |
| Web site did not do anything beyond trade-out ads that required splitting costs and revenues | 10 |
| Web site was profitable independent of their other media operation | 10 |
| Did not respond/other | 26 |

## Economics and Promotion in Converged Operations

Since convergence is as much an economic issue as a news content issue, promotion and profitability factors were examined. News managers were asked, "If you have a converged operation, what is your opinion of your company's success in promoting it?" Over 70 percent rated it positive. Asked to give the single most common style of selling advertising on their various news outlets, the news managers reported a variety of models, summarized in Table 12.4.

## Final Pros and Cons

Why should companies continue or consider doing some form of news convergence? Those with converged operations were asked to choose the most important reason why the company adopted convergent journalism practices. Most of the responses reflected the view that "good journalism is good business." Practicing convergent journalism made them more competitive, more attractive to the audience, and more efficient (Table 12.5).

**Table 12.5** ▰ Why Converge?

| Primary Reason for Adopting Convergence | Percentage |
|---|---|
| "Better Positioning of Ourselves in the Market" | 28 |
| "Better Service to Audience" | 22 |
| "Economy of Scale/Cost Savings" | 11 |
| "Corporate/Management Policy or Directive" | 6 |
| "Other" or "No Response" | 33 |

**Table 12.6** ■ Why You Do Not Have a Converged News Operation

| Reason for No Convergence | Percentage |
|---|---|
| "No Other Media to Work With" | 17 |
| "It Would Be Helping the Competition" | 6 |
| "Current Employees Could Not Adapt" | 6 |
| "Resistance of Other Media We Might Attempt Doing Convergence With" | 6 |
| "Other" or "No Response" | 65 |

Those who were not doing convergence were asked to provide the most important reason why. Although more than half selected 'no opinion," the reasons that were given reflect a traditional management mind-set (see Table 12.6).

For years it has been standard for media companies to rarely if ever even mention a competitor, much less work with them. Also, there has been anecdotal evidence that some operations have at times had a patronizing view that it is the employees who cannot adapt (the euphemism for "they're not good enough" and/or "we won't pay for training"). Finally, when asked, "If you did start convergence, what is your opinion of how the quality of news coverage would change?" almost three-quarters (73 percent) of respondents thought it would be positive or very positive.

Respondents were asked about the most important factor determining a person's ability to work in a converged newsroom. Three primary reasons emerged, involving reluctance to adopt convergence (see Table 12.7). These reflected the individuals themselves rather than any environmental constraints or lack or resources. Fifty percent said, "Attitude about his/her work expectation versus what the company wants"; 17 percent thought, "Turf battles in the workplace over who is print and who is broadcast"; and eleven percent said, "Age of the person/number of years in the news business."

Others added their own thoughts about potential difficulties. One said, "Their experience in the particular field—no experience with broadcast equipment is bad

**Table 12.7** ■ Barriers Affecting Ability to Work in a Converged Newsroom

| Description of Barriers | Percentage |
|---|---|
| Attitude about his/her work expectation versus what the company wants | 50 |
| Turf battles in the workplace over who is print and who is broadcast | 17 |
| Age of the person/number of years in the news business | 11 |
| Other | 22 |

for the print people, no experience writing is bad for the broadcasters." Another suggested it could be "Lack of time to do quality work for both print and broadcast or another combination." One newspaper manager stated, "Our cooperative efforts with a local affiliate station are limited to weather news coverage and cross-promotion. We don't have print reporters going on-air. And so far we've done no team-reporting project, which is a step we discussed at the beginning of the relationship." This is one example on the low end of the convergence continuum identified by Dailey et al. (2005).

## Study Two: A Case Study of News Convergence

The survey results provide some insight into the views of those working in smaller markets about convergence efforts and their benefits. To understand more of the dynamics of a converged news environment and to apply Dailey et al.'s convergence continuum, it is helpful to see how one such operation has worked. In Quincy, Illinois, at the time of this study, one company owned the local newspaper (the *Herald-Whig*), WGEM—the local NBC-affiliated TV station, a local cable-only Fox-affiliated channel running on the Quincy cable system (CGEM), local broadcast radio stations, and a Web site. Their news environment was built around employees performing multiple duties to provide content across the various platforms. Researching how this facility operates was accomplished by a site visit and interviews with employees and news managers.

The opportunity to study this operation occurred during the summer of 2002, as the researcher was working in the Hannibal-Quincy-Keokuk television market as a participant in the Radio Television News Directors Association (RTNDA; n.d.) Educator in the Newsroom fellowship program. Although the researcher was working at a competing TV station in the market, an on-site visit was arranged. Just as convergence efforts between competing companies often become a matter of trust, attempting a site visit to "collect data" from "the competition" also required a measure of trust from managers at the site being visited, analyzed and documented.

The researcher spent two days interviewing staff members and touring the media facilities in July 2002, with individual interviews lasting between five and 30 minutes. Several weeks afterward, a final full interview was conducted with Managing Editor Mark Magliari. The researcher's approach was to observe objectively and allow "truth to emerge," rather than the observer placing his own preconceived ideas upon the phenomenon being observed. Since the researcher is also the instrument in ethnographic research, it is also important to note that the researcher was an experienced broadcaster, which may have introduced some bias of broadcast versus print.

One limitation of this approach was the conflict of interest inherent in getting staff to voice honest reactions and then bringing these anonymously before management for comment and reaction. The researcher spoke with many staff members, from the lead anchor to reporters, radio air talent and TV and newspaper photographers. To maintain respondent anonymity, the researcher purposefully did not identify or attribute any specific opinions expressed by staff members, but instead discussed the ideas in the aggregate with Magliari. This approach allowed a number of issues to be discussed with management, both the positive and negative outcomes associated with convergence.

## Challenges and Drawbacks

The Quincy operation, as it was judged in the 2002 study, demonstrates how new employees in broadcast or print news careers may need to be prepared to work in a multitasking environment. Television photographers might be expected to shoot video for a news story but also take digital still images for the newspaper or the Web site. The photographers might also post news content on the Web site. Television reporters might have to create a television news story as well as a newspaper story. Having such a broad set of technical skills may be problematic.

Perhaps the greatest challenge in merging print and broadcast is the difference in the two cultures. Magliari said potential problems exist when a new television reporter, asked to also write a newspaper article on a story, might think, "Well, if I'd wanted to work for a paper..." (personal communication, November 2002). Likewise, a newspaper reporter asked to provide a story for both print and broadcast might hesitate because he or she was "not a TV person." Redding (2002) quotes Al Tompkins of the Poynter Institute, who states. "The stereotypes are that TV is not serious enough and print is too damned serious. None of these are accurate, but it's very difficult to convince each other. And that's the reason that really serious news operations have serious conversations before convergence" (p. E1). While these cultures may be different, sometimes differences can be beneficial. Magliari said having the relationship with the newspaper also gave a strong journalistic base to the broadcast side (personal communication, November 2002).

The broadcast and newspapers' newsrooms also had technical difficulties when "talking to each other"—they used different newsroom computer systems, with the broadcast system PC-based and the newspaper's system Macintosh-based. Sharing stories was simple enough via e-mail; the sharing became complicated, however, when it came to accessing archives or assignments/futures files. The newspaper typically had much better planning for future news, allowing both print and broadcast to have a better assignments process—maintaining those files for both to review was not working.

Finding a way to merge the software would depend upon commercial software developers recognizing a need for cross-platform/cross-distribution compatibility, or for the companies to have specially written software. Given the complexity of television news–producing software that allows interaction with wire service feeds, access to atlas and other search data, creation of networked work stations, and the like, the former option seems more practical that the latter. However, the former may not yet be realistic for commercial software developers to consider.

In this particular case, the broadcast and print newsrooms also were not in the same building. There were some discussions of eventually combining all operations in one facility, but technical and cost questions were unresolved. Even if they were to be under one roof, the newspaper has essentially two deadlines a day, which do not match or correspond to the WGEM-TV deadlines of four big newscasts per day. Print and broadcast newsgathering also works differently. Print journalists can cover a story from 60 or 70 miles away by making phone calls. Radio journalists can call and "go live" or get sound bites over the phone. But television crews must go to the story location to record video and interviews.

The term "local" has different meanings for different media. Cable system distribution is only on the Quincy cable system—the 24 other counties listed as the WGEM service area did not get CGEM. The radio station signals covered most—but not all—the counties in the area. Each of the cities in the television market also has its own local newspaper; many of those cities also have their own local radio stations. When covering a story where the live truck was not available or beyond signal range, video could be fed via broadband connection. But getting a signal the other way, from CGEM to other cable systems, was not possible. The WGEM digital(DT) signal was seen as a potential means of transmitting a signal to other cable systems.

Is there a threat to the viability of a newspaper in this kind of environment, where consumers get news throughout the day from radio, broadcast TV, cable TV, and Internet distribution? Magliari points to the "random access" to specific information a newspaper reader wants and the portability of a newspaper as supporting factors favoring the paper's continued existence (personal communication, November 2002). As is generally recognized, newspapers changed in reaction to the introduction of radio; radio changed once faced with television competition; and broadcast television changed in response to cable TV and the Internet. Therefore, the *Herald-Whig* benefits from this partnership and is strengthened rather than weakened.

Multitasking is not a new concept in smaller-market television, and may be more applicable to larger-market and even national network environments. As is common in smaller markets, the WGEM-TV news director was also the anchor for the 6 p.m. newscast and the afternoon radio news anchor. Several employees noted that no one in their building had only one job. Often one of the evening anchors in such local small markets is also the producer, while the other anchor may also be a daytime reporter. In relation to this typical workload of small-market television, WGEM-TV made an effort to have the on-air look of a station in a larger market. The company sought to pay employees enough to minimize the turnover typical in smaller markets. Achieving the look and feel of a larger-market operation requires financial investment, so those who were expected to do multiple jobs also could expect to be paid a competitive wage.

The challenge, however, was not necessarily finding employees who could do multiple jobs. In this converged operation, the difficulty was one of overcoming cultural differences between print and broadcast and coordinating the differences in specific skills and tasks. For example, there are significant differences between print and broadcast news writing, video recording versus still-image photography, audio versus video production of news, or on-camera versus radio- or print-oriented interviewing. At WGEM, the newspaper managers were involved in the hiring of the broadcast employees, since they had to find college graduates who could write a newspaper story and a television story (Magliari, personal communication, July 2002).

## The Successes and Benefits of Convergence

Perhaps the greatest benefit for operations like the one in Quincy is that they can focus on gathering and reporting news and are able to cover a greater range of stories. Rather than reducing costs, convergence more likely provides real financial benefit through cost containment. Costs of otherwise duplicate news crews are eliminated. For example, rather than sending two vehicles with two crews on a trip to cover a story several counties away from the main operation, one vehicle could take both groups. Rather than sending a television news photographer and reporter along with a newspaper reporter and photographer to a story, one pair could often cover the story for television and the newspaper, and that content could then be repurposed to the radio stations and the Web site. Despite the standard expectation of multitasking, the primary responsibilities of WGEM-TV news employees were to the TV station, while *Herald-Whig* reporters' primary duties were toward reporting newspaper stories.

The philosophy of WGEM seemed to be contrary to how many local-market converged media operate. While most such operations are more promotional than editorial (Redding, 2002), at WGEM they tried to place stories in all outlets, based particularly on fitting the content to the distribution. Typically, for example, Web-posted stories are designed essentially to build viewers toward the television newscast. But even as content is often repurposed, some stories work better for a particular medium. Stories with strong visual elements, for example, work well on television. Some stories are extremely complex and may rely on a lot of numbers or financial data and may be better presented in a newspaper.

Repurposing is growing more commonplace nationally, as broadcast network–delivered programs increasingly rerun within the same week on a cable network. In the Quincy operation, it is a necessary process in providing content to all outlets. An interview done for a Sunday morning WGEM-TV public affairs show would also run on CGEM and the audio portion would be played on both radio stations. Sound bites from the interview would be used in both radio and television news stories. Content from the interview could find its way into the newspaper and onto the Web site. The morning radio show also ran live on the cable TV channel, CGEM. A live broadcast of a sporting event produced locally would air on radio and CGEM, with the CGEM video being supported by the same play-by-play commentary as aired on the radio station.

The local managers preferred to use the term "synergy" rather than "convergence," and focused on newsgathering and distribution rather than the corporate bottom line. Still, the business model developed in Quincy, Illinois, appeared to be working effectively. While there were some cultural and technical limitations, cross-ownership of multiple media allowed Quincy Newspapers Incorporated (QNI) and the Quincy Broadcasting Company (QBC) to extend their ability to serve their audience and to do so in a way that was economically sound.

Can a local market converged television, radio, newspaper, and Internet news-delivery business operate effectively? This study demonstrated that QNI and QBC were successful during the two years' time they had worked together with the print, broadcast, and Internet operations. The benefits seen from this relationship included newsgathering, planning, and broad topic coverage as well as economic efficiencies. The greatest difficulties included some technical challenges, such as electronic sharing of planning and archive files, along with the inevitable differences that exist in deadlines and the culture of doing print-style versus broadcast-style news.

Additionally, although the WGEM managing editor noted that the company being locally owned allowed them to care about the local community and to be partners in community development, making a profit for the company clearly was still vital. The FM radio station's change to all-news boosted the station's advertising dollars due to an influx of political advertising dollars that now fit the station format better than it had for music radio. Internal research also showed that news programming on the FM was helping make the WGEM TV morning news show and 6 p.m. news more dominant in the market (Magliari, personal communication, November 2002).

Is the public better served by this arrangement, or is it more beneficial to the owners? Several WGEM staff members were questioned regarding the station's efforts and success at offering alternative points of view and ability to do so with multiple means of distribution. The homogeneity of the city of Quincy was noted as a limitation in allowing this question to be fully explored or answered. It seems in a very real way that the successes of the synergy are bound within the general city limits of Quincy. The city of Hannibal, Missouri—which is only 15 miles away from Quincy, but has its own local paper and radio—is much more ethnically and socially diverse than Quincy. Particularly if QNI/QBC could get CGEM on the Hannibal cable system, future research could compare the impact there versus what has been accomplished in Quincy.

Reflecting back to the convergence continuum noted earlier, this case study shows that a converged operation can engage in several stages simultaneously. While QNI and QBC attempted to achieve full convergence, on any given day one part of the convergence operation might do simple cross-promotion, another part might clone content, some work might reflect coopetition, and some stories would demonstrate full content sharing. The Quincy case study helps support Dailey et al.'s (2005) convergence continuum as a means of studying converged news operations.

## Conclusion and Implications

As mentioned at the beginning of this chapter, the danger is that media convergence could be a euphemism for corporate media to expand their control of information as well as enable profit maximization. But a kinder, gentler form of convergence is taking shape and being practiced in smaller markets throughout the United States, and it seems the public is being well served. Will the synergy evidenced by the survey of news managers and supported by the successes in Quincy continue to work? The findings suggest that it will, barring unforeseen regulatory or technical changes. This is a business model ideally suited to the current marketplace philosophy of broadcast (de)regulation, and educators must prepare college graduates interested in news careers. Connecting the Quincy case study with the survey of Texas media markets, a picture begins to emerge. In Quincy, the reduced competitive pressure seemed to have enabled social networking and relationships that overcame professional biases so often found in larger markets. Co-ownership versus partnerships also seems to be an important variable.

There are important trends to consider today when studying convergence. Three in particular are the following: (1) the introduction of mobile media, such as cell phones and iPods, to the converged media mix, (2) new competition from the likes of Yahoo! and Google as each tries to create and distribute localized content, and (3) the continued growth and potential challenges of citizen journalism efforts through Internet Web sites, podcasts, and blogs.

The case study presented here shows one example where a co-owned news convergence effort has been reasonably successful. The survey of Texas media markets shows that both print and broadcast news managers tend to agree that convergence is good for their news efforts, and those who do not have converged operations also believe it would be beneficial. While the news industry continues to develop converged newsrooms and educators continue to struggle with how, or whether, to prepare college graduates for a new media environment, many questions remain unanswered. Debate among practitioners and scholars shown in studies cited here as well as others suggests convergence is here to stay, but the hope of some commentators for a "super journalist" or of more insightful news content remains elusive. More audience-centered studies seem to be the logical next step.

## References

Birge, E. (2004, May 1). Teaching convergence—But what is it? *Quill, 92*(4), 10.

Castaneda, L., Murphy, S., & Hether, H. J. (2005). Teaching print, broadcast, and online journalism concurrently: A case study assessing a convergence curriculum. *Journalism and Mass Communication Educator, 60*(1), 57–70.

Corrigan, D. (2004, November). Convergence works for media owner but not for news consumer. *St. Louis Journalism Review, pp*. 14–15.

Dailey, L., Demo, L., & Spillman. M. (2005). The convergence continuum: A model for studying collaboration between media newsrooms, *Atlantic Journal of Communication, 13*(3), 150–168.

Filak, V. F. (2004). Cultural convergence: Intergroup bias among journalists and its impact on convergence. *Atlantic Journal of Communication, 12*, 216–23.

Kaufhold, G. (1998, June 22). It's not convergence—it's diffusion! *Electronic News, 44*(2224), 8.

Kraeplin, C., & Criado, C. A. (2005). Building a case for convergence journalism curriculum. *Journalism and Mass Communication Educator, 60*(1), 47–56.

Landler, M., & Fabricant, G. (2002, August 4). A tortoise, many hares and a web of convergence. *The New York Times,* C1.

Lowrey, W., Daniels, G.L., & Becker, L.B. (2005). Predictors of convergence curricula in journalism and mass communication programs. *Journalism and Mass Communication Educator, 60*(1), 32–46.

Mohl, J. (2004, May 1). Planning for a changing profession. *Quill, 92 (4)*, 3–8.

Nielsen Media Research. (n.d.). Local market universe estimates. Retrieved February 8, 2007, from http://www.nielsenmedia.com/nc/portal/site/Public/

Nunnally, J.C., & Bernstein, I.H. (1994). *Psychometric theory* (3rd ed.). New York: McGraw-Hill.

Radio Television News Directors Association. (n.d.). *Educator in the newsroom fellowships*. Retrieved February 15, 2007 from http://www.rtnda.org/resources/excel.shtml

Redding, A. (2002, August 25). Convergence in name only without trust. *South Bend Tribune,* p. E1.

Robins, W. (2001, March 5). If Eminem and Elton can do it, why not old and new media? *Editor and Publisher,* 134, 15–17.

Silcock, B. W., & Keith, S. (2006). Translating the Tower of Babel: Issues of definition, language and culture in converged newsrooms. *Journalism Studies, 7*(4), 610–627.

Singer, J. B. (2004). Strange bedfellows? The diffusion of convergence in four news organizations. *Journalism Studies, 5*(1), 3–18.

Tatro, N. (2002, May 27). Urge to converge changes media companies' mission. *Crain's Chicago Business,* p. 11.

Tanner, A., and Duhé, S. (2005). Trends in mass media education in the age of media convergence: Preparing students for careers in a converging news environment. *Simile,* 5(3), 1–12.

Thelen, G., Kaplan, J., & Bradley, D. (2003). Debate. *Journalism Studies, 4*(4), 513–521.

Wimmer, R.D., & Dominick, J.R. (2006). *Mass media research: An introduction* (8th ed.). Belmont, CA: Thomson Wadsworth.

CHAPTER 13

# Beyond the "Tower of Babel"

## Ideas for Future Research in Media Convergence

Susan Keith
B. William Silcock

Newsroom convergence has become a growing subfield for journalism and mass communication researchers, as this volume indicates. Using both quantitative and qualitative methods, researchers are beginning to piece together a picture of how the industry and academy *are* and *should be* approaching forays across platform lines. Several survey-based studies have measured the reach of convergence practice across the United States (Criado & Kraeplin, 2003; Dailey, Demo, & Spillman, 2005b; Duhé, Mortimer, & Chow, 2004; Kraeplin & Criado, 2006; Lowrey, 2005), while a variety of qualitative studies have illuminated the convergence practices of specific professional organizations (Dupagne & Garrison, 2006; Huang et al., 2006b; Ketterer et al., 2004; Silcock & Keith, 2006; Singer, 2004). Other work has described how the academy has responded to the industry's convergence practices (Castaneda, Murphy, & Hether, 2005; Hammond & Petersen, 2000; Kraeplin & Criado, 2005; Lowrey, Daniels, & Becker, 2005). Still other studies have offered research-based normative advice for how professors should teach convergence journalism (Filak, 2006; Huang et al., 2006a).

Some research in this rising subfield has raised questions about whether convergence—especially cross-platform partnerships between newspapers and television stations—functions as its early proponents envisioned. Several authors have presented evidence that convergence partnerships, including some of the most famous, are not as integrated as industry rhetoric sometimes suggests (Curkan-Flanagan & Hardenbergh, 2002; Dailey et. al, 2005b; Dupagne & Garrison, 2006; Lowrey, 2005; Silcock & Keith, 2006). One study has even raised questions about whether audiences accept the idea of print-broadcast-Web content sharing (Ketterer, Smethers, & Bressers, 2005).

Nevertheless, gaps remain in our understanding of convergence, the coming together of media platforms—print, broadcast and online—in technological, economic, and cultural ways. The goal of this chapter is to point out some of those holes, drawing on the authors' research on cross-platform partnerships on opposite sides of the country, published in an article titled "Translating the Tower of Babel? Issues of Definition, Language and Culture in Converged Newsrooms" (Silcock & Keith, 2006). Our aim is to suggest questions that future studies need to ask to complete our understanding of the continuing transformation of journalistic practice. We attempt to meet this goal by presenting findings from our study in two sections, each of which is followed immediately by discussion of the new or incompletely answered questions those sections raise.

## Tracking Convergence in Tampa and Phoenix

In 2002, the authors of this chapter began conducting interview-based research with journalists at two highly publicized convergence partnerships. One, based in Phoenix, Arizona, paired two news outlets located several blocks apart: KPNX-TV, the market-leading NBC affiliate, long owned by Gannett Co. Inc., and the *Arizona Republic* newspaper, acquired by Gannett in 2000. The other involved the Tampa, Florida, partners sometimes referred to as the "poster children for convergence": WFLA-TV, the market-leading NBC affiliate, and *The Tampa Tribune* newspaper, both owned by Richmond, Virginia–based Media General and located in a facility built with convergence in mind. Initially, we sought to determine whether journalists experienced the sort of language-based confusion we had noticed in our own conversations about teaching journalism across platform lines. One of us had spent 16 years as a newspaper reporter and editor and taught primarily print courses; the other had worked as a television reporter and news director for 21 years and taught primarily broadcast courses. As a result, when the school where we then both taught, Arizona State University, attempted to introduce some elements of convergence into beginning journalism courses (Foote, 2002), we found we each of us viewed news with a distinct "vocabulary of precedents" (Ericson, Baranek, & Chan, 1987). We had to explain to each other terminology particular to our separate fields, such as "vo-sot" in television and "slot," "rim," and "spike" in print. We had to learn from each other that the daily list of stories is called the "rundown" by television news directors and the "budget" by newspaper editors. In addition, we found that our fields assigned very different meanings to some terms they had in common, including "editor," "deck," and "package."

Drawing on Gurevitch and Levy's contention (1990) that a common language helps foster a shared professional culture in newsrooms and Zelizer's (1993) notion that "shared discourse" (p. 221) helps unite journalists into interpretive communities, we wondered whether journalists working in converged environments might face substantive struggles over language. Our research (Silcock & Keith, 2006) revealed, however, that language challenges were minor and easily overcome. As *Tampa Tribune* sports reporter Roy Cummings put it when he described doing reports for WFLA-TV:

> [There have] been a couple of times where they would tell me I'm going to do something and I wouldn't know what it was and I would have to ask. But I've done it enough now that I know how the whole thing works. Like the earpiece or whatever. They told me, "Do you have your own . . . ?"—I don't know what it is—IUB, or whatever they call it. It's been awhile since I've done one, so I've forgotten the phrase. But I say, I say, "No, I don't think I have one of those." (personal communication, Spring 2002)

Although Cummings was unable to recall the proper terminology for an earpiece—IFB, for "interruptible feedback"—he did correctly use the television terms "cut-in," "stand-up," and "package" in other parts of an interview, indicating that he had absorbed some of the vocabulary of television news. Other journalists interviewed pointed out that it was not difficult for print reporters to learn new television terminology because they typically learned newsroom-specific jargon at each newspaper they worked for. Television reporters were rarely forced to learn print jargon, because television news staffs were so small, compared with newspaper staffs, that few television reporters had time to write for newspapers. One who did, WFLA consumer reporter Victoria Lim, a columnist for *The Tampa Tribune,* struggled with the different grammar rules employed by print. "Tense was a real problem for me," she said, "because in the beginning I wrote some of my copy in the present active tense like you would for television, which is not an appropriate tense for a newspaper when reporting on something that has already happened" (personal communication, Spring 2005). However, in the grander scheme of convergence, commented Gil Thelen, then the publisher of *The Tampa Tribune,* language confusion was not "deep and profound"; it was "more operational."

The more significant challenges for the convergence partners were economic and cultural, a finding that was clarified when we did a second round of interviews in 2005, talking with some original and some new sources. By then, although the Tampa convergence operation seemed little changed, the Phoenix partnership, which had been in full flower in 2002 when we began the study, had devolved substantially. At its peak, the partnership had featured regular appearances by the *Arizona Republic*'s business editor and about a dozen other newspaper staff members on KPNX. In addition there was a KPNX video camera installed in the

*Republic* newsroom; two veteran *Republic* journalists who cross-trained at KPNX were serving as "multimedia editors"; a videographer was working out of the *Republic* newsroom;, and the *Republic* was offering KPNX enterprise stories daily. In 2004, however, the partners reduced these efforts largely because they found that convergence was not profitable. The business editor joined KPNX full-time as a weekend anchor, the multimedia editors took print positions, and the *Republic* videographer was retrained as a still photographer, As Tracy Collins, deputy managing editor for presentation at the *Arizona Republic,* told us:

> "We found there was no measurable difference in our circulation numbers and no measurable difference in their overnight ratings based on the stuff that we were doing.... We started to realize through market research and everything that the newspaper and the broadcast audiences were fairly different and that sharing between was not necessarily going to drive one to the other. (Silcock & Keith, 2006, p. 615).[1]

Instead, the partners began to focus more heavily on feeding content to their common Web site, www.azcentral.com, with KPNX launching in April 2005 three-minute Web newscasts updated several times each day. Some links did remain between the station and newspaper. For example, KPNX's consumer reporter continued writing a column for the *Republic,* the *Republic*'s weather was "branded" with the name of a KPNX meteorologist, the video camera remained in the *Republic* newsroom, and KPNX staff members had access to the *Republic*'s archive. But those things, a former KPNX producer told us, "could be happening in a market where you don't really have what someone might term convergence" (Silcock & Keith, 2006, p. 616).

---

1. Profit is not the only consideration for Gannett's Phoenix convergence operation. Whether it continues long term in any form will depend on the outcome of what happens with the Federal Communications Commission's (FCC) disputed cross-ownership rule. Until 2003, an FCC rule adopted in 1975 prevented a company from owning a television station and newspaper in the same market. (Media General's Tampa partnership was grandfathered in because they already owned both *The Tampa Tribune* and WFLA-TV when the 1975 rules were written.) That rule would have prevented Gannett Co. Inc. from renewing KPNX's license when it expired in 2006 if Gannett owned the *Arizona Republic* then—unless the company obtained a waiver exempting the Phoenix partners from the rule. In July 2003, however, FCC commissioners voted 3–2 to relax several media ownership rules, including the cross-ownership rule (Ahrens, 2003). A year later, however, a three-judge panel of the U.S. Court of Appeals for the Third Circuit ordered the FCC to reconsider the rules changes, which it said the Commission had failed to adequately justify (Labaton, 2004). Several large media companies, including Gannett and Media General—which had print-broadcast partnerships in cities other than Tampa that would have been affected by the old rule—appealed the Third Circuit's ruling. But in June 2005, the U.S. Supreme Court refused to hear their appeal. A year later, Gannett applied for renewal of KPNX's broadcast license and a permanent waiver of the cross-ownership rule, as the FCC announced it would begin reviewing ownership rules, as ordered by the court. Gannett reported in 2007 that it did not expect the FCC to rule on its license application until the cross-ownership rules were settled (Gannett Co. Inc., 2007).

Nevertheless, Collins referred to the station and newspaper's focus on their joint Web site as just another form of convergence. It was a type of convergence, however, that did not seem to be covered by either of the two major convergence models proposed by scholars. The earlier, proposed by Dailey, Demo, and Spillman (2005a), suggested that there is a "convergence continuum" of five overlapping stages. These range from cross-promotion, in which partners merely advertise each other's journalism; through cloning, in which partners republish each other's reporting with few changes; to "coopetition," in which partners simultaneously cooperate and compete to report the news first; to content sharing, in which partners share, then repackage each other's content; and finally to full convergence, in which teams made up of staff members from both partners work together to report news. Later, Lowrey's work (2005), derived from the Dailey et al. (2005a) model, suggested that convergence relationships can be characterized by whether they involve content partnering (the simplest level, at which journalistic content is shared), procedural partnering (which involves shared planning, scheduling, and story preparation) or structural partnering (which includes such attributes as a shared assignment desk, multimedia editors who make assignments across platform lines, and beats staffed with both newspaper and television reporters).

The devolved *Republic*-KPNX partnership seemed to contain some elements of the cloning stage of the Dailey et al. (2005a) model. However, the fact that the media outlets were feeding a passive third platform—a Web site that did not share content with the station or newspaper—and creating new content especially for that platform suggested that the practice went beyond cloning. The Phoenix practice also seemed to exceed the content-partnering level proposed by Lowrey (2005), without reaching his procedural or structural partnering stages. Determining that none of the existing models exactly fit what we had observed, we proposed the name "co-(re)creating" to describe the type of partnership—a newspaper and television station retrenched from day-to-day convergence and feeding a common Web site—that existed in Phoenix in 2005 (Silcock & Keith, 2006, p. 623). In this new term, "co" indicated a level of cooperation—though a somewhat reduced one—between the convergence partners; "(re)" referred to the fact that both the television and newspaper partners were repurposing their content for the Web site; and "creating" referred to the fact that the partners, especially KPNX, were creating for the Web site some new content that did not appear on news shows or in the newspaper.

## Research Gaps: Convergence Economics and Structures

These findings about the structure and economics of partnerships suggest several opportunities for future research that might fill gaps in our understanding of convergence. First, consider the movement in Phoenix away from attempts to integrate newspaper and television operations on a daily basis because those attempts did not stop falling newspaper circulation or raise news ratings for the market's top-rated station. It confirms the assertion by Quinn (2004a, 2004b, 2005) and others that convergence has been as much a result of economic pressure as technological advances and altruistic attempts to provide news in multiple ways to time-starved audiences. If the nation's largest newspaper company, Gannett, cannot figure a way to make money from substantial, daily acts of convergence by a television station and newspaper under common ownership—one of the factors that usually promotes convergence (Quinn, 2005)—what does that suggest about the longevity of convergence partnerships between partners owned by different groups? The current fate of those partnerships, which were announced by the dozens in the early years of the century, would seem ripe for study.

Second, the movement in Phoenix to what managers termed a different form of convergence suggests that the definition of "convergence"—always a contentious topic in convergence research—will continue to be fluid. As Dailey et al. (2005b) wrote, "the definition of convergence is evolving even as newsroom partnerships evolve" (p. 37). That evolution suggests that scholars need to continue to produce case studies of convergence operations, both to record the evolution, or devolution, of industry practice and to educate journalism professors on what they should be preparing students to face. There is a particular need for qualitative studies based on in-depth, in-newsroom observation like that practiced by Singer (2004, 2006), Klinenberg (2005), and Boczkowski (2004). This type of research has rarely been conducted in recent years by journalism and mass communications researchers—Klinenberg and Boczkowski are sociologists—perhaps because it is viewed as too time-consuming or because there is the perception that access is difficult to negotiate. Yet it is valuable because it can delve below the rhetoric of convergence. There also is a place for further quantitative surveys tracking convergence practice. Although a fairly large number of studies have measured the diffusion of convergence (Criado & Kraeplin, 2003; Dailey et al., 2005b; Duhé et al., 2004; Kraeplin & Criado, 2006; Lowrey, 2005), the identification of a new form of convergence in Phoenix also suggests that more national surveys will be needed to follow the evolution of convergence practice.

Third, our identification of a form of convergence that did not fit existing models suggests that there is a need for more theorizing about convergence structures. The models that have received the most attention—Dailey et al.'s five-stage "convergence continuum" and Lowrey's three-type partnering-basis model—seem to be based on the notion that convergence involves only two entities, functioning to some extent as equals. Although it is indeed the television-newspaper dimension of convergence partnerships that has received the most scholarly attention, most convergence partnerships also involve a third platform: at least one Web site. How those Web sites figure into convergence is not fully articulated by any existing full-scale models, though we attempted to describe one version with our "co-(re)creating" label (Silcock & Keith, 2006).

Fourth, convergence research concentrated on structures needs to further examine what might be called "intra-organization convergence," the internal practice of multimedia journalism by print or broadcast entities acting without partners. A growing number of newspapers, for example, are rejecting the notion that they need to partner with a television station to harness the power of video and the audiences it attracts—a notion that was once at the very heart of convergence. They are instead producing their own Web video newscasts, with varying levels of professionalism (Malone, 2007). These innovations need to be examined closely for they seem to signal that the future of journalism, at least for print-based outlets, is not convergence but migration—to the Web. As Wasserman (2006) writes, "The technological superiority of online distribution for multimedia presentation and its vast potential for interactivity will make the Internet the principal venue for news and topical commentary" (p. 24). Although some convergence studies, such as Singer (2004), have looked closely at online operations that are part of print-broadcast partnerships, more work is needed.

## Cultural Challenges in Tampa and Phoenix

Other gaps in the current understanding of convergence were suggested by our findings about clashes between print and television cultures in the Tampa and Phoenix convergence operations. Some of these were not surprising. "Anybody knows in journalism TV people and print people fight like cats and dogs. . . . We still encounter it and deal with it," Forrest Carr, then the news director of WFLA-TV said in 2002, when the station and *The Tampa Tribune* had been practicing convergence in some form for 16 to 18 years, depending on which event is used as the starting point. Sometimes print and television staffers' distrust of each other resulted only in fairly inconsequential stereotyping, such as that which former *Arizona Republic* business editor Brahm Resnick experienced when he first visited KPNX. He told us: "I came in and was meeting people, and one of the morning reporters said, 'Oh, you're Brahm Resnik? I was kind of picturing somebody with, I don't know, with a beard and a pipe and a tweed coat.' "

Other times, distrust had what would appear to be serious implications for the success of integration. For example, even at the height of the Phoenix convergence project, the *Arizona Republic* deliberately limited what KPNX could learn about upcoming *Republic* stories—even though the newspaper could access the television station's rundown, the list of stories that would air on future broadcasts. As former multimedia producer Mike Stephens explained:

> We tried to control as much as we could the information that we gave them and . . . how it was presented. . . . We wanted to retain a fair amount of control over it. . . . [Y]ou don't just want to give them everything, because TV has a tendency to kind of run with it and, and it doesn't check facts as quickly and

> so we were kind of, we would be the people who put on the brakes a lot of the times. You know we never gave them the access to go into our budgeting system...just to have free rein to see everything themselves that we were doing. It kind of went through us. (Silcock & Keith, 2006, p. 617)

This suggested, we noted, "an underlying assumption that the newspaper was more likely than the TV news operation to gather news that was valuable and, thus, should be closely guarded" (Silcock & Keith, 2006, p. 617).

That assumption was rooted, in part, in the differences in work routines between print and broadcast journalists. Under pressure to fill several broadcasts a day despite small staffs, broadcast journalists were accustomed to employing practices—such as using single-sourced stories, focusing on breaking law-enforcement news, and doing little advance planning—that their print partners found odd, if not downright unacceptable. Print journalists also sometimes did things that made life difficult for their television partners. For example, Stephens said, the *Arizona Republic* occasionally would successfully pitch a story to KPNX early in the day, then decide to hold the story because too little space was available in the next day's newspaper. In that situation, Stephens explained in 2002, the agreement between the *Arizona Republic* and KPNX was "if we are giving them something they wouldn't have gotten on their own—which we do every single day—they can't run with it." He said the *Republic* would never wait until 4:45 p.m. to tell the station that it was holding a story scheduled to be mentioned on the 5 p.m. broadcast, "but even when it's a couple of hours out from a newscast, if they start thinking that this is what they're going to have (and a story is held)...they get a little annoyed."

Some cultural clashes, however, were more serious than annoyance. Journalists from both partnerships said that there had been issues surrounding differing ethical standards between print and broadcast related to visual journalism. Michelle Bearden, religion reporter at *The Tampa Tribune,* said the clash of rules about what constituted visual truth-telling became especially clear when a print photographer and broadcast videographer made images for the same story at the same time. If the videographer missed a shot, such as someone walking through a door, she said, "They'll go, 'Can you walk back in the door again?' Ethically, that makes the print guys crazy. They don't like that, and of course they won't shoot that" (Silcock & Keith, 2006, p. 619). Because of similar conflicts, KPNX news director Mark Casey said, the Phoenix convergence partners dropped a plan to have *Republic* still photographers shoot video.

## Research Gaps: Newsroom Cultures

Such cultural conflicts have already been the subject of some convergence research. Filak (see Chapter 7, this volume), for example, found that journalists considering hypothetical convergence plans exhibited the sort of intergroup bias that might be predicted from former WFLA news director Forrest Carr's comment that "TV people and print people fight like cats and dogs." Both print and broadcast journalists were more likely to approve of a convergence plan if they believe it had been written by people who worked in their medium. Cultural conflict has also been the subject of a series of studies drawn from Singer's observation and interviews in four newsrooms. In a study based on diffusion of innovations theory, Singer (2004) reported that cultural clashes existed among early and late adopters of convergence within the same organizations. In another article based on the same field research, Singer (2006) reported that print and broadcast journalists in convergence partnerships distrusted each other's reporting methods, news judgments, and news values. In addition, some journalists interviewed thought that convergence violated the journalistic norm of editorial independence, either because of what they perceived as erosion in the wall between news and advertising online or because television sometimes was constrained to merely teasing stories that would be in the next day's newspaper, rather than reporting immediately the news those stories contained. Although these examinations of cultural clashes in converged organizations have been extremely important, there remains room for more research on the culture(s) of convergence. Our study suggested research is particularly needed in three closely entwined areas.

First, while the fact that journalists from different media do stereotype each other has been adequately documented, more research is needed to examine the serious ethical conflicts that lie at the heart of some stereotypes. Thoughtful interviews with journalists in convergence partnerships might turn up examples of conflicts of ethical standards other than those found by Singer (2006) and the authors of this chapter (Silcock & Keith, 2006). Survey research could determine how widespread such conflicts are and to what degree journalists perceive them as problematic. This would lay groundwork for media ethics scholars to provide normative guidance on whether and how organizations employing different standards of truth-telling—involving, for example, the ethics of recreating reality for visuals—should move toward a single standard, either in all their work or in work on a common platform, such as a Web site.

Second, in a related area, there is a need for research focusing narrowly on how open or closed convergence partnerships are, that is, to what degree partners shield information from each other. Such information would be valuable in determining to what degree partners are engaged in "coopetition," the word coined by Dailey et al. (2005a) to describe partners who simultaneously cooperate and compete.

Third, the effects of the very different work routines of print and broadcast journalists seen in our study suggest that it would be useful to study which, if either, of those routines are being embraced by online journalists. A few studies of online journalists have suggested that they have values (Arant & Anderson, 2001) and gatekeeping orientations (Cassidy, 2006) similar to print journalists. But work describing online journalists' routines is still in its infancy, and there is a particular lack of research on online work routines, which probably are still solidifying.

As Singer (1998) has noted, this is an area in which approaches based on the sociology of news work would be especially welcome.

## Other Research Gaps

Other significant areas not suggested by our study also remain to be explored. For example, it appears that only a single published study (Huang, Rademakers, Fayemiwo, & Dunlap, 2004) has attempted to determine the truth of the frequent criticism (e.g., Corrigan, 2002; Kaplan, 2003) that the practice of convergence produces lower-quality journalism. Huang et al. (2004) found that was not the case at *The Tampa Tribune,* but more work in this area is needed. In addition, little, if any, research has considered the contention, voiced by focus-group members in one study (Ketterer et al., 2005), that convergence produces homogenized journalism, with the result that news consumers who pay attention to more than one convergence partner merely see the same stories over and over again. Although a few studies have compared content across media platforms (e.g., Choi, 2004; Lee, 2004), that research has been largely limited to comparing print and online content and has specifically sought to test the effects of convergence.

Researchers also need to examine newsroom convergence on a larger scale to determine how it fits into other work on journalism and mass communication and, on a broader scale, into work on communication. This chapter, the work of two former journalists, has focused on convergence as other journalism and mass communication scholars have: as an issue of bringing together media platforms, largely print and broadcast, with disparate journalistic traditions. That orientation can be useful because it mirrors many journalists' experiences of convergence and because it reflects the realities of platform-based categorization still used in many journalism and mass communication programs. This perspective can be limiting if it leads researchers who consider their field to be newsroom convergence to ignore the work of scholars who study "new" media or online journalism. As newspapers and television stations increasingly depend on their online presences, each literature is crucial to the other. It would also be unfortunate if too tight a focus on "convergence" as denoting a print-broadcast partnership led journalism and mass communication researchers to ignore the work of scholars who see convergence as embracing information delivery on a variety of devices—such as cell phones and MP3 players—or as another name for media consolidation. These views of convergence, often based on different outlooks than those shared by the journalists-turned-professors who have produced much of the newsroom convergence scholarship deserve attention. Although it remains to be seen whether most newsroom convergence projects will achieve full integration, that remains a noble goal for the scholarship of convergence.

## References

Ahrens, F. (2003, June 3). FCC eases media ownership rules. *The Washington Post,* p. A1.

Arant, D., & Anderson, J. Q. (2001). Newspaper online editors support traditional ethics. *Newspaper Research Journal, 22*(4), 57–69.

Boczkowski, P. J. (2004). *Digitizing the news: Innovation in online newspapers.* Cambridge, MA: MIT Press.

Cassidy, W. P. (2006). Gatekeeping similar for online, print journalists. *Newspaper Research Journal, 27*(2), 6–23.

Castaneda, L., Murphy, S., & Hether, H. J. (2005). Teaching print, broadcast, and online journalism concurrently: A case study assessing a convergence curriculum. *Journalism and Mass Communication Educator, 60*(1), 57–70.

Choi, Y. (2004). Study examines daily public journalism at six newspapers. *Newspaper Research Journal, 25*(2), 12–27.

Corrigan, D. (2002, October). Convergence: Over-worked reporters with less news. *St. Louis Journalism Review,* pp. 20–21, 23.

Criado, C.A., & Kraeplin, C. (2003, August). *The state of convergence journalism: United States media and university study*. Paper presented to the Association for Education in Journalism and Mass Communication, Kansas City.

Curkan-Flanagan, M., & Hardenbergh, M. (2002, April). *Central Florida's convergence triangle: A qualitative analysis of two major converged local television news operations*. Paper presented to the Broadcast Education Association, Las Vegas.

Dailey, L., Demo, L., & Spillman, M. (2005a). The convergence continuum: A model for studying collaboration between media newsrooms. *Atlantic Journal of Communication, 13*(3), 150–168.

Dailey, L., Demo, L., & Spillman, M. (2005b). Most TV/newspaper partners at cross promotion stage. *Newspaper Research Journal, 26* (4) 36–49.

Duhé, S.F., Mortimer, M.M., & Chow, S.S. (2004) Convergence in TV newsrooms: A nationwide look. *Convergence: The International Journal of Research into New Media Technologies, 10*(2), 81–104.

Dupagne, M., & Garrison, B. (2006). The meaning and influence of convergence: A qualitative case study of newsroom work at the Tampa News Center. *Journalism Studies, 7*(2), 237–255.

Ericson, R., Baranek, P., & Chan, J. (1987). *Visualising deviance: A study of news organization*. Toronto: University of Toronto Press.

Filak, V. F. (2006). The impact of instructional methods on medium-based bias and convergence approval. *Journalism and Mass Communication Educator, 61*(1), 48–61.

Foote, J. (2002). Convergence challenge. *Feedback, 43*(1), 7–13.

Gannett Co. Inc. (2007, March 1). *2006 annual report*. Retrieved July 2, 2007, from http://library.corporate-ir.net/library/84/846/84662/items/233865/06AnnualReport.pdf.

Gurevitch, M., & Levy, M. (1990, November). Universal meanings from the global newsroom? *Television: The Journal of the Royal Television Society,* pp. 19–22.

Hammond, S. C., & Petersen, D. (2000). Print, broadcast and online convergence in the newsroom. *Journalism and Mass Communication Educator, 55*(2), 16–26.

Huang, E., Davison, K., Shreve, S., Davis, T., Bettendorf, E., & Nair, A. (2006a). Bridging newsrooms and classrooms: Preparing the next generation of journalists for converged media. *Journalism and Mass Communication Monographs, 8*(3), 221–262.

Huang, E., Davison, K., Shreve, S., Davis, T., Bettendorf, E., & Nair, A. (2006b). Facing the challenges of convergence. *Convergence: The International Journal of Research into New Media Technologies, 12*(1), 83–98.

Huang, E., Rademakers, L., Fayemiwo, M. A., Dunlap, L. (2004). Converged journalism and quality: A case study of the Tampa Tribune news stories. *Convergence: The Journal of Research into New Media Technologies,* 10(4), 73–91.

Kaplan, J. (2003). Convergence: Not a panacea. In G. Thelen, J. Kaplan, & D. Bradley, Debate. *Journalism Studies, 4*(4), 513–521.

Ketterer, S., Smethers S., & Bressers, B. (2005, August). *But, will it play in Lawrence? Audience perceptions of convergence on the* Lawrence Journal-World, *News 6 and LJWorld.com.* Paper presented to the Association for Education in Journalism and Mass Communication, San Antonio.

Ketterer, S., Weir, T., Smethers, J. S., & Beck, J. (2004). Case study shows limited benefits of convergence. *Newspaper Research Journal, 25*(3), 52–65.

Klinenberg, E. (2005). Convergence: News production in a digital age. *Annals of the American Academy of Political and Social Science, 597,* 48–64.

Kraeplin, C., & Criado, C.A. (2005). Building a case for a convergence journalism curriculum. *Journalism and Mass Communication Educator, 60*(1), 47–56.

Kraeplin, C., & Criado, C.A. (2006). Surveys show TV/newspapers maintaining partnerships. *Newspaper Research Journal, 27*(4), 52–65.

Labaton, S. (2004, June 25). Court orders F.C.C. to rethink new rules on growth of media. *The New York Times,* pp. A1, C4.

Lee, G. (2004, August). *Salience transfer between online and offline media in Korea: Content analysis of four traditional papers and their online siblings.* Paper presented to the Association for Education in Journalism and Mass Communication conference, Toronto.

Lowrey, W. (2005). Commitment to newspaper-TV partnering: A test of the impact of institutional isomorphism. *Journalism and Mass Communication Quarterl, 82*(3), 495–515.

Lowrey, W., Daniels, G. L., & Becker, L. B. (2005). Predictors of convergence curricula in journalism and mass communication programs. *Journalism and Mass Communication Educator, 60*(1), 32–46.

Malone, M. (2007, June 11). Newspapers find online video niche. *Broadcasting and Cable*. Retrieved July 7, 2007, from http://www.broadcastingcable.com/article/CA6450358.html

Quinn, S. (2004a). Better journalism or better profits? A key convergence issue in an age of concentrated ownership. *Pacific Journalism Review, 10*(2), 111–129.

Quinn, S. (2004b). An intersection of ideals: Journalism, profits, technology and convergence. *Convergence: The Journal of Research into New Media Technologies 10*(4) 109–123.

Quinn, S. (2005). *Convergent journalism: The fundamentals of multimedia reporting*. New York: Peter Lang.

Silcock, B. W., & Keith, S. (2006). Translating the Tower of Babel? Issues of definition, language, and culture in converged newsrooms. *Journalism Studies, 7*(4), 610–627.

Singer, J. B. (1998). Online journalists: Foundations for research into their changing roles. *Journal of Computer-Mediated Communication 4*(1). Retrieved July 5, 2007, from http://jcmc.indiana.edu/vol4/issue1/singer.html.

Singer, J. B. (2004). Strange bedfellows: Diffusion of convergence in four news organizations. *Journalism Studies, 5*(1), 3–18.

Singer, J. B. (2006). Partnerships and public service: Normative issues for journalists in converged newsrooms. *Journal of Mass Media Ethics, 21*(1) 30–53.

Wasserman, E. (2006). Looking past the rush into convergence. *Nieman Reports,* Winter, pp. 34–35.

Zelizer, B. (1993). Journalists as interpretive communities. *Critical Studies in Mass Communication 10,* 219–237.

CHAPTER 14

# Web Logs

## Democratizing Media Production

Bryan Murley

U.S. mid-term election night, 2006—close Senate races in Tennessee, Virginia, and Missouri pointed to the potential for a shift in power for the U.S. Congress from Republican to Democratic control. Meanwhile, in Washington, D.C., a major news network was acknowledging another shift in power by hosting a party for top Web log authors at an Internet café. Authors of Web logs—frequently updated Web pages whose contents appear in reverse chronological order—had gained center stage. Throughout election night coverage, CNN reporters mingled with the bloggers, interviewing them and reporting from the event. In so doing, CNN was illustrating a testy relationship that was growing warmer between bloggers and mainstream media outlets.

> "Bloggers are leading the conversation," said David Bohrman, CNN's Washington bureau chief. "You could argue that most of the political dialogue in this country is happening online, so if you don't incorporate that into your coverage, you're missing a major element." (Gold, 2006, p. 17)

If convergence is the changing face of audiences, industries, content and technologies, as Kolodzy (2006) points out in *Convergence Journalism,* then certainly Web logs have been at the forefront of such convergence. The platform has enabled audience members to "grab the microphone," contributed to shifts in the media industry, provided content using multiple media, and fostered technological developments that have benefited media practitioners as well as lone-wolf bloggers.

## A Brief Historical Overview

> *Personal journalism is also not a new invention. People have been stirring the pot since before the nation's founding. (Gillmor, 2004, p. 1)*

The Internet is not the first medium to foster "outside the establishment" commentary and opinion, and so the phenomenon of Web logs falls within a historical understanding of mass communication. Indeed, blogs and assorted other forms of "user-generated content" are only the latest means for people to voice their beliefs, feelings, and opinions. From the invention of the alphabet, humans have set down their opinions and thoughts on the universe.

Religious and philosophical texts, especially, led to expressed opinions upon the nature of life (e.g., the Bible and Plato's *The Republic*). Such "opinion" was for the most part limited to the elites of societies until the 1400s, when Johannes Gutenberg helped introduce the movable-type printing press to Europe. This invention is credited with helping change the fabric of cultural, social, and religious life in Europe. Undoubtedly, it aided the spread of the Protestant Reformation, beginning with the rapid printing and distribution of Martin Luther's *95 Theses,* which was distributed around the world after it was first hammered to a church door in 1517 (Emery & Emery, 1992).

The 18th Century saw an explosion in the power and prominence of the press as an agent for change within society. Nowhere was this more evident than in the American colonies of Britain. Printers and journalists of the colonial era including James and Ben Franklin, Thomas Paine, and William Bradford used their presses to print pamphlets and broadsheets with a mixture of news, gossip, and propaganda critical of authority figures (Emery & Emery, 1992).

> The greatest stimulus to the development of the American press of this period was the rising political tension that was to culminate in the War of Independence. The press had an essential role in the drama about to unfold. The newspaper thrives on controversy, provided it is able to take part in the discussions with any degree of freedom. (p. 33)

The spirit of the 18th-century pamphleteers lived on through the 20th century but barely, being mostly swallowed by the institutionalization of journalism into media corporations and the professional development of journalism. Flames of the "do-it-yourself" opinionated attitude of these rebels began to burn again in the 1980s and 1990s, as desktop publishing and the rapid spread of the personal computer put the physical layout and design of printed materials within reach of middle-class Americans. Then came the Internet, which (among other things) offers a means of mass distribution of opinion at little or no cost.

Beginning in the late 1990s, some technically savvy Web users began updating their personal Web sites frequently, signaling the change in the page with a "date stamp." These Web pages were hand-coded by "Web enthusiasts" who linked to other pages they found around the Internet, according to Rebecca Blood (2005), a historian of the Web log phenomenon. She noted that an early Web logger's page of links to other Web logs "lists the 23 known to be in existence at the beginning of 1999." (p. 129)

## September 11, 2001

The growth in Web logs took a decided shift in 2001. When hijackers commandeered four commercial airplanes on September 11, 2001, their actions spurred one of the most dramatic expansions in publishing in history. Blogger and tech journalist Dan Gillmor (2004) pegged the rise of blogs to the events of that day. While frequently updated personal Web pages had become a small but growing phenomenon, after 9/11, blogs began to grow at a rapid rate.

As Web logs began to evolve with the introduction of easy, free Web-based software, most notably the Blogger platform, the format began to develop its own conventions of behavior. To properly understand Web logs, it is necessary to understand "blog"—the shorthand for "Web log"—as both a noun and a verb.

Drezner and Farrell (2004) offer a concise definition of a Web log (noun) as

> a Web page with minimal to no external editing, providing on-line commentary, periodically updated and presented in reverse chronological order, with hyperlinks to other online sources. Blogs can function as personal diaries, technical advice columns, sports chat, celebrity gossip, political commentary, or all of the above. (p. 5)

This definition certainly catches the physical characteristics of Web logs. However, the verb form, "to blog," expands into social aspects of the activity. Blogging involves updating a Web log frequently, often writing in a personal style, using hyperlinks to cite other Web-based sources, and engaging with others who blog. The Web log form of writing varies from a traditional journalistic form, and journalists who blog learn to use the format in a different manner from other media. In "A Blogger's Blog: Exploring the Definition of a Medium," boyd (2006) echoes this nuanced definitional approach, arguing that "blogs must be conceptualized as both a medium and a bi-product [*sic*] of expression."

## Technology Enables Blogging to Grow

As the Web log platform has grown, technological advancements have helped categorize and connect Web log information from various blogs. An early example of this was the "trackback" or "pingback," which would allow Web log A to display hyperlinks to Web logs B and C when those Web logs referenced a particular post on Web log A. RSS (Really Simple Syndication) allowed blog readers to "subscribe" to a Web log, and receive updates as the Web log author posted new material without having to actually surf to the Web log through a Web browser. "Tagging," a service popularized by Technorati, allowed Web log authors to categorize their posts according to keywords for easier retrieval and association with other related Web log information. Each of these innovations showed the desire of those in the Web logging community to enable easy dissemination and discovery of relevant content by information consumers.

## Content and Controversies

While Web logs most closely resemble diaries or journals in their format, Web log authors saw natural opportunities in the form for an expansion of the craft of journalist. Web log authors have been quick to claim an intellectual heritage in the footsteps of Luther's Protestant Reformation and the pamphleteers of the 18th century (e.g., Gillmor, 2004; Hewitt, 2005), arguing that they return journalism to a form that is more opinionated and less driven by the standards of the profession and concerns of corporate media owners (Gillmor, 2004).

Since 9/11, bloggers participated in the shaping of news coverage related to at least four major news events in the United States: the fall of Sen. Trent Lott, the Iraq War, the 2004 presidential election, and the downfall of CBS News anchor Dan Rather.

In late 2002, new Senate majority leader Lott made controversial remarks at the 100th birthday party for Sen. Strom Thurmond, R-S.C. The remarks[1] hardly merited a mention in mainstream media outlets in the days after the event (Kurtz, 2002). However, several bloggers picked up on the remarks and commented on what Lott said. The blogospheric reaction helped convince traditional media outlets to reexamine Lott's remarks and his personal history. At the end of the controversy, Lott stepped down as Senate majority leader. Wright's (2003) study of the timeline of this controversy showed that blogs contributed to the eventual mainstream media return to this story.

As the Iraq War began in 2003, bloggers again found a place in the media spotlight. During the Iraq conflict, Web loggers used the format to sound off on contemporary news reports. Sometimes, the Web log reports conflicted with news reports. Other times, the Web log reports confirmed and expanded on news media information. "Salam Pax" was the pseudonym of a blog publisher who lived in Baghdad, Iraq, and posted his comments in the weeks leading up to the

1. Said Lott: "I want to say this about my state: When Strom Thurmond ran for president, we voted for him. We're proud of it. And if the rest of the country had followed our lead, we wouldn't have had all these problems over all these years, either." (p. C01)

invasion of Baghdad (http://dear_raed.blogspot.com). "L.T. Smash" was a soldier who published a Web log from his post in the Middle East (http://www.lt-smash.us). These were examples of "public" Web logs. Both Pax and Smash represented uncensored "real" voices that otherwise would not have received an audience.

Soon, journalists reporting on the war began to publish Web logs. At least two embedded reporters, Kevin Sites from *CNN* and Joshua Kucera from *Time* magazine, began to use blogs, which were promptly shut down by their employers (Dube, 2003; Sites, 2003). Other journalists who had Web logs maintained an anonymous presence by posting under a pseudonym (e.g., Media Minder, mediaminded.blogspot.com, "a copy editor at an American newspaper"). A few news organizations embraced this new technology and hosted blogs on their news sites.

In August 2003, Daniel Weintraub, a political reporter for the *Sacramento Bee,* posted an item on his Web log that led to controversy during the California gubernatorial recall election. Weintraub wrote a Web log entry about Lt. Gov. Cruz Bustamente that upset members of the Hispanic community. The incident led *Sacramento Bee* management to bring Weintraub's Web log under the editing umbrella of the newspaper, a decision that did not sit well with some bloggers (Glaser, 2003).

The 2004 presidential election focused a great deal of attention on Web logs as, for the first time ever, bloggers were credentialed at both major political conventions (Walton, 2004). During the 2004 campaign, Web loggers were seen as having a part in both the rise and the fall of presidential candidate Howard Dean. In a sense, each of these skirmishes displayed the testy relationship between traditional journalistic businesses and Web logs, the medium and the authors.

The "perfect storm" for blogs as an agenda-setting medium arrived on September 8, 2004. CBS News anchor Dan Rather reported on President George Bush's Texas Air National Guard service for *60 Minutes Wednesday*. Within an hour of the broadcast, bloggers had raised questions about the authenticity of one of the memos used in the *60 Minutes Wednesday* report. The bloggers were right: The memos were fake. The controversy led to an independent panel investigation into the newsgathering processes used in the story (Thornburgh & Boccardi, 2005). Eventually, four CBS staffers were fired, and Dan Rather stepped down as *CBS Evening News* anchor in March 2005 (Associated Press, 2005).

Drezner and Farrell (2004) noted the ability of blogs to play an agenda-setting role: "The rapidity of blogger interactions affects political communication in the mainstream media through agenda setting and framing effects. The agenda-setting role is clear—if a critical number of elite blogs raise a particular story, it can pique the interest of mainstream media outlets" (p. 17). This agenda-setting capability is enhanced when considered in light of research that shows blog readers consider blogs more credible than traditional media (Johnson & Kaye, 2004). Murley and Roberts (2005) found that top-ranked Web logs acted as "second-level" agenda setters by citing and then reframing content from traditional media outlets.

Bloggers tend to link to traditional media sites frequently, although linking patterns differ according to political orientation. Also, top Web logs of different political leanings tend to link to other Web logs that share those leanings (Adamic & Glana, 2005). Within the blogosphere, a number of bloggers have risen to the top in terms of links and traffic, which hints that these top blogs have a greater agenda-setting ability than lesser-known bloggers (Shirky, 2003). The tendency of Web loggers and Web log readers to orient themselves to like-minded individuals has led some to caution that the Web log genre encourages a form of "cyber-balkanization"—a term coined by sociologist Robert Putnam (2000)—diminishing the common ground people within a culture need to operate. This is especially true in the arena of political blogs, where researchers from the Institute for Politics, Democracy and the Internet found that daily political blog readers were more likely to be involved in partisan political efforts—both online and offline—than other blog readers or the general population (Graf, 2006). However, these concerns are not unique to the Web log format, and have been a part of the debate about the effects of the Internet upon society since early in its existence (Van Alstyne & Brynjolfsson, 1995).

Blogs have been distinguished as a new genre of journalism by their personalization, participation with the audience, and connection with other blogs.

> Here, this works on two levels: the bloggers have turned consumption of mainstream news into a production of a new news product, while the blogs' audiences also participate in content construction and meaning production by participating in the comments section. In this way, blogs are not a closed text with their intended meaning already fully inscribed but instead come into being through this performance between the blogger and the audience. (Wall, 2005, p. 166)

Trammell and Keshelashvili (2005) confirmed this interactive nature as a major component of so-called "A-list" bloggers—those who publish Web logs with high traffic and/or high numbers of hyperlinks to their blogs from other bloggers—when examining their self-presentation and impression management strategies. Despite these attributes exhibited by A-list bloggers, Singer (2005) found that—at least in their early stages—most political journalist bloggers were maintaining traditional roles as gatekeepers of information, against the overall participatory nature of the Web log format. Carpenter and Buis (2006) confirmed Singer's findings about linking among journalist bloggers, but also found a "willingness of bloggers to engage with users creating a product constructed by both the journalist and the news user" (p. 17). This "co-production" of the news product is a major shift in the practice of journalism. As Robinson (2006) noted in her textual analysis of journalistic Web logs: "Slowly, the online medium is forcing journalism to define a new identity" (p. 79).

A great disservice would be done if one were to focus solely on blogs at their intersection with journalism and so-called "mainstream media." Indeed, one might argue that Web logs have had more influence the farther from journalism's main topical focuses they have strayed.

As just one example, Web logs have been instrumental in building and maintaining the emerging church movement, a movement that challenges traditional Protestant Christianity along a postmodern paradigm. By using Web logs and other

Internet-related methods of communication, the leaders of the emerging church movement have confronted misconceptions about their theology and church practices and shaped the debate about their movement on their own terms.

Evangelical D.A. Carson wrote a book in 2005 that took a critical view of the movement. Carson argued that the movement could not be reconciled with conservative theology. Even as the book was being released, Andrew Jones, a prominent blogger in the emerging church movement (www.tallskinnykiwi.com) responded to Carson's charges. Jones' first response included this passage, which is emblematic of the strength of the Web logging phenomenon:

> I also write this as a blogger with a loud voice on the internet. When people type "Don Carson, emerging church" into their Google search engines, the postings on my blog (tallskinnykiwi.com) have been coming up in the Top 3 listings—well ahead of your tapes or book, making me somewhat of a doorway to your material. This was not intentional. Just a perk of being a new media publisher rather than an old media book person. (You won't find my writings in a book store). (Jones, 2005)

Jones' discourse points to one of the features that has helped Web logs gain entreé into national and international debates—the power of search. Relatively early in their practice, Web loggers began to discern that hyperlinks to other news sites and Web logs would increase their "rank" within search engine results. This led to much debate as to whether blogs should be excluded from search results. However, blogs have not been removed from search results, and some blogs have been indexed in news search results by Yahoo! News and Google News.

In other areas, as well, Web logs have demonstrated a potential for greater participation in media creation on both personal and cultural levels. Deuze (2006) argued that Web logs and other participatory media tools helped to explain the rise in ethnic or minority media around the world, another example of the broad appeal of the instant publishing format.

## Who's Reading?

With all these people writing and creating content, someone has to be reading it, right? Here it should be qualified that accurate numbers of readers of Web logs are extremely difficult to come by.

Despite the difficulties, researchers are attempting to piece together some parts of the readership portrait. The Pew Internet and American Life Project reported that "by the end of 2004, 32 million Americans were blog readers" (Rainie, 2005, p. 1). That total works out to 27 percent of Internet users. Pew's research showed that blog readers were "more likely to be young, male, well educated, Internet veterans" (Rainie, 2005, p. 2). Most of those figures were supported by a blog readers' survey conducted by Blogads, an advertising firm that represents Web

**Table 14.1** ▰ Blog Readership by Country—Edelman Survey

| Country | Read Blogs (%) | Do Not Read Blogs (%) |
|---|---|---|
| Japan | 74 | 26 |
| South Korea | 43 | 57 |
| China | 39 | 53 |
| United States | 27 | 69 |
| United Kingdom | 23 | 75 |
| France | 22 | 68 |

*Source*: (Murray, 2006)

logs, in March 2005, except that Blogads found more readers over 30 years of age (Copeland, 2005).

In February 2006, the Gallup Poll indicated that "one in five Web users read Web-logs, or 'blogs,' either frequently or occasionally." This is 40 million readers, a number which Gallup maintains has not increased since 2005.

While it might be tempting to dismiss the low readership of Web logs, the numbers mentioned are only U.S. readership numbers. Studies regarding Web logs in other parts of the world show that there is growing readership—especially in Asian countries.

A 2006 study by Edelman, the public relations firm, showed high rates of blog readership in Asian markets. (Table 14.1)

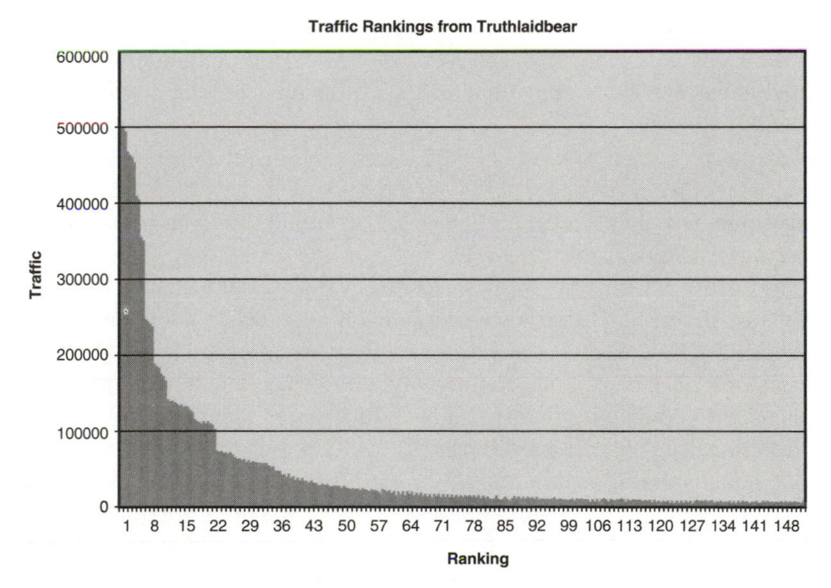


**Figure 14.1**

Other countries in the study—Italy, Poland, Germany and Belgium—showed readership numbers in the teens. In all the countries, Edelman found that politically active "influencers" were more likely than the general Internet population to read Web logs (Murray, 2006).

Despite the increasing numbers of readers, most of the readership remains concentrated among a handful of Web logs, as Shirky's (2003) power law predicted. To illustrate this point, the traffic rankings for the top 200 Web logs according to the traffic rankings at the Truth Laid Bear weblog (http://truthlaidbear.com/TrafficRanking.php) were input into a spreadsheet, producing a graph illustrating the power law in effect (Figure 14.1). The curve mirrors the curve Shirky produced in 2003. (Note: Shirky used linking patterns instead of traffic.)

# Where Things Stand in 2007

## Incorporation?

News media outlets have not left the terrain of Web logs alone. Some forward-thinking media companies have incorporated staff-written Web logs into their offerings, while others have pushed the envelope by seeking local readers to write Web logs hosted on the news organization's Web site.

For news organizations, the Web log has been a tool to expand and explain their coverage to readers. John Robinson, the editor of the Greensboro *News and Record* in North Carolina, has used his Web log to engage readers in a discussion about the community aspects of the newspaper's Web site and also explain coverage decisions along the way. The *News and Record* hosts numerous staff Web logs and is but one example of how media companies are attempting to use the platform.

Lawrence.com, the Web site of the Lawrence (Kansas) *Journal-World,* was an early adopter of the reader-produced Web log, and others have followed suit. Blufftontoday.com, a widely cited example of a community-building news site, offers Web logs to readers as well.

The quality and quantity of such Web logs varies widely. Some readers post to the Web logs infrequently, and some gain followings and elicit numerous comments on their blogs. The subject matter varies widely, as well. From a traditional publishing standpoint, such content uncertainty might be seen as a liability. For new media publishers, the content uncertainty is more a function of the platform.

Additionally, some media companies are attempting to build closer ties to the Web log community by highlighting Web log posts that link to the media company's articles. Washingtonpost.com was an early leader in this effort, partnering with Technorati to offer a place on each story that highlights Web log articles that link to the story (Easter, 2005).

However, media companies—both U.S. and European—still struggle with the basics of blogging, such as hyperlinks and ease of use. The *Guardian* (U.K.) began a "group blog" in March 2006 that was modeled after a U.S. group blog, *The Huffington Post,* but the *Guardian*'s blog has been plagued by growing pains (Macdonald, 2006). Washingtonpost.com had to shut down reader comments on a *Post* Web log after a particularly abusive exchange regarding an article by ombudsman Deborah Howell. Executive Editor Jim Brady wrote in a blog post announcing the comment embargo: "Transparency and reasoned debate are crucial parts of the Web culture, and it's a disappointment to us that we have not been able to maintain a civil conversation, especially about issues that people feel strongly (and differently) about" (Brady, 2006).

## Blogging as Brand

As the media establishment has attempted to incorporate Web logs into their product, media entrepreneurs have struck out from the media establishment as well. The aforementioned *Huffington Post* was an early example of a Web log that leveraged the "star-power" of media celebrities to increase readership and advertising income. Furthermore, journalists have moved to the Web log format as well. Om Malik, who wrote a popular Web log, named *GigaOm,* for Business 2.0, left the company in 2006 to strike out on his own (Marshall, 2006). New Media maven Jeff Jarvis argued that more journalists would become independent and build a personal "brand" for their abilities in this new media environment—probably using Web log software and techniques (Jarvis, 2006). There are also efforts by groups of bloggers and advertising agencies to monetize the growing trend through advertising sales. Some bloggers have also made efforts to aggregate content and act as a "portal" to the Web log world, as is the case of *Pajamas Media,* a consortium of highly trafficked Web logs founded in 2005 (Lamb, 2005).

Here it should be noted that the Web log format has suffered some damage due to its open, conversational ethos, mainly from "spammers." As noted in this quote from Six Apart (2004), the maker of several types of Web log software, "Whether we want to admit it or not, our little corner of the Web has blipped onto the radar screens of spammers. Gone are the days where the ultimate simplicity of a comment form and a submit button immediately and frictionlessly connected you with hundreds or thousands of your readers."

Spammers attempt to use unmoderated comments or trackbacks (links to their own Web content) to Web log posts as a means of increasing the prominence of their products in search engines. Web logs faced monumental assaults from spammers who developed automated scripts that would insert thousands of comments into a Web log's comment system. These attacks led some authors to close comments entirely, while they also spurred innovations within the Web log software industry, including the use of Turing tests,[2] comment registration, scripts that would close comments after a set period of time, and "blacklists" that would delete any comment containing a list of key spam words (Six Apart, 2004).

## The Future?

After five years of uncommon growth in influence and number, Web logs appeared to be the "granddaddy" of Web-based personal publishing enterprises. With podcasting, vlogging (video weblogs), social networking, and media sharing (à la YouTube) competing for user-generated content, one could reasonably wonder whether the venerable date-stamped journaling that is the staple of the blogosphere was beyond its utility.

Yet, as Dave Sifry noted in his "State of the Blogosphere" report in October 2006, the blogosphere continued to grow—doubling in size about every 236 days, with approximately 100,000 new Web logs created every day. Currently, the service tracks 57 million Web logs, with 55 percent of those updated within the last three months (Sifry, 2006).

Some would even wonder whether blogging was doing anything to democratize media production at all, as Shirky's (2003) power curve implied the concentration of attention among a few top blogs while many others were left out in the cold. In addition, concerns have been raised about the "cyber-balkanization" of the blogosphere, where bloggers and blog readers only interact with others of like mind.

Such concerns are certainly worth considering, when one looks only at the process of producing a Web log from the viewpoint of growing an audience for content. It is a paraphrase of the philosophical question about a tree falling in the woods: "If a person blogs in the blogosphere, and no one is there to read it, does it make a difference?"

While self-publishing can be a lonely endeavor, with few readers beyond family and friends, there are signs that the blogosphere is a nuanced ground for the development of individuals who focus on niche communities. While much communication research has focused on "political" Web logs, there are fruitful avenues of exploration yet to be delved into in these niche self-publications.

Web logs are also "morphing" as social network sites like MySpace and Facebook have incorporated Web log–style "notes" into the software of the sites. This incorporation allows a social network–based "blogosphere" to emerge among

---

2. A Turing test is a means used to determine whether the comment being submitted to a Web log is a result of human activity, or the work of a computer software program. Most frequently, blogging Turing tests ask the commenter to enter a series of randomly generated letters and numbers into a text box. A computer program could not respond properly to the test.

younger Internet users, who can follow the journal writings of their online "friends." Additionally, in 2006, Six Apart, the company that developed the early Web log software Movable Type, introduced a new blogging platform called Vox, which exhibited some of the features of social-networking sites.

Additionally, blogs have become much more international in scope, as individuals in different countries have begun to use this publishing tool to comment on their sociopolitical milieu. Bloggers have gained a voice in repressive environments such as Iran and China. Sifry noted in his October 2006 "State of the Blogosphere" report that English is the primary language for 39 percent of Web logs. Japanese is the second most used language at 33 percent, and Chinese is third at 10 percent. It is important not to view the impact of Web logs solely through the lens of Western capitalist societies. Rebecca MacKinnon (2007) summed this up: "People in different countries and language groups use blogs in different ways for different purposes. You really can't take generalizations about how the U.S. blogosphere works and apply them to any other country or region's blogosphere."

Neither can we assume that Web logs alone will foster a democratization of political reality in disparate regions of the world. For instance, while Web logs have been popular for following dissident voices within Iran, as of this writing there were few indications that such dissidents were effective in influencing the political realities of that country.

However these concerns play out, the most important contribution of Web logs to the media landscape has been their ability to infuse new and different voices into the mediated discussion. The blogosphere has promoted such voices, and—coupled with the ubiquity of search as a tool for accessing information–given those voices a larger stage. While traditional publishing and broadcasting outlets continue to maintain their sizable head start in brand presence, not to mention resource allocation, Web log authors have shown that sometimes David can best Goliath in the mediasphere.

In the U.S., media, public relations, advertising, and technology businesses are all trying to figure out what to do with Web logs. For journalists, the platform provides both opportunity and danger. The opportunities include learning new means of practicing the craft of journalism, promoting transparency, and building closer relationships with readers. The dangers include miscalculating the effect and reach of Web logs, or dismissing them as a passing fad or "unserious" hobby. The bloggers who joined CNN in Washington on a cold November night in 2006 did not look like they were engaged in a passing fad.

## References

Adamic, L. A., & Glance, N. (2005). *The political blogosphere and the 2004 U.S. Election: Divided they blog*. Paper presented at the 2nd Annual Workshop on the Weblogging Ecosystem: Aggregation, Analysis and Dynamics, Japan.

Associated Press. (2005). CBS ousts four for roles in Bush guard story. Retrieved April 22, 2005, from http://msnbc.msn.com/id/6807825/

Blood, R. (2005). Weblogs: A history and perspective. In E. P. Bucy (Ed.), *Living in the information age: A new media reader* (2nd ed., pp. 129–133). Belmont, CA: Thomson Wadsworth.

boyd, d. (2006). A blogger's blog: Exploring the definition of a medium [Electronic Version]. *Reconstruction: Studies in Contemporary Culture,* 6. Retrieved January 4, 2007 from http://reconstruction.eserver.org/064/boyd.shtml.

Brady, J. (2006, November 10). Comments turned off. from http://blog.washingtonpost.com/washpostblog/2006/01/comments_turned_off.html

Carpenter, S., & Buis, L. (2006, October 21). *J-bloggers and the blogosphere: A study of how j-bloggers affect social capital and content*. Paper presented at the Convergence and Society: Ethics, Religion, and New Media, University of South Carolina.

Carson, D.A. (2005). *Becoming conversant with the emerging church : Understanding a movement and its implications*. Grand Rapids, Mich.: Zondervan.

Copeland, H. (2005). *Blogads blog reader survey*. Retrieved January 2, 2007, from http://www.blogads.com/survey/2005_blog_reader_survey.html

Deuze, M. (2006). Ethnic media, community media and participatory culture. *Journalism, 7*(3), 262–280.

Drezner, D., & Farrell, H. (2004). *The power and politics of blogs*. Paper presented at the American Political Science Association, Chicago.

Dube, J. (2003). Another journalist blogger shuts down. Retrieved December 5, 2003, from http://www.cyberjournalist.net/news/000240.php

Easter, E. (2005). *Washingtonpost.com partners with Technorati to deliver content and comments from blogs*. Retrieved November 10, 2006, from http://www.washingtonpost.com/wp-adv/mediacenter/html/release_technorati_091505.html

Emery, M., & Emery, E. (1992). *The press in America: An interpretive history of the mass media* (7 ed.). Englewood Cliffs, N.J: Prentice Hall.

Gallup. (2006). Blog readership bogged down. Retrieved February 12, 2007, from http://www.galluppoll.com/content/?ci=21397

Gillmor, D. (2004). *We the media: Grassroots journalism for the people*. Sebastapol, CA: O'Reilly Media.

Glaser, M. (2003). *Weblogs are pushing the newsroom envelope on writers' sponaneity*. Retrieved September 24, 2003, from http://www.ojr.org/ojr/glaser/1064357109.php

Gold, M. (2006, October 30). CNN hopes blogging is election-night blessing. *Los Angeles Times*.Retrieved July 14, 2008 from http://articles.latimes.com/2006/10/30/calendar/et-news30

Graf, J. (2006). *The audience for political blogs: New research on blog readers.* Retrieved November 10, 2006 from http://www.ipdi.org/uploadedfiles/Blog%20Readership%20-%20October%202006.pdf

Hewitt, H. (2005). *Blog! Understanding the information reformation that's changing your world.* Nashville, TN: Thomas Nelson.

Jarvis, J. (2006). *Independent journalist as brand.* Retrieved November 10, 2006, from http://www.buzzmachine.com/index.php/2006/10/13/independent-journalist-as-brand/

Johnson, T. A., & Kaye, B. K. (2004). Wag the blog: How reliance on traditional media and the Internet influence credibility perceptions of weblogs among blog users. *Journalism and Mass Communication Quarterly, 81*(3), 622–642.

Jones, A. (2005). An open blog post for don carson 1.1 [Electronic Version]. *Tall Skinny Kiwi.* Retrieved September 11, 2005, from http://tallskinnykiwi.typepad.com/tallskinnykiwi/2005/04/an_open_blog_po.html.

Kolodzy, J. (2006). *Convergence journalism: Writing and reporting across the news media.* Lanham, MD: Rowman & Littlefield.

Kurtz, H. (2002, December 16). A hundred-candle story and how to blow it. *Washington Post,* p. C01.

Lamb, G. M. (2005, November 30). A one-stop shop for the "best" blogs. *Christian Science Monitor,* p. 14.

Macdonald, N. (2006). *"Comment is free," but designing communities is hard* [Electronic Version]. *Online Journalism Review.* Retrieved November 10, 2006, from http://www.ojr.org/ojr/stories/060817macdonald/

MacKinnon, R. (2007). *Asia leads the world in blog readership.* Retrieved February 12, 2007, from http://rconversation.blogs.com/rconversation/2007/01/asia_leads_the_.html

Marshall, M. (2006). *Om Malik quits biz 2.0, raises cash to build out broadband news site.* Retrieved November 10, 2006, from http://www.siliconbeat.com/entries/2006/06/12/om_malik_quits_biz_20_raises_cash_to_build_out_broadband_news_site.html

Murley, B., & Roberts, C. (2005, October). *Biting the hand that feeds: Blogs and second-level agenda setting.* Paper presented at the Conference on Media Convergence: Cooperation, Collisions and Change, Provo, Utah.

Murray, R. (2006). *A corporate guide to the global blogosphere.* Retrieved February 12, 2007, from http://www.nxtbook.com/nxtbooks/edelman/whitepaper010907/index.php

Putnam, R. (2000). *The other pin drops, communicating online.* Retrieved July 15, 2007, from http://www.inc.com/magazine/20000515/18987.html

Rainie, L. (2005). *The state of blogging.* Retrieved January 2, 2007, from http://www.pewinternet.org/PPF/r/144/report_display.asp

Robinson, S. (2006). The mission of the j-blog: Recapturing journalistic authority online. *Journalism, 7*(1), 65–83.

Shirky, C. (2003). *Power laws, weblogs, and inequality*. Retrieved April 22, 2005, from http://www.shirky.com/writings/powerlaw_weblog.html

Sifry, D. (2006). *State of the blogosphere, October, 2006*. Retrieved November 10, 2006, from http://technorati.com/weblog/2006/11/161.html

Singer, J. B. (2005). The political j-blogger: "Normalizing" a new media form to fit old norms and practices. *Journalism, 6*(2), 173–198.

Sites, K. (2003). *Pausing the warblog, for now*. Retrieved December 7, 2003, from http://www.kevinsites.net/2003_03_16_archive.php

Six Apart. (2004). *Six Apart guide to comment spam*. Retrieved November 10, 2006, from http://www.sixapart.com/pronet/comment_spam.html

Thornburgh, D., & Boccardi, L. D. (2005). *Report of the independent review panel:* CBS. Retrieved July 14, 2008 from http://wwwimage.cbsnews.com/htdocs/pdf/complete_report/CBS_Report.pdf

Trammell, K. D., & Keshelashvili, A. (2005). Examining the new influencers: A self-presentation study of a-list blogs. *Journalism and Mass Communication Quarterly, 82*(4), 968–982.

Van Alstyne, M., & Brynjolfsson, E. (1995, December 15–18). *Electronic communities: Global village or cyberbalkans?* Paper presented at the International Conference on Information Systems, Cleveland, Ohio.

Wall, M. (2005). "Blogs of war": Weblogs as news. *Journalism, 6*(2), 153–172.

Walton, M. (2004). *Bloggers get convention credentials: Instant, interactive communication making a political mark*. Retrieved April 22, 2005, from http://www.cnn.com/2004/TECH/internet/07/23/conventionbloggers/

Wright, C. (2003). Parking Lott: The role of web logs in the fall of Sen. Trent Lott. *GNOVIS, 3*(Fall 2003), 30.

CHAPTER 15

# Global Aspects of Convergence

Kenneth C. Killebrew

## Introduction

Globally, convergence is a central theme when discussing the new media environment. Convergence occurs at many levels, in many ways, but news organizations in particular are evolving in unique ways depending upon the governmental and regulatory forms or structures. The United States has a fairly libertarian approach (O'Siochru & Girard 2002) compared to the paternalism of several European systems. In Africa there is relatively little experience with new media for the masses, so media convergence remains in its infancy, save for a few isolated cases. In Asia, countries like Korea and Japan are at the forefront of change, having been transformed by the combination of a younger generation adopting—and adapting—almost all new forms of communication.

It is difficult to summarize global trends in media convergence in a single chapter. But in general, media convergence propels each country and region forward in ways that are impossible to predict. Every government seeks to improve itself and the conditions of its citizenry, and communication technologies are tantalizing in their promise. Therefore, regulations are drafted and passed to harness the potential threats of converging media rather than geared toward slowing or thwarting its development. Convergence, for many, is not the problem; it is a solution to many of the information ills of society.

For example, in the fall of 2006, academics and professionals met at Keio University in Japan to investigate digital media convergence in Asia. Representatives from media organizations and universities around the world reported on attempts to improve the vastly growing enterprises of the Internet and its concomitant Digital Age activities. Ultimately, they created a manifesto dedicated to the premise that technology is the handmaiden of tomorrow in what is becoming a global "Creative Society" (Keio University, 2006).

Experiments in new approaches to modern information dissemination extend to both formal and informal groups in this new society. Traditionally trained journalists as well as citizen journalists, bloggers, Web designers, and others were included as defining entities in this creative society. The question then was "How would this new society inform, educate, and entertain itself?" Rather than having a single answer, globally the question appears to have several, depending on a host of factors.

In Europe, there is a decided commercial tinge to media convergence. Groups meet to discuss effective marketing strategies for the Internet and digital media so as to develop branding opportunities for their products. In late 2006, David Currie, Chairman of the Office of Communication in the United Kingdom said, "The reality of convergence—and the sweeping transition from analog to digital technologies—is radically changing the communications sector" (Currie & Richards, 2006, p 19). His audience of Asian and European scholars and media leaders was focused mostly on the needs of the next generation of information providers rather than the needs of the audience for that information.

At the same time, half a world away, factions in Brazil were working to determine who should control the tide of new media information. Government, business, and activist leaders split over whether new media would hasten the nation's development or create threats to Brazil's national identity. The result was to create a regulatory council that would bring all the elements together in its own vision of a new world information order. That vision is still evolving as leaders grapple with the true underpinning of the issues: uncertainty in times of change.

Unlike Brazil, China's government is attempting to find ways to implement communication innovations while shutting off the spigot of external information. Despite having the most elaborate system of censorship in the world (Walton, 2001), the "great firewall" and the "golden shield" have not been able to prevent the blogosphere from calling attention to corrupt or inept party officials. The Chinese government has contracted with major software developers like Cisco and search engine companies including Google to stem the tide of outside information deemed to be counterproductive to the established culture.

The Internet and the free flow of information are also bringing important benefits to government officials. In an environment that encourages legitimate profit, revenues from "keyword" sources brought more than $150 million into the coffers of government-controlled companies in the third quarter of 2006 alone (Zhang, 2006). The government is looking to improve those revenues further for 2007 and 2008. In China the goal seems to be controlling undesirable or disruptive information while at the same time taking advantage of the marketing tools commonly used by the rest of the world (Zhang, 2006).

Throughout 2008, Chinese officials tried to put on a more open face and create the image of a progressive culture. To facilitate that openness, non-Chinese journalists now only need the permission of individuals to interview them. Previously journalists needed state approval as well. But in other ways, things remain the same; officials have reiterated that Web sites operating within the borders of the country but not sanctioned by the government are forbidden from reporting news. By 2008, Reporters Without Borders indicated that China ranked 163 out of 169 nations on its list of repressive nations (Reporters sans frontières, 2008).

The changes for the Olympics likely will have little effect on the nation's traditional media systems, but the evolution in new media may find ways to frustrate the old order. The Chinese are learning that for every electronic plug in the information dike, a new player will crop up to let the information flow. University of Toronto researchers have developed new software, called "Psiphon" which creates workarounds to the government's attempts at information control. Early reports are that the software works very well (Beach, 2006).

Regardless of the approach, technological growth and innovation seem to be outpacing the ability of governments to control the spread of information. This chapter attempts to explore these differences and similarities on a global scale. It will first examine technological change, then the "social solutions" to technological change and how news organizations are positioning themselves for the future.

The importance of understanding the evolution of convergence is seeing that it is not simply all about the technology. Nor is it just about finding and developing information resources. The larger issues are those that reveal how and why we can all be participants, and that what was once prescribed to be a one-way communication process can now be a dialogue. Control of that dialogue now routinely shifts from the source to the receiver and back again, sometimes quite rapidly. Thus, one implication of media convergence is that the number of participants discussing a given topic can rapidly expand or contract depending on the interests of the audience (the participants), rather than on an institution such as government, business, or even the media. Therefore, the starting point for this analysis begins with the technology and the companies that enable convergence to happen.

## Technological Convergence Unchecked

Few nations are truly attempting to curtail the evolution of communications and information technologies. Most are seeking to understand what is happening to information and where the technology is taking them. Cellular telephones are a case in point. Last year's cell phone is this year's multi-functional mobile information center, with access to e-mail, the Web, music, text messaging, and now live mobile TV. The cell phone is not convergence, but it is an important component that enables convergent applications that bring Global Positioning System (GPS) locators and satellite mapping into the palm of your hand. A Blackberry® or Treo™ is routinely used to track the standings of fantasy soccer, football, basketball, and baseball leagues. The tracking is instantaneous, accurate and can become an emotional roller-coaster for those deeply invested in the game. Thus, any discussion of media convergence has to include cell phones and other technologies that will enable new channels of distribution for media messages.

The portability of cheap communications devices is leveling the field between the so-called "haves" and "have-not" nations. As these costs continue to decline and resources continue to rise, the traditional north-south divide could be erased, at least in terms of communications. New models of media convergence and convergent journalism could be expected to take root and prosper.

## Europe

Media convergence in European nations has more of a managed feel than in the United States. Especially in regard to telecommunications, the Office of Communication (Ofcom) in the United Kingdom, for example, along with the European Union (EU), holds annual meetings to discuss what is happening and to plan for changes both in economic and regulatory terms. David Currie, Chairman of Ofcom and member of the House of Lords noted, "In New Europe and the major emerging economies, fixed wireline infrastructures are being leap-frogged in favour of a greater predominance of mobile and wireless infrastructure" (Currie & Richards, 2006). France, Spain, Germany, Ireland, and Italy also grapple with widespread diffusion of mobile media. Mobile telephone use has increased dramatically the past five years and now accounts for more than half of all telephone business in those countries (Office of Communication, 2006). Data gathered by Ofcom shows that the evolution to a wireless society is ongoing throughout the world. Reports from the EU, Australia, Taiwan, Hong Kong, and a host of universities attest to the onslaught of technological innovation.

While fixed-wire and wireless telephone revenues compete across international economies, other areas of convergence also are moving rapidly. Computer use, and more specifically Web use, continues to grow at a breakneck speed, with less-developed nations often growing in their online use faster than developed nations. There is an expectation that developing nations will ultimately create a new marketplace for the rapidly evolving wireless products.

Juniper Research, a U.K. business research firm, notes that the total global mobile entertainment market—including gambling, adult content, mobile games, mobile music, mobile TV, and infotainment—was worth over $17 billion in 2006

(Entertainment Mobile, 2006). Revenues are expected to top $47 billion (U.S.) by the end of 2009. Much of that revenue (Head, 2006), amazingly, is in downloads to telephones of ringtones. With the shift toward mobile TV well on its way globally and in the United States in 2007, those mobile revenues are expected to shift to downloaded videos.

Spain's national telephone company, Telefónica, is an example of how convergence is transforming media, telecommunications, and related businesses. Telefónica has moved its marketing focus from national to global, and wireline to wireless. Many of its South America holdings, including Brazil, Peru, and Argentina, are eager to embrace mobile television. While Telefónica has transitioned to a global force in the telecommunications industry, the rest of the nation also has been moving forward technologically, moving ever closer to total media convergence. By the end of 2006, Telefónica's revenues from wireless operations in Spain were more than half of total revenues (*Telefónica, S.A., annual report 2006* ).

The Milken Institute sponsored a Global Conference on Expanding Opportunities in the global marketplace in 2006. A particular focus of the conference was on the development of new markets through media convergence. In this context, convergence meant building branding techniques in a wireless world. The fragmentation of traditional markets was seen as a difficult challenge for those seeking to extend their branding information (Bronfman, Cappello, & Kelly 2006).

Social-networking sites such as Facebook and MySpace are some of the fastest growing sites for information between and about individuals (Ewing, 2007). In early 2007, Vodafone Group, one of Europe's largest mobile telephone companies, had signed an exclusive partnership deal with the Web site MySpace, allowing customers to access and update their MySpace pages from their cell phones (Ewing, 2007). MySpace, a unit of Fox Interactive Media, said that the deal marked its first foray into the European cell phone sector.

By spring 2007, Vodafone also was offering access to Facebook, but when Vodafone advertising turned up on a right-wing political Web site in August 2007, through Facebook, Vodafone quickly pulled out of its agreement with the social-networking group. Caution is now the word at Vodafone Group headquarters (Kiss, 2007).

## Asia

While Europe is rapidly moving into the world of converged media technology, some Asian nations remain at the forefront in the development and use of technology. A number of Asian countries have embraced lifestyles involving broadband Internet and mobile media devices. With cell phone usage in Hong Kong reaching 157 percent in April, 2008 (OFTA, 2008) and mainland China signing up more than 200,000 new Internet users each day (China Internet Network Information Center, 2008), the continent is fully committed to technological convergence and the communications revolution (Lloyd, 2004). Multiple economic reports from Beijing indicate that 37 percent of the Chinese population now use cell phones (Wireless World Forum, 2006), and the world's most densely populated areas are showing skyrocketing use of mobile technology. The number of cell phone subscribers in China topped 500 million in the summer of 2007 (RNCOS, Mobile Industry Research, 2007).

That mobile technology is no longer just for telephone calls and text messages, and in China people are subscribing to television broadcasts on their mobile phones (Wireless World Forum, 2006). The integration of the technologies is reaching such a high level in China and other parts of Asia that the term "telephone" is being replaced with more inclusive terms as UMD (ultra mobile device) (Intel Software Network, 2008).

Online convergence of technologies continues unabated. Internet telephone service (VoIP) has grown rapidly in a number of countries, essentially doubling from 19 million in 2005 to 40 million worldwide in 2006 (Wireless Federation, 2007). In France, more than five percent subscribe to VoIP service (*Telefónica, S.A., annual report 2006*).

The unprecedented growth in mobile markets as well as the Internet is creating shifts in advertising budgets away from traditional platforms. Over the next five years, Internet advertising is expected to double from US$25.5 billion in 2007 to US$51 billion by 2012 (Boston Business Journal, 2008). Furthermore, this projected growth will push the Internet from being the fifth most popular advertising medium to number two. While much of this growth is simply due to the creation of a new marketplace for advertising placement, it is drawing attention from all quarters. From a somewhat different perspective, Paul Budde Communication, an Australian firm, notes that video-on-demand (VoD) in all forms will reach 350 million households by 2010 ("2006 Global Digital Media," 2006).

In Korea, Disney's ABC Television network already has licensed more than 200 hours of popular U.S. programs, including episodes of *Lost* and *Desperate Housewives* for mobile TV viewing. Disney officials said that global positioning of programming will be among the fastest growing markets for the reuse of material originally developed for traditional television (Adegoke, 2006). Disney has leased the shows to TU Media Group in Korea and expects that even stronger program deals will be made in the near future.

In terms of technological convergence, both TU and Disney believe that mobile television will be a critical growth area for mobile technology in the near future. Disney and TU are not the only companies watching and pushing the launch of mobile television in 2007. Sony Ericsson, a joint venture between the two technology giants in cell phone production, now predict that mobile television will be a global phenomenon by 2008, and they expect nearly one- third of the world's population will be watching television via mobile technology by 2010 (Mannion, 2006).

Mobile devices are not limited to traditional cell phones. Home access networks, provided through the existing and developing capacities of the mobile networks, will allow families the opportunity to watch television both at home and on the move. The oft-maligned delivery of telecommunications signals through common power lines is changing in areas of Europe and Asia. The new systems combine traditional landline usage with Wi-Fi mobile technology, and newly built homes are increasingly media-ready and not bound by hard wiring.

Competition, a willingness to take risks, and ever-evolving technological change are moving global convergence from abstract theory to practical reality. Connectivity rates, particularly for broadband Internet use and mobile "devices," have rapidly risen. Politics and culture still need to catch up. Traditional journalists and editors are caught between the two, raising questions about how to reach audiences, maintain profitability, and still maintain a strong ethical system for generating news. Across the world, answers are being sought, ideas being introduced, and experiments undertaken, all in the hope that the right questions have been asked.

## Global Developments and "New News"

The Keio University program mentioned earlier in this chapter is part of the university's Research Institute for Digital Media and Content. Its fifth annual convergence symposium examined everything from government policies and regulations to the creation of new technologies that would enhance the human experience in gathering information by dissipating the "noise" of the information society (Keio University, 2006). The research is based on two notions. The first premise is that today's media system is outdated because the information provided to the public is too simple and presented out of context (Tsuchiya, 2006). The second notion is that, although traditional media oversimplify information, the Web presents information in an unorganized fashion that provides great diversity, but little direction to people seeking information/knowledge (Tsuchiya, 2006). The result is that members of the public are often distracted by the volume or "noise" of the Web environment.

News managers across the globe are struggling with how to cut through the noise and reach new audiences and retain market share. One problem is in defining what online market share means. So far there is no perfect means of controlling the dissemination of unwanted information across geographical borders. The spread and reach of online news is but one of the issues that governments and news providers are struggling with in the news and information age.

For example, online newspaper readership in Spain has jumped 183 percent in the past five years (Varela, 2005b). The skyrocketing shift has left many newspaper editors confused about how to handle the rapid shift in information access. Coupled with the explosive shift to online readership, there has also been the development of "citizen journalists."

Editors at *Periodistas 21* believe the trend is part of the evolving nature of journalism (Varela, 2005b). In early 2006, the newspaper launched an online citizen's journalism project, and by mid-year began evaluating how information received and posted by the citizen journalists was tracking with traditional journalism characteristics.

In an article posted to the Editor's Web log, the writers noted that citizen journalism needs to be carefully vetted (Varela, 2005a). They cited at least five other areas of continuing concern with citizen journalism, including the following:

- Editors must divide testimonies that help understand information from those that only serve to sensationalize an event.
- Because journalism is an act of verification, editors must take care to ensure that images and text are not manipulated.
- Mainstream media must take strides to differentiate between first-hand accounts of news and tragedies and expressions of feelings and opinions about an event.
- Mainstream media organizations know that photographic galleries increase Web site traffic; editors must ensure that online photography meets the same quality standards as traditional print photographs.
- Editors also must control the postings. Too many citizen contributions will only create "noise" about the event. Editors must be willing to edit the site so that the audience does not become confused by an avalanche of information.

Citizen journalism, according to Enrique Dans, a professor at the Instituto del Empresa in Spain, typifies what he calls third-level journalism, or simply "Journalism 3" (Varela, 2005a). Dans defines "Journalism 1" as traditional journalism, a level where many newspapers across the world remain. He calls interactive or Web-based journalism and the conversion of traditional newspaper journalism to the Web as "Journalism 2." Newspapers participating in Journalism 2 are those which recognize that readers have moved to the Internet, but still rely on traditional means of reporting and delivering information to the masses. The emphasis remains on traditionally trained reporters and editors in fixed media organizations. Journalism 3 takes the reporting out of the hands of traditional journalists and places information dissemination squarely in the hands of the people.

How citizen journalism will evolve depends on a host of factors. Citizen-created content on the Web competes with and complements traditional media news product. With the popularity of individual Web sites, media must continue to attract readers and viewers to their vetted Web sites. Issues of credibility, accuracy, and ethics come into play, and some media engage in educational and marketing programs aimed at showing the public how they can participate in news gathering. As individuals routinely contribute to traditional media sites, both the media and the citizen journalist will regain the public's trust.

## The Conundrum: Giving the Public What It Wants

For all the concerns about accuracy and truth, the world still enjoys a good conspiracy. The love triangle of three American astronauts that led to the arrest of one astronaut was front page news in most of the major media around the globe. The death of Anna Nicole Smith in early 2007 led to more speculation. Blogs and chat rooms exploded with discussions about "who did what to whom and under what circumstances." In 2008 the Hong Kong entertainment world made global headlines with the saga of nude and sexual photos of young stars that found their way onto the Internet (Magnier, 2008). Sifting fact from fiction has always been the tough part of good journalistic editing. It is likely to remain that way regardless of the country where the information originates.

In Europe, Ofcom reports that television viewing is down while Internet usage is exploding in a number of countries (Office of Communication, 2006). The expectations for global viewership of wireless television are high, with some predicting the television market will be even more lucrative than other wireless endeavors, whether via the Internet or mobile telephone technology (Mattias, 2006).

To cope with these changes in audience consumption patterns, the *Guardian* Newspapers in London, Gannett Corporation in the United States and others have adopted a Web-first approach to news dissemination. The *Guardian* was one source in a major study on technological and journalistic convergence published in late 2006 by the European Union. In part, the report noted a variety of innovations the *Guardian* would introduce, including the following:

- "Web-first policy"—Noted earlier, this value-decision by the editors allows online reader access to major news by foreign correspondents and business journalists before it appears in the paper.
- "Customized print"—In late 2006, the *Guardian* launched G24, a free "print and read" service for news content, updated every 15 minutes. Users log onto Guardian Unlimited and download an 8–12 page pdf file featuring the latest news. They can select from any of five news streams: general news, international, economics, sports and media stories.
- "Comment Is Free"—This collective group blog brings together regular columnists from the *Guardian* and *Observer* newspapers with other writers and commentators; readers are invited to comment on anything they read with the aim of creating a space for open debate. The blog is updated regularly through the day. The site also carries all the editorials from the *Guardian* and *Observer* newspapers, giving readers the chance to comment on these articles directly for the first time. There is also a photo-blog from one of the paper's photographers.
- "Been There"—This is a travel site that relies on readers writing in with their experiences and their recommendations for hotels, sights, bars,

clubs, beaches, restaurants, and so forth, and even, submitting pictures. Extracts form a regular feature in the printed paper's Travel supplement on Saturdays.

- "Podcasts"—Podcasts include political and cultural debates, a regular Media Guardian podcast and a comedy podcast featuring the comedian Ricky Gervais which topped the *iTunes* charts in January 2006, having been downloaded more than two million times in a month ("Interactive Content and Convergence," 2006).

While the newspaper has concentrated on the broadband online platform, top editors of the newspapers believe mobile devices will play a strong role in shaping the future content of the newspapers ("Interactive Content and Convergence," 2006, pp.177–178). As bandwidth reaches appropriate levels, the newspapers will shift to provide content that takes advantage of mobile technology.

## Digital Development and Other Forms of Media Use

Digital "convergence" appears to be coming of age in Europe. A number of nations are already participating in mobile television projects, and more are expected to get into the mobile television market in the next few years (Walko, 2006). Broadband Internet and mobile networks now make it possible to broadcast, stream or download digitized content from a variety of platforms to a number of devices, often on an on-demand, interactive basis. "Interactivity" now relates not only to accessing content itself (as in interactive computer games) but to the many options left to consumers in the digital environment: navigation and search modes and multiple ways of accessing content in "pull" business models, as opposed to traditional "push" models.

Europe has indeed witnessed an impressive array of new media developments over the past few years in terms of supply (launch of online and mobile content services, new media deals) as well as demand (usage and technology adoption). At the end of 2006, EU officials confirmed the overall trend as "very positive" despite a variety of challenges that still need to be addressed for digital distribution to become a mass market service more quickly and more widely ("Interactive Content and Convergence," 2006, p. 263).

One of the challenges facing European markets is that they are not always at the forefront of digital distribution of content and lag behind more advanced markets in some aspects. Several independent reports indicate that Europe is second behind Japan and Korea (but before North America) for mobile content distribution and mobile television, and second behind the United States for broadband content distribution (Office of Communication, 2006, p. 308).

In March, 2008, a press release by the European Union noted that eight member states had greater broadband penetration than the United States (Europa,

2008). The European Commission's 13th progress report on the single telecoms market noted that Denmark, Finland, the Netherlands, and Sweden had penetration rates above 30% compared with the U.S. rate of 22.1%. Other European countries with penetration higher than the U.S. include the United Kingdom, Luxembourg, Belgium, and France (Europa, 2008).

In mobile device usage Europe lags behind only Japan. Japan and several other Asian nations remain at least 18 months ahead of Europe in the development and distribution of mobile television, a gap that is likely to continue for at least several more years.

By the end of 2005, digital on-demand movie distribution and music downloads in Europe also remained behind Japan and to a lesser extent the United States. Mobile digital radio is predicted to reach about five per cent of the European population by 2010 ("Interactive Content and Convergence," 2006, p. 32). Podcasting is expected to reach about half of that mark (two to three percent of the public), but by 2010 those levels will be only slightly behind usage levels in the United States.

Like elsewhere in the world, online and mobile digital game distribution has been adopted rapidly in Europe. The report estimates that the total value of the European "digital" games market was €698 million in 2005, of which 48 percent (€334 million) was contributed by the mobile sector, or about 11 percent of the total digital game market. By 2010, the forecast for the digital games market is to reach €2.3 billion, or about 33 percent of the total game revenues ("Interactive Content and Convergence," 2006, pp. 96–97).

To sum up, just how is the world generally "connecting up"? According to Nielsen Online (2007), many of the most developed nations are still increasing in Internet growth, but at rates of only one to two percent in a given month. The volatility of the Internet marketplace, the impact of technology, and the need to show continued growth still results in huge increases in certain other parts of the world. From October to November 2007, for example, Australia and Brazil still showed Internet growth of between seven and eight percent (Nielsen Online, 2007). With a world population of more than eight billion people, there is a lot of room for that growth, and it can be expected to remain as precarious as it has been for the past decade.

## Nations Transforming: Regulatory Action Versus Media Access

With the development of more future devices the end user should have all of the information he or she needs in the new global media environment. However, the new environment is worrisome to many. Some governments, including those mentioned in the introduction, are concerned with ensuring content conforms to appropriate cultural standards.

Various nations on a number of continents find themselves at cross- purposes. In China, for example, the Communist Party tries to control information by limiting access to the Internet and monitoring the activity of its citizens. Yet the more moderate forces in the government have been less willing to enforce bans on certain activities. The government is concerned with opening China's borders to the free flow of information—good and bad—with the rest of the world. As China awakens and takes its place on the world stage, the leaders are striving to find a "middle road" between freedom and censorship.

The governments in the Middle East have also had to grapple with the limits of free expression. No nation encourages the free transfer of all types of information. National security (especially military secrets) and morality (child pornography and other taboos) have always been subject to controls and violators prosecuted. But traditionally closed societies must now determine the degree of various kinds of information that are permissible. Once information begins to trickle, it is difficult to keep it from becoming a raging torrent.

Another factor influencing how governments react is that information flow runs in both directions. Increasingly, governments are also turning to the Internet to get their information out to people. The European Union tried for years to present its story to a diverse and sometimes unsympathetic audience. In 2006, after The Netherlands and France moved to undermine the union's constitutional vote and urged a pullout from the organization, confusion on various issues became epidemic. National television stations carried reports on numerous stories that took positions based on their individual national and cultural identities. The EU went on the offensive.

In the fall of 2006, the European Union launched EUX.TV to broadcast on the Internet. It avoids traditional terrestrial media and broadcasting and moves to educate the entire union about union issues from the organization's international perspective. Raymond Frenken, a well-known Dutch journalist, was tapped to launch the new endeavor. EUX.TV purports to "provide a range of programmes, interviews, debates, and news stories about European politics at an international, national, and local level" (R. Frenken, personal communication, September 2006).

The channel launched from Brussels, the official capital of the European Union, but its primary operations are based in Maastricht, The Netherlands. The programs air in real time and are readily accessible. All of the stories carry at least some video, often including interviews with proponents and opponents of various issues. A recent count showed stories on environmental issues, EU politics, Turkey's bid to join the union and The Netherlands' attempt to scrap parts of the EU constitution.

In essence, the programming could be viewed as a government channel in action, but its placement on the Internet assures it of an international audience. It is too early to determine how or even if the new online network will influence the workings of the European Union.

On another front, the EU has been pondering just how to view the future of the regulatory landscape. A report issued in late 2006 posed a variety of questions for regulators based on the fundamental tenets of regulatory behavior.

> The three traditional public policy goals have been quality, plurality and standards. There is rightly a debate about how far, with the proliferation of content and delivery systems, public policy and regulatory intervention remains necessary to deliver those objectives; and, if so, what the right form of intervention should be and where it should apply. (Currie & Richards, 2006, p. 21)

Similar to the United States and other Western nations, the focus is on when to protect or "dole out" the spectrum, how to ensure that the public interest is maintained and whether there is a role for regulators beyond the basic limits of structure. The report poses a variety of questions and reflects the frustration of some governments, noting that "as someone has remarked, on the Internet even the First Amendment is merely a local ordinance" (Currie & Richards, 2006, p. 21).

One of the questions raised deals with traditional broadcasters. The authors point out that, by continuing to regulate traditional broadcasting in much the same way as it has been regulated in the past, they are ignoring the fact that the Internet is fully protected with its right to publish and can simply become the next arena for broadcast content (Foster & Kiedrowski, 2006).

Still other nations in Europe have worked to ensure that the traditional forms of journalistic convergence exclude broadcasters from the playing field. Austria, Germany, the United Kingdom, and others have cast a wary eye toward media companies that want to own both broadcasting and print media. Much like in the United States, the nations have created roadblocks to cross-media ownership. The reasons are varied. Some countries are just getting into the realm of allowing broadcasters free-market status. Others simply worry that the number of voices allowed a nation or region will be limited when media companies are allowed to own both broadcast and print outlets.

While the reasons on the surface appear logical, in the long run the restrictions will simply hasten the adoption of Internet-based television programming and news transmission. The effect likely will be to harm the broadcast industry rather than protect the public. Where regulation creates a vacuum, technology will find ways to fill the void.

The past few years have seen numerous conferences examining the potential impact of the unfolding technologies on people. The results provide some suggestions and point to various broad trends. Some technologies are expanding, such as wireless broadband, while others seem to be rejected or fade away, like the compact disk and the videotape recorder. The debate on preserving national and regional identities continues, and governments have shown some sympathy to preserving culture. Even as convergence technology favors certain practices and perspectives, the ever-evolving nature of people and technologies continually forces us to revisit the issue.

## State of "Global" Journalism Integration

In 2002, Stone and Bierhoff predicted how organizations could move forward by moving through convergence development until reaching an integrated state of truly online-broadcast-print newsrooms. Five years later, some of the world had moved toward their model by bringing communications and journalism activities into an integrated stream. But many other nations, often encumbered by long-standing rules and regulations regarding broadcasting, have failed to begin the process, and the technological changes in the future may make the Stone-Bierhoff model less likely.

Stone and Bierhoff (2002) examined the state of convergence in four European nations; the United Kingdom, France, Spain, and Sweden. At the time, multiple corporations were moving toward integrating news operations into converged media. By 2007, some of those companies had been successful; others had either reached a cooling point in their march toward integration or had backed off the notion of attempting to converge.

The reasons behind the slowdown are as diverse as the attempts to integrate. In some instances, the rapid development of high-speed Internet access and the development of Internet Protocol TV (IPTV) have made integration with traditional broadcasters somewhat unnecessary. Some nations, fearing that too much media power would be aggregated into a single news source, blocked attempts by media companies to own multiple platform operations in the same geographical areas. Cultural issues of convergence, which were only beginning to be recognized in 2002, also have led to infighting and resistance from those working in traditional print, traditional over-the-air-broadcast, and online.

Many news organizations have had difficulty in retaining journalists in order to assemble a set of full-time cross-media platform specialists. This has led to more intense training efforts outside of the organizations and in universities or trade schools to bring the needed skills into the evolving journalistic workplace. Meanwhile, because the broadcast leg of the journalism industry is less integrated into the convergence system, the move between online and traditional print journalism is moving relatively well, although only after protracted false starts.

The regulatory blockade isolates broadcasters from the benefits of integration. This has had various levels of effect on broadcasters. In order for broadcast media to "engage" in online activity, it must create its own online unit. But traditionally, broadcast media have fewer resources and archives to bring to most journalistic stories. Therefore, the quality of broadcast journalism may suffer, since broadcast media will not have direct access to much of the information available to competing news organizations with a Web presence.

In some instances, broadcasters may be able to readily go around the regulatory arena and join in the information venture by creating working relationships (or partnerships) with the online side of the integrating media. Attempts to create lasting partnerships with print media have met with mixed results in a variety of ownership contexts. Generally, these partnerships are short-lived, as the organizations behind the media outlets seek more strategic (and profitable) outcomes.

## Conclusions

The term "sea change" may actually understate the advances in technology and information convergence that are taking place globally. A few years ago hardly anyone thought we would be downloading MySpace information to our cell phones from the Internet, or even that a Web site like MySpace would exist. These developments pose enormous challenges to the traditional information providers. Newspapers and broadcasters have had to change rapidly to meet these challenges, and in some instances engrained practices and regulatory controls choke the efforts to change. Governments and leaders are charged with overseeing the health, safety, and welfare of the people and the nation they serve. It is in every nation's interest to seek rational ways of channeling information within time, space, and manner limitations in order to help its citizenry. In the United States, Viacom brought legal action against YouTube because the site allows users to post video clips without compensation. China is also working with media companies to bring copyright action against Web sites inside its borders. Every nation is confronting the flow of various kinds of information on a case-by-case basis.

On the upside, high-quality technology, provided relatively cheaply, is flooding the marketplace. If you need to communicate, the options are seemingly unlimited. News and information, programming, even gossip and silliness are easily accessed in an open and converged media network.

The European Union study noted that there six areas where convergence efforts were being hindered, or where those efforts would face hindering factors in the near term. Those areas included the following:

1. **Technology issues**—Mainly involving consumer access to enabling technologies.
2. **Copyright issues**—Includes difficulties in accessing content, due to the definitions of new media exploitation rights, terms of trade and collective management of rights.
3. **Digital piracy issues**—Includes the disparity of legal means to fight piracy in the different EU member states (or globally).
4. **Legal and regulatory issues**—Including the regulation of new media services and non–linear content services.
5. **Competition issues**—Including gatekeeping issues in the value chains.
6. **Various economic issues**—Including access to funding, skills in converting traditional to converged media, the cost of digitization and consumer acceptance (though lately this has seemed to be a lesser issue).

Governments and media professionals as well as individuals, parents, and each of us as citizens will need to watch these developments carefully. Governments have shown they are not ready for the aggregated issues involving convergence. Media professionals, while making progress, are approaching convergence in an uneven way that is keeping media access blurry in some parts of the world. Individuals need to watch changes carefully or they may risk investing in products that are here one day and gone the next. Parents will need to continue to be watchful as new types of content, some unwanted, find their way to their children's mobile devices. As citizens we need to be prepared to inform ourselves about the evolution of media.

The days of simply thinking about telephones as a two-way communication instrument are over. Television sets are now media centers. Geographical characteristics that once defined us are fading swiftly as we become global citizens. Governments will need to work together to solve the issues that will influence both technological and journalistic convergence. Cooperation will be a required commodity as the new borderless media world takes hold. How well and how soon governments or regions develop appropriate ways of guiding and controlling this new environment is anyone's guess. The frontier is upon us. It is time to saddle up our intellectual horses and ride into the future.

## References

2006 Global digital media—Convergence, triple-play report (2006, November). International Markets, executive summary. Retrieved July 11, 2008 from html://www.budde.com.au/publications/annual/global-market/digital-media

Adegoke, Y. (2006, December 1). Big media sees opportunities in Web user content. Reuters.com. Retrieved July 11, 2008 from http://www.reuters.com/article/MediaMarketing06/idUSN3017420020061201

Beach, S. (2006, February 16). Scaling the firewall of digital censorship—Oliver Moore. *China Digital Times*; reprinted from *Globe and Mail*. Retrieved February 2007 from http://chinadigitaltimes.net/2006/02/scaling_the_firewall_of_digital_censorship.html

Boston Business Journal (2008, May 30). Researchers: Internet advertising growth will continue to surge. Retrieved July 11, 2008 from http://www.bizjournals.com/boston/stories/2008/05/26/daily35.html.

Bronfman, E., Cappello, R., & Kelly, M. (2006, April 21–24) *Media convergence and the revolution in marketing and brand building*. Panel proceedings before Milken Institute Global Conference on "Expanding Opportunities in the Global Marketplace," Los Angeles.

China Internet Network Information Center (January, 2008) Statistical survey report on the Internet development in China. Retrieved July 2, 2008 from http://www.cnnic.cn/uploadfiles/pdf/2008/2/29/104126.pdf

Currie, D., & Richards, E. (2006, November). *Communications the next decade: A collection of essays prepared for the UK Office of Communications*. Available through the U.K. Office of Communications Web site at http://www.ofcom.org.uk/research/commsdecade/

Entertainment Mobile. (2006). *Juniper Research White Paper*. Juniper Research Limited, Century House, Hampshire, UK. Available at http://www.juniperresearch.com/shop/products/whitepaper/pdf/White_Paper_Mobile%20Entertainment.pdf

Europa (2008, March 19). 8 EU Member States ahead of the US in broadband deployment says Commission's Telecoms. Europa Press Releases Rapid. Retrieved July 12, 2008 from http://europa.eu/rapid/pressReleasesAction.do?reference=IP/08/460&format=HTML&aged=0&language=EN&guiLanguage=en

Ewing, J. (2007, February 13). Vodafone and MySpace connect to conquer. *Business Week* [Electronic version]. Retrieved July 12, 2008 from http://www.businessweek.com/globalbiz/content/feb2007/gb20070213_212313.htm

Foster, R., & Kiedrowski, T. (2006). Overview. Opening report on *Communications: The Next Decade,* a report commissioned by the Office of Communications, U.K.

Head, W. (2006, November 30). TV and games will push mobile entertainment market to $47 billion by 2009. *ITnews,* London. Retrieved July 12, 2008 from http://www.itnews.com.au/News/42911,tv-and-games-will-push-mobile-entertainment-market-us47bn-by-2009.aspx

Intel Software Network. (2008, February 26). UMPC Portal. Available at http: www.softwarecommunities.intel.com/articles/eng/3754.htm

*Interactive content and convergence: Implications for the information society.* (2006, October). White Paper commissioned by the European Commission. By Screen Digest Ltd, CMS Hasche Sigle, Goldmedia Gmbh, Rightscom Ltd. Retrieved July 12, 2008 from http://ec.europa.eu/information_society/eeurope/i2010/docs/studies/interactive_content_ec2006.pdf

Keio University. (2006, August 28–29). *Manifesto towards the creative society*. Paper presented at the Digital Media and Content 5th International Symposium, Keio, Japan. Retrieved March 13, 2008, from http://www.dmc.keio.ac.jp/symposium/2006aug/manifesto_e.html

Kiss, J. (2007, August 2). Facebook ads pulled in BNP row. *The Guardian,* London [Electronic version]. Available at http://www.guardian.co.uk/media/2007/aug/02/digitalmedia.facebook

Lloyd, E. (2004, November 30). *Hong Kong's Digital Potential. I-media connection*. Available at http://www.imediaconnection.com/content/4697.asp

Magnier, M. (2008, February 27). Hong Kong ogles, blushes. *Los Angeles Times* [Electronic version]. Retrieved March 7, 2008, from http://www.latimes.com/news/nationworld/world/la-fg-hksex27feb27,1,6805735.story

Mannion, P. (2006, December 1). Ericsson and Intel collaborate to push mobile broadband. *EETimes,* United Business Media, London. Retrieved March 13, 2008, from http://eetimes.eu/showArticle.jhtml?articleID=196601090

Mattias, K. (2006, December 1). Cell-phone TV to reach mass audience in 2008, Ericsson. Retrieved February 23, 2007, from www.post-gazette.com/pg/06335/741032–96.stm).9

Nielsen Online (2007). Selected nations active home Internet users, June-July 2007. Retrieved January 16, 2008, from http://www.clickz.com/showPage.html?page=3626923

Office of Communications (2006, November 29–30). *Communications - the next decade*. Presentation before the Office of Communication's 2006 International Conference on Global Communications Technology, London. Retrieved July 12, 2008 from http://www.ofcom.org.uk/research/commsdecade/

Office of the Telecommunications Authority (OFTA), the government of the Hong Kong Special Administrative Region (2008). Key telecommunication statistics. Retrieved July 4, 2008 from http://www.ofta.gov.hk/en/datastat/key_stat.html.

O' Siochru, S., & Girard, B. (2002). *Global media governance: A beginner's guide*. Lanham, MD: Rowman & Littlefield.

Reporters sans frontières (2008). *2008 Annual report on press freedom*. Report available at http://www.rsf.org/article.php3?id_article=24025

RNCOS, Mobile Industry Research (2007, June 13). Chinese mobile industry reaching new highs. Retrieved July 11, 2008 from http://www.rncos.com/Blog/2007/06/Chinese-mobile-industry-reaching-new-highs

Stone, M., & Bierhoff, J. (2002). The state of multimedia newsrooms in Europe. In *The European multimedia landscape*. International Institute of Infonomics, European Centre for Digital Communication. Available at http://www.ecdc.info/documents/reports/0707171438The+European+Multimedia+landscape.pdf or at http:www.mudia.org or at http:///www.ecdc.info or http://www.infonomics.nl. Heerlen, The Netherlands.

*Telefónica, S.A., annual report 2006* (2006). Available at www.telefonica.com/eng

Tsuchiya, Y. (2006, August 28–29). *Digital communication tools in perspective of socio-media studies*. Paper presented at the 5th Annual Research Institute for Digital Media and Content, Keio University, Keio, Japan.

Varela, J. (2005a, June 20). AEDE(3): Blogs challenge to the press. *Periodistas 21*. Retrieved July 11, 2008 from http://periodistas21.blogspot.com/2005/06/aede-3-los-blogs-desafan-la-prensa.html

Varela, J. (2005b, June 22). Spain: Newspapers need to adapt to the Web and citizens journalism. *Periodistas, 2*. Retrieved July 11, 2008 from http://wef.blogs.com/editors/2005/06/spain_newspaper.htlml

Walko, J. (2006, November 27). Ericsson, Sony collaborate on mobile TV. *CommsDesign,* CMP Media. Retrieved July 12, 2008 from http://www.commsdesign.com/news/tech_beat/showArticle.jhtml?articleID=196513760

Walton, G. (2001). *China's golden shield: Corporations and the development of surveillance technology in the People's Republic of China.* Montreal: Rights and Democracy.

Wireless Federation. (2007, May 17). Retain VoIP users up twofold to 40 million in 2006. Retrieved July 7, 2008 from http://wirelessfederation.com/news/retail-voip-users-up-twofold-to-40-million-in-2006/

Wireless World Forum. (2006, June) China's booming market to total 500 million people. Retrieved July 11, 2008 from Wirelessinsightasia.com

Zhang, L. (2006, August). Behind the Great Firewall: Decoding China's Internet media policies from the inside. *Convergence: The International Journal of Research into New Media Technologies,12*(3).

CHAPTER 16

# Tracks, Silos, and Elevators

## Postsecondary Convergence Journalism Education in the United States

Timothy E. Bajkiewicz

But it can be an unwieldy concept.

*South and Nicholson (2002), on convergence*

Convergence may be a boat-car.

It was an enduring image from Bob Haiman, President Emeritus of The Poynter Institute in St. Petersburg, Florida, during his presentation at their 2004 seminar "Convergence Journalism for College Educators." The Amphicar was a 1960s German-imported hybrid that was both a car, with speeds up to 70 mph, and a boat, with twin propellers out the back and a watertight body (International Amphicar Owners Club, 2006). Only 3,000 made it to the United States before the company went under (so to speak), citing highway and environmental regulatory issues. The vehicle was also expensive to buy and operate, and its steel frame tended to rust. It tried to be too many things; it was a novelty; and it was a failure. In an earlier column, Haiman said the same fate may befall convergence:

> People want a specific device or service to function brilliantly.... They want the best of the breed, not a hybrid that tries to do two different things and adds up not doing either of them particularly well. I get nervous about news staffs trying to be boat-cars. (Haiman, 2002, n.p.)

Imagine if, during those few years, staffs and faculty from schools across the country retooled for boat-cars and abandoned proven, seemingly dependable, and separate programs on making and repairing cars and boats. Instead, they combined them to potentially create multifunctional, maybe even "super," mechanics. "The industry demands we stay ahead of the wave," educators would have explained. "Consumers are becoming platform-independent. We'll provide it however they want it." (Sound familiar?) Entire curricula, entire programs would have been, well, all washed up. The United States may now have a European-like train system, if only because other transportation manufacturing options no longer existed or never sufficiently rebounded.

Will the same be true for convergence journalism, and with it journalism education? The stakes are high as educators and administrators react to what they see as 21st-century industry and audience media-use changes. Michael Parks, director of the University of Southern California's School of Journalism, said, "Readers, viewers, and listeners got up and moved. They're making different sets of decisions on when, where, and how they will get their news" (Birge, 2004, p. 12). Safe to say, journalism schools and their faculty did not move with them, given the glacial pace of the academy. It is probably no coincidence that some of today's terms for journalism program curricula—tracks, silos, and even elevators—hail from agrarian times and journalism education's late-19th-century beginnings in the United States (Dickson, 2000). For convergence advocates these terms represent the old order, a medium-centric approach that does not recognize new realities. Hampden H. Smith III, then-head of the department of Journalism and Mass Communication at Washington and Lee University, commented, "Too many of our faculty teach what they did in the newsroom 20 or more years ago as though Gutenberg were still working in the composing room. Our [students] deserve a better education than that" (Birge, 2004, p. 13). With a record high enrollment of more than 195,000 undergraduate students in U.S. journalism and mass communications programs in 2005–2006 and growth in enrollment for the past seven years (Becker, Vlad, Tucker, & Pelton, 2006), the stakes are high, indeed.

Regardless of whether convergence "floats," it arguably brings to journalism education some of the most fundamental changes in curriculum and perspective ever witnessed, changes advocates say are needed to best position students, faculty, and programs for media's moving-target future. This chapter explores the recent research and conversations surrounding convergence and their implications for programs in schools, colleges, and universities around the United States, with specific attention to news-related studies in journalism and telecommunications. This discussion and analysis utilizes several of the latest studies about convergence education, a newer area for communication scholarship (e.g., Criado & Kraeplin, 2003;

Huang et al., 2006; Tanner & Duhé, 2005; Wenger, 2005a; Wenger & VanSlyke Turk, 2005). It begins with a brief history of journalism education in the United States, which set the cornerstones for contemporary efforts. Next, it examines the expressed need by educators for convergence training and what is happening in journalism programs around the country. Then it explores the kinds of skills, traditional and convergence-related, that professionals say they seek in journalism and mass communication graduates, and how those skills compare with faculty perceptions of needs. Finally, the convergence education conversation comes full circle as the case is made that it is actually a continuation of the age-old theory-practice debate between the academy and industry, and that journalism educators must hold their ground and demand industry share in its responsibilities for heading into what is considered journalism's next great frontier.

## Building Blocks: A Brief History of Journalism Education

The realities of 21st-century audiences and media industries may bring what is considered "new convergence" to journalism and mass communication education, but the related issues and concerns have been hotly debated since formal journalism education began in the mid-19th century. Just as convergence brings disparate concepts and players together to create original and different content, societal and cultural factors merged to help facilitate journalism's foothold in modern American education. Paraphrasing the philosopher Santayana's (1905) oft-quoted phrase, being condemned to repeat or even continue the unlearned lessons of journalism education could spell disaster not only for convergence, but for journalism education's future vitality and relevance.

Discussion of training future newspaper editors came as early as 1789, when John Ward Fenno, publisher of *The Gazette of the United States,* "called the newspapers of that day the 'most base, false, servile, and venal publications that ever polluted society,' and suggested that the evil might be removed by the appointment of college-trained editors" (cited in O'Dell, 1935, p. 1). Journalism education's genesis, and opportunity, came with the post–Civil War South. Universities responded to the war's dire economic and employment aftermath with new curricula to train and empower their citizens for the reconstruction effort, including the University of North Carolina, the University of Virginia, and Washington College (O'Dell, 1935), now Washington and Lee University, in Lexington. The latter's second namesake, Confederate General Robert E. Lee, accepted that school's presidency in 1865 to help, among others, young Southern men prosper in the still-struggling young United States. Four years later he proposed the first journalism scholarships for 50 such men, marking the first attempt to merge professional journalism training and a college education (Dickson,

2000). Professional criticism hastened the program's demise in 1878, after it had offered only 17 scholarships. Lee's death in 1870 removed the scholarships' primary architect and advocate, but newspaper publishers and tradesmen expressed harsh skepticism that a college education could replace the difficult apprentice program they completed (Sloan, 1990)—to this day a sore spot in the relationship between the academy and industry.

Journalism education grew as it rode the vocational movement of the 1870s, spurred on by land grant state universities and their edict to promote practical education. "Progressive education" was another buzz-term of the day, with its grand vision of social service and opportunity, as administrators established the elective system in higher education to, among other things, keep students and faculty more interested. The vocational movement moved to many universities in the Midwest and West, including Kansas State College, which offered the first journalism course in 1873 (O'Dell, 1935). Between 1873 and 1903, 14 colleges and universities offered such courses, including Indiana, Iowa, Michigan, and Missouri. Cornell strived to establish a professional program model in 1875 and offered a "Certificate of Journalism." Required coursework included philosophy, Latin, and two classes that could be seen as promoting an early convergence perspective: phonography and telegraphy.

Joseph Pulitzer had hoped to establish the first true journalism school, based on a curriculum he proposed in 1892 (Dickson, 2000). "He believed with a passion that at its best, journalism was a great intellectual profession of vital importance to the nation's welfare" (Brian, 2001, p. 277). Pulitzer's plan stressed liberal arts and news editorial concepts, and remains one of the blueprints for journalism curriculum today: law, ethics, history, "truth and accuracy," principles of journalism and, of course, news (Dickson, 2000, p. 10). Pulitzer thought associating journalism with education might help restore some credibility to the craft after it was tarnished by the "yellow journalism" of the late 19th century, from which Pulitzer profited handsomely.

In 1903 Pulitzer offered a $2 million endowment to either Columbia University or Harvard University to implement his ideas. When his offer to the schools became public:

> Skeptics ridiculed the idea of teaching journalism, citing the well-worn maxim that journalists are born, not made. Pulitzer challenged them to "name some great editor born full-winged like Mercury, the messenger of the gods. The only position that occurs to me which a man in our Republic can successfully fill by the simple fact of birth, is that of an idiot. (Brian, 2001, p. 283)

In the meantime, President Charles Eliot of Harvard proposed what is now considered the model for a vocational approach to journalism education, with not only traditional courses, but training in newspaper administration and manufacture, business, and advertising. Pulitzer, like many modern journalism educators, demanded the program enforce strict separation between the editorial and business sides. His famed comment was "If my wishes are to be considered, business instruction of any sort should not, would not, and must not form any part of the work of the college of journalism" (Pulitzer, 1904, p. 656).

At around the same time, the School of Journalism at the University of Missouri had earned the distinction of being first when it opened in 1908 (O'Dell, 1935). Although Missouri's effort predated Harvard's, with Missouri publishing their first curriculum in 1898, Eliot's ideas strongly influenced the program. Today, Missouri is recognized as the model of the vocational, practical approach. Ultimately, Pulitzer did not live to see his wish to open a school fulfilled. Columbia University received Pulitzer's money after he died in 1911, and their journalism school opened the next year. It was front page news for *The New York Times* and, of course, Pulitzer's *New York World* (Brian, 2001).

Journalism education was becoming well established by the time broadcasting and radio entered American living rooms in the 1920s, so schools often paired the two. Washington State University and the University of Southern California offered the first broadcast journalism courses in 1929 (Dickson, 2000), the latter recognized today as a leader in convergence education (Thelen, 2002). By the end of the 1930s, schools offering a bachelor's degree in broadcasting included the University of Wisconsin, Brigham Young Univerisity, Iowa State University, Syracuse University, and the University of Maryland. Radio engineering was the most common course; others included radio speaking and scriptwriting. Television entered broadcast curricula in the 1940s, although few schools included it. By the late 1980s, broadcast news was the most common sequence for broadcast education programs, followed by television, production, radio, film, and management (Warner & Liu, 1990).

Journalism education's growth skyrocketed through the 20th century, with 33 universities and colleges offering classes in 1912, and by 1929, 190 schools teaching journalism, including 56 professional schools with more than 5,000 students; by 1940, 542 four-year schools (Dickson, 2000). The growth for broadcasting was not as steep, but still impressive, with 81 programs by 1954 and 147 by 1967 (Kittross, 1989). In 2005, 461 degree-granting senior colleges and universities offered at least 10 news-editorial journalism courses (Becker, Vlad, Tucker, & Pelton, 2006).

As this brief summary illustrates, the first part of the 20th century saw pioneers and visionaries lay a firm foundation for contemporary journalism education. This growth spoke to more than Pulitzer's dreams or Eliot's pragmatism—it reflected a growing acceptance of journalism in higher education. But prominence in the academy is tied to broader societal and economic placement, and there journalism still struggles. As many journalists do today, Pulitzer envisioned journalism not as a trade, but as a profession. A pamphlet he commissioned about the need for such education stated, "All other high pursuits have become learned professions; but journalism lingers in the primitive condition" (O'Dell, 1935, p. 100). Convergence could change that perception by providing the opportunity for journalism to show the world that effective and essential media communication is just as complex, difficult, and vital as any traditional profession.

## Convergence in Journalism Education

"To converge, or not to. . . ." Like any good writer, Shakespeare would have adapted to his time. Journalism education faces the same poetic question, although today's script is a bit smudged. For more than a century, journalism schools have served two masters, each with a very different sense of time: the academy, with its sometimes maddeningly slow pace, and industry, with its faster, more intense momentum (Dickson, 2000). For several decades this had not been an issue, since journalism teaching and techniques changed little. Then audiences took notice of developing technologies while the Internet blossomed into the largest communication platform yet designed, which together created a very complicated situation for journalism practitioners and educators. Despite these changes, the tempo of the academy remained unhurried, but deliberate; however, this new reality prompted journalism faculty to reconsider their mission within their academic mandate. This section examines how journalism educators have specifically addressed convergence in their programs and curricula.

### Now in Newsrooms and Classrooms

One way to begin the discussion is with a contrast of perceptions from industry professionals and journalism educators regarding the prevalence of convergence in American newsrooms. Huang et al. (2006) found that almost half of editors from all media forms said they produced news content for multiple media on a regular basis, while Tanner and Duhé (2005) found this true for close to 90 percent of television news directors. Meanwhile, more educators than professionals have said convergence will be journalism's next big thing. Of lead administrators of programs accredited by the Accrediting Council for Education in Journalism and Mass Communication, 72 percent said they believe convergence to be the future of the industry, with 79 percent highly influenced or influenced to include convergence in their curricula because they believe the industry is moving in that direction (Tanner & Duhé, 2005). For other administrators, 83 percent said it is very important or important to train students to work in converged newsrooms (Wenger, 2005a). Using the same data, Wenger and VanSlyke Turk (2005) placed education administrators and industry professionals into two attitude groups, convergers or traditionalists. They labeled more educators as convergers, 76 percent, than online editors, 73 percent and television news directors, 69 percent. They placed the majority of newspaper editors, 57 percent, in the traditionalist camp.

Next, a vision of the future: In 2001 the leading organization of educators in journalism education, the Association for Education in Journalism and Mass Communication (AEJMC), released "Journalism and Mass Communication Education: 2001 and Beyond," a volume of final reports from subcommittees and task forces charged with seeing into the unknown. As part of this work, Pavlik, Morgan, and

Henderson's (2001) vision regarding technology change included several observations and recommendations that directly impact the prospects for, and realization of, convergence education. They noted changes regarding information and the tools of journalism in five broad areas: acquisition, storage, production, distribution, and presentation. This led to "four dramatic sets of implications" (p. 17) for changes in journalism education: storytelling and content, professional workflows, management and culture of the industry, and the relationships between journalism organizations and their publics. They concluded that journalism education "has a unique opportunity to take a leadership role in how [journalism and mass communications] is conceptualized and taught and how the industry operates in the 21st century and beyond" (p. 20).

Convergence advocates see a primary leadership opportunity with the playbill of higher education: curricula. By the time of the 2001 AEJMC report, several studies agreed that a majority of journalism schools were already implementing convergence, but often in small ways. Between 1998 and 2002, around 60 percent had changed curricula or added courses to address convergence (Huang et al., 2006). Less than one-third of journalism deans and department heads surveyed in fall 2002 said convergence curricula of some type had been in place for three or more years, 37 percent said one to two years, and 19 percent said less than a year (Criado & Kraeplin, 2003). Of that 85 percent total, two-thirds described the curricular changes as minor, and a quarter described them as major. Lowery, Daniels, and Becker (2005) found similar results with the most comprehensive sample studied: 463 journalism program administrators surveyed as part of the University of Georgia's annual enrollment study. As with previous findings, 85 percent said they are somehow converged.

In other research, when a converged curriculum was defined as "one that teaches all journalism students how to generate news content for print, broadcast, and online" (Wenger, 2005a, p. 10), 13 percent of administrators said their curricula are highly converged, 65 percent labeled their program as somewhat to moderately converged, and 22 percent said their program is not converged. However, in 2005, fewer of those administrators described their curricula as highly or moderately converged than they did in 2004, down to 42 percent from 57 percent (Wenger & VanSlyke Turk, 2005). For those administrators whose programs were already committed to convergence, half described their overall experience as positive or somewhat positive, while 81 percent said they would increase their convergence commitment in the next three to five years (Tanner & Duhé, 2005).

The momentum of so many educators and programs headed down some path toward convergence could result in a self-fulfilling prophecy with broad, hopefully positive, implications. As journalism schools produce more convergence-trained (or even convergence-minded) graduates, newsrooms could feel the urge to converge from within. Convergence could become a commonplace, grass-roots perspective instead of a new-fangled, management-implemented strategy. For this to occur, educators must demonstrate the vision and discipline many discuss. They would need to steer a steady curricular course on convergence while resisting industry's often short-sighted criticisms.

## Sequence Swansong?

Soon after program administrators and faculty decide to pursue convergence (if not before) everyone gets down to curricular brass tacks: How should convergence be implemented in our curriculum? For the vast majority of journalism schools, the more relevant question is What do we do with our sequence/track/silo/elevator? Sequences emerged in journalism schools during the 1950s as the demand increased for specialized study in not only print journalism, but broadcasting, public relations, and advertising (Weaver, 2003). Today, sequences are the primary organizational tool for curricula. Although they could be arranged to emphasize any subject or concept, in modern journalism schools "sequence" is shorthand for medium-specific training: For example, journalism students learn reporting for use in print newspapers, and not what a television story may need.

In most programs, students in different sequences like journalism, television news, and maybe even online only meet in the halls. James Gentry, labeled a "pioneer in cross-media preparation" for redesigning the curriculum at the University of Kansas, said, "The kids called it an elevator. They'd get in, and the doors would close," and they rode it to graduation, never stopping at another sequence's floor (Birge, 2004, p. 11). Programs at the University of Kansas (Outing, 1999) and the University of Southern California (South & Nicholson, 2002) were among the first to boldly go with convergence where no journalism curricula had gone before. The AEJMC (2001) millennial report recommended cross-media education and content concentrations, instead of media-specific efforts—in other words, a completely new paradigm for teaching journalism in the new media age. Huang et al. (2006) found that well over half of faculty agreed or strongly agreed that their traditional sequences should be reorganized for convergence, while less than a quarter disagreed or strongly disagreed. One faculty member noted, "Sequences in many J-schools are regarded as dinosaurs because it no long makes sense to teach broadcast and newspaper, for instance, as two unrelated bodies of knowledge in the era of media convergence" (p. 255). However, 63 percent of those same faculty, as well as media professionals, also said students should still study a specialization.

It seems just as difficult to define what constitutes a converged curriculum as it does to define convergence itself. Lowery et al. (2005) proposed three stages of curricular reform that effectively rank a program's amount and type of convergence. Their schema also provided categories for other research findings and began a common metric for this aspect of convergence, something which the field now lacks and desperately needs (Dailey, Demo, & Spillman, 2005). When classifying current journalism programs into these stages, a bell curve results, with few at the ends and the majority somewhere in the middle. In the first stage, "static" programs remain with traditional curricula and emphasize medium-specific training. Findings for "static" programs vary widely, from Lowery et al.'s (2005) 15 percent and Wenger's (2005a) 22 percent of nonconverged schools, to Kraeplin and Criado's (2005) finding of more than half. For the second stage, "supplementary" programs also stay with their traditional curricula, but with added courses or sequences that address convergence. Most journalism programs that tried to implement convergence fall into this category. For this middle ground, research is in agreement that it comprises about half of programs. In the third stage of curricular reform, "realigned" programs "reconceptualize and reshape their overall curricula structures" (p. 35) and would be considered truly converged. Again figures vary, from Lowery et al.'s (2005) finding of less than one-third of programs, to 13 percent (Wenger, 2005a), to Kraeplin and Criado's (2005) finding that only one of 46 administrators described their program this way.

Given these studies, a majority of programs described themselves as converged in some way and had some kind of converged program in place since about 2000. On the scale of academic time, this is but a breath—ancient manuscripts have much older layers of dust. Still, considering the tectonic shifts necessary to move modern university faculty and curricula, it demonstrated an impressive and nimble response by programs around the country. Practitioners may not appreciate these efforts (they have likely never sat through a curriculum faculty meeting), but they should take heart that educators worked diligently to motivate the academy to respond to the changing media landscape. Journalism education is now better positioned for changes with convergence, or whatever may be next.

Administrator altruism may not be the primary motivating factor—there is also the money (Phipps, 2000). Convergence deals with new media, a sexy buzzterm for "the future," which has led universities to magically increase budgets and find updated facilities, and foundations to lavish multimillion donations and endowments. This has been especially true for broadcast education, which is equipment-intensive and budget-straining. More importantly, journalism schools may be getting more than just new media toys out of convergence—they have earned new respect. "Improved status means media education is no longer being relegated to what some universities regarded disdainfully as vocational training. Instead, journalism schools are being invited to share facilities and opportunities with other departments" (p. 12). These programs include engineering and computer science, whose students would not typically be seen near the journalism building. But journalism generates content, and in our modern, creativity-starved, every-interest-needs-an-outlet universe, content on campus translates to academic influence.

## The Accreditation Factor

Accreditation is a hotly debated indicator of respect and quality from academic peers around the nation, and its standards directly affect if and how some journalism programs approach convergence. More than 40 percent of program administrators saw the restrictions of accreditation as a challenge for implementing convergence curricula (Tanner & Duhé, 2005). Accreditation is part of a very old discussion. When journalism education first began at the turn of the century, newspaper professionals became increasingly interested in the process as more college-trained journalists entered the job market (Weaver, 2003). Educators wanted to share information and establish standards for the growing number of journalism programs. In 1912 these concerns coalesced into the American Association of Teachers of Journalism, the forerunner of today's Association for Education in Journalism and Mass Communication (AEJMC), the principal organization of journalism educators. In 1917 the largest schools of the time formed the American Association of Schools and Departments of Journalism (ASSDJ), which still exists today. Both organizations began to set curriculum standards in 1923 with the Council on Education for Journalism. Formal accrediting efforts followed in 1939 when the AASDJ and newspaper groups formed the National Council on Professional Education for Journalism, which is today's Accrediting Council on Education in Journalism and Mass Communications (ACEJMC).

Ultimately these organizations changed the landscape of journalism education and formalized the media industry's influence in journalism schools, since the ACEJMC includes representatives from professional media organizations (Weaver, 2003). By mid-2007, 109 journalism and mass communication college and university programs in the United States were ACEJMC-accredited, with one in Chile (ACEJMC, 2007). Lowery et al. (2005) found at least 463 specific journalism programs around the country, while the U.S. Department of Labor's Bureau of Labor Statistics (2006) estimated more than 1,200 colleges offer journalism, communication, and related programs. Among the journalism programs, around 25 percent are accredited.

Whether or not a journalism program is accredited, or seeks such status, greatly influences how or even if its faculty may approach convergence in its curriculum. ACEJMC currently evaluates programs based on nine standards, including curriculum and instruction, diversity and inclusiveness, faculty, scholarship, student services and other resources, professional service, and assessment. Of course, ACEJMC's (2001) millennial report complemented these standards. The report labeled its program recommendations, which essentially outlined convergence education, as "advantages" when compared to accreditation standards (p. 18). By the current standards, those especially relevant to convergence include "certain core values and competencies to be able to" (1) understand concepts and apply theories in the use and presentation of images and information, (2) write correctly and clearly in forms and styles appropriate for the communication professions, audiences, and purposes they serve, and (3) apply tools and technologies appropriate for the communication professions in which they work (n.p.).

Accreditation standards have fanned the flames of the classic liberal arts versus vocational training debate, as some journalists demand skills classes while educators traditionally advocate a core of liberal arts studies (Sloan, 1990). Yet almost everyone in both groups wants students with a broad liberal arts background (Kraeplin

& Criado, 2005). The balance of liberal arts and skills classes involves the "80/65" rule, as outlined in ACEJMC's (2004) "Standard 2, Curriculum and Instruction." This standard requires that students take a minimum of 80 semesters hours (or 116 quarter hours) outside of journalism and mass communications, and a minimum of 65 semester hours (94 quarter hours) in the liberal arts and sciences. For 2002 this requirement was reduced from a minimum of 90 outside semester hours, and programs were able to exempt some classes (ACEJMC, 2001). The Council adopted the 90/65 standard in 1985, which led to discussion about how accredited programs defined liberal arts courses. One study of 87 ACEJMC-accredited schools (Hoskins, 1985) found that identifying liberal arts was a subjective process, and pointed out that philosophers had been debating the role of liberal arts in education and society for millennia. A 1999 *Editor and Publisher* editorial staunchly defended these journalism accreditation efforts, saying "Too many j-schools had become glorified trade schools as journalism majors heavied up on technical courses that blunted a well-rounded liberal arts education. ACEJMC insists on bringing that mix of career and culture into a more proper balance" (Editorial Staff, 1999, p. 18).

National university accrediting standards typically require at least 120 semester hours for a baccalaureate degree (e.g., Southern Association of Colleges and Schools; Commission on Colleges, 2001). With ACEJMC's 80/65 rule, around 40 semester hours remain for a journalism curriculum. This is the difficult math for administrators and faculty who are considering redesigning their program to somehow address convergence, or for programs that already have done so and are taking another look. As South and Nicholson (2002) observe, "Curriculum revision is a zero-sum game; if schools add courses in multimedia journalism, they may have to subtract courses in traditional reporting" (p. 15). They mention how some schools "offload" technology-intensive courses from their journalism curriculum to outside departments to meet accrediting standards, and then make those courses requirements for program entry.

The 2001 ACEJMC report also recommended content specializations, which translate to coverage areas (e.g., science or health reporting), instead of media specializations (e.g., print or broadcast). Besides better preparing students for a cross-media future, the report said this would encourage courses outside the department and those deemed "professional," which would not conflict with the 80/65 rule. One example of content specialization could be Ohio State's program in public affairs journalism (Fitzgerald, 2003). However, Ohio State and other big schools, including Michigan and Wisconsin, actually dropped accreditation to "take their programs in non-traditional directions" (Fitzgerald, 2003, n.p.).

For many programs, that direction could be convergence, with or without the rigors and limitations brought by accreditation. Lowery et al. (2005) found not being ACEJMC-accredited predictive of whether a school pursued a converged curriculum. In the majority of programs with both specialized sequences and convergence courses, faculty perceived industry trends toward convergence, which discriminated among those that emphasized only convergence or only sequences. Faculty interest was also predictive of converged programs, as was more faculty members per student, which often occurs in smaller programs. Surprisingly, they found smaller programs with fewer attachments to industry more likely to be fully converged. This seems counterintuitive, since connections to industry should correlate with faculty perceiving industry convergence. Citing institutional theory as a possible explanation, the researchers noted that professional environmental pressures, which should come from industry attachments, often influence change; however, since few news outlets are actually converged, no such pressure exists for convergence curricula. Instead, larger, more professionally connected programs realize "the perceived need to maintain legitimacy" (p. 35), and so "make a show of pursuing convergence" (p. 43).

Lowery et al. (2005) concluded that larger, mainstream programs should study smaller, converged programs. While it is true that smaller programs tend toward faster and more agile curriculum reactions, in this situation they also act as potentially unwilling, and definitely incomparable, convergence test beds for their larger academic colleagues. In addition, smaller programs do not have the same industry leverage as their larger counterparts, which may slow convergence's potential adoption, and perhaps even its eventual success. Mindful of the importance faculty perception plays in this process, the researchers suggested increased study of industry convergence trends. However, that kind of current information relies heavily on cooperation from journalism professionals, who may be reluctant to share what they consider proprietary information and may be weary of research attention. Instead, Lowery et al. suggest, industry should welcome such curiosity, since "these programs provide the skills and expertise that news organizations need in their labor force" (p. 44). Many consider skills the currency of journalism education, and they are especially important and complex with convergence. The next section addresses the classes and skills educators and professionals see as most important to put convergence into professional practice.

## Where It Happens: Convergence Classes and Skills

Skills act as the most significant bridge between journalism education and the news industry. For professionals, the amount and quality of skills in recent graduates act as a continually updated report card of journalism school performance. According to Huang et al. (2006), journalism education will be most impacted by what they termed role convergence, where media professionals will need multiple skills to competently perform more and different newsroom tasks. But educators have expressed concerns that students in convergence programs are learning too many skills, and that they do not achieve mastery in one area (Tanner & Duhé, 2005). Professionals share this fear, but were divided about the need or relevance of multimedia skills for new hires in the near or far future (Birge, 2006).

A quick look at if, and how, professionals produce for multiple media provides convergence educators with good indicators for possible needed skill sets. In short, complementary media have stuck together, while television and print rarely contribute content to each other (Criado & Kraeplin, 2003). Online news was the clear content winner in these relationships, with about half of television and newspaper reporters very frequently providing stories, especially breaking news, considered a staple for online journalism (Online News Association, 2001). Television reporters helped their radio partners, with about one-third very frequently writing radio news and being interviewed on-air. For television and newspaper partners, convergence often meant additional sources: About half of newspaper reporters and/or editors were at least occasionally interviewed for television; likewise, almost three-fourths of television reporters would at least occasionally be quoted in the newspaper. However, for these traditionally warring groups, convergence did not mean generating more cross-platform content: More than two-thirds of newspaper reporters never wrote radio news, nor did they write for the television newscast; similarly, two-thirds of television reporters never contributed to the newspaper. In other research with television news directors practicing convergence, three-quarters said they produced content for online, about half included radio, and a little more than one-third mentioned content for newspapers (Tanner & Duhé, 2005).

When it comes to where convergence is taught, of the 65 percent of administrators who described their program as somewhat to moderately converged—the vast middle populated by most journalism programs—about half said their programs "provide opportunities . . . to develop competencies" (Wenger, 2005a, p. 10) in convergence while still in a specialized sequence. About one-third (Tanner & Duhé, 2005; Wenger, 2005a) to more than half (Kraeplin & Criado, 2005) of programs used a specific class or two for convergence skills, while around three-quarters of programs featured at least one course specifically for online journalism (Lowery et al., 2005). Meanwhile, almost 90 percent of individual faculty incorporated convergence into their classes, along with other curricular approaches (Tanner & Duhé, 2005). When convergence was part of the curriculum, it was sprinkled throughout, including lecture courses (71 percent), introductory skills courses (67 percent), and advanced skills courses (84 percent).

## Basic Skills Basically Still Count

Given industry's lukewarm reception to convergence, educators should tread cautiously in emphasizing convergence skills for every print journalism and broadcast news student. This touches upon another of journalism education's great debates. As some scholars said, "It is hard to find a skill that journalism educators or practitioners do not deem important for students to acquire before entering the job market" (Martin et al., 2004, p. 14). Almost all educators and professionals are in agreement about the importance of basic journalism skills, including writing and reporting, interviewing, news judgment, copy editing, media law, and ethics

(Kraeplin & Criado, 2005; Wenger & VanSlyke Turk, 2005). Pavlik et al. (2001) agreed that journalism programs should always stress basics like teaching news values, relying on good sources, checking facts, using balance and fairness, developing good interviewing skills, and adhering to high ethical standards. Under the heading "The Basics Still Matter," Pryor (2006) wrote, "The journalism applicant with both a depth and a breadth of skills and a willingness to work at engaging the audience has the edge" (n.p.).

If stressing the basics were not enough (and there are plenty of basics), more than three-quarters of educators, editors, and news production professionals said students needed multiple skill sets, like writing, editing, video production, and online (Huang et al., 2006; Wilkinson, this volume). However, few convergence programs train students with all the skills needed for cross-platform journalism (Kraeplin & Criado, 2005). Almost all students were trained to write for print media, but only around 60 percent trained most or some of their students to write for online and broadcast, and about 70 percent trained only some or few of their students for Web language and design skills and on-air presentation for radio or television.

An earlier discussion mentioned how more educators than practitioners believed convergence was prevalent; this also held true with the perceived need for convergence skills. Around 85 percent of professors, editors, and news professionals agreed or strongly agreed that students should learn to write for multiple media, and that students with a visual emphasis should learn about photography and video (Huang et al., 2006). In both cases, significantly more educators agreed. In another study with more specific findings, almost all educators said convergence skills were very or moderately important to media managers when hiring, while the same was true for less than three-quarters of television news directors and newspaper editors (Kraeplin & Criado, 2005). Also, Wenger and VanSlyke Turk (2005) found that educators and professionals considered collaboration, cross-platform writing, and multiplatform storytelling as the most important convergence skills, with more educators than practitioners believing students need to write across multiple platforms and have multimedia story planning skills, while more professionals emphasized collaboration skills.

## Teaching and Training Technical Skills

Technical skills deserve separate discussion, considering they often spark heated dialogue among faculty members (and did long before the arrival of convergence), and that, as Wenger (2005b) pointed out in *Quill,* many converged programs stress online projects and the accompanying skills. On the upside, both educators and practitioners agreed that skills for the Web and for capturing audio and video were important (Wenger & VanSlyke Turk, 2005). Interestingly, between 2004 and 2005 the gap narrowed between those groups' responses about the importance of Web skills, such that they were essentially equal: Professionals increased their perceived importance, while educators decreased theirs. This disagrees with findings from

Kraeplin and Criado (2005) that more than twice as many educators (66 percent) than television news directors (31 percent) and newspaper editors (32 percent) thought Web language and design skills were important. In other research, fewer than 2 percent of television news directors saw Web design as important, although 18 percent included the skill of adapting visual content for multiple media (Tanner & Duhé, 2005).

Online professionals are in the best position to comment about online matters. In Paul's (2001) skills survey of 55 online news managers in connection with the ACEJMC millennial report, around three-quarters included the importance of updating time-sensitive material and editing text for online, while around half included creating multimedia projects, hand-coding raw HTML, and server-side file management. Most said new hires did not need to know scripting languages like JavaScript, or about original Web page design or Web editors (e.g., Macromedia's Dreamweaver). Online managers stressed the availability of training, with one commenting, "The core journalistic skills are still the most highly required and technical skills training can, in many cases, be done on an 'on the job' basis" (p. 26–27). They would rather see new journalists "who have no fear of new technology" (p. 29).

Training, and the lack of it, has been a sore spot with journalists. According to a 2002 Knight Foundation study (John S. and James L. Knight Foundation, 2002), journalists cited lack of training as their top source of job dissatisfaction, even after salary, and more than two-thirds said they received no regular training. During that time, news organizations had not increased training budgets for more than a decade. Of course, hiring managers would like to see fully trained, turnkey journalism school graduates. As one professional-turned-academic wrote, "There's no avoiding the reality that editors want beginners who already have experience, gained at someone else's expense" (McClelland, 2004, p. 16). About one-third of television news directors said they offer such training for their employees, although very few taught convergence skills (Tanner & Duhé, 2005). A significant majority of newspaper editors would like to see the kinds of technical skills needed for convergence learned in journalism schools, while even more nonmanagement newspaper professionals would rather have on-the-job training (Huang et al., 2006).

Skills make journalism happen, and they are the meat and potatoes of every journalism school. Educators who teach more convergence skills than practitioners think necessary probably create a more-rounded journalism graduate, but at the cost of a shallow, albeit wide, knowledge pool. Thelen (2002) cautioned, "Journalism schools must continue to produce graduates who are competent in one craft area: reporting, design, producing, directing, editing" (p. 16). If professors believe these skills so important, then curriculum needs to catch up with perception so they are offered in more than one or two classes, the current norm for most programs integrating convergence. Unfortunately, it is a no-win situation: Students' entire college careers could be dedicated to skills training, and then on their first job they might be criticized for not knowing some organization-specific need. News managers want "eager learners" (p. 16) in newsrooms, but learning cannot be one-sided. Organizations that do not make skills teaching and training regularly available, especially if they expect convergence activities, are not investing in the future of their employees, nor the craft or business of journalism.

## Faculty: The Face of Convergence in the Classroom

Journalism faculties carry the frontline burden of teaching, training and mentoring the next generation of graduates to work in converged newsrooms—a responsibility that directly impacts industry. In 2005, newspapers hired one out of every four new employees directly from colleges, with 85 percent of those straight out of journalism schools; in radio and TV news, freshly minted journalists accounted for 90 percent of college hires (Becker, Vlad, Pelton, & Papper, 2006). As convergence enters more newsrooms, faculty will be under increasing pressure to continue these trends and supply quality graduates. Discussions of definition, perception, and accreditation will mean precious little without the classroom leadership necessary to make convergence an educational reality.

Convergence curricula will potentially affect a lot of educators—more than 10,000 full- and part-time faculty members work in the country's journalism and mass communication programs (Becker, Vlad, Tucker, & Pelton, 2006). In 2001, 61 percent of all journalism school faculty members were male, and 85 percent were ethnically European, with 9 percent African-American and 3 percent Hispanic (Becker, Huh, Vlad, & Mace, 2003). That analysis of 12 annual surveys found faculties more diverse than in the recent past, both in terms of gender and ethnicity, although gains in ethnic diversity were negligible. Professors were evenly distributed by rank, with about 30 percent at each level of assistant, associate, and full, and 12 percent at instructor. Specifically, those who teach in journalism, broadcast news, and photojournalism sequences will be most affected by convergence—in 2001 they accounted for about 40 percent of all full-time journalism school faculty members. Huang et al. (2006) found 71 percent of specifically journalism educators typically to be male, between 46 and 55 years old; to hold a Ph.D.; to do academic research; to have worked in news media between one and 10 years; and probably to still practice news professionally.

Convergence classrooms, in terms of faculty and their students, will probably look as they do in most journalism schools—uneven in terms of gender and ethnicity. Despite slow but steady growth in terms of gender diversity, in 2001 women comprised less than 40 percent of faculties, but more than 60 percent of students (Becker et al., 2003). Ethnically, minorities made up 26 percent of students in journalism classes, leaving a gap of 11 percent with faculty, with that gap expected to widen with continued student diversification. One silver lining: Despite a 15 percent increase in the number of journalism programs and 17 percent more students from 1989 to 2001, faculty growth has outpaced these gains, with 27 percent more full-time faculty and a shocking 49 percent more part-time faculty.

This demographic picture of journalism programs has several implications for successfully implementing convergence curricula. Although administrators are often in the best position to affect issues relating to faculty gender and ethnicity, individual faculty members in converged programs should encourage their classes to cover these issues while taking advantage of the multiplatform storytelling tools at their disposal. Since about 60 percent of faculty members hold ranks of associate or full professor, the majority of potential convergence educators are likely several years removed from the industry, meaning their technical skills may be a bit rusty. This does not bode well, since convergence is easily the most technically intensive and diverse kind of journalism ever practiced.

Faculty convergence skills, especially technical skills, may be convergence education's Achilles' heel. Although more than three-quarters of educators say they are theoretically equipped to teach convergence (Huang et al., 2006), and administrators assert nearly all their faculty want to teach it, more than a third cannot decide how best to do so (Tanner & Duhé, 2005). Their own technical competencies and comfort zones will surely impact their implementation strategies. Administrators say that of their faculties, one-third (Wenger, 2005a) to more than half (Tanner & Duhé, 2005) lack the proper training to teach convergence. More than half of journalism professors had not taught a course in the last five years that required skills outside of their own expertise, and only half feel technologically ready for the demands of teaching convergence (Huang et al., 2006).

Technology is crucial to practicing and teaching convergence, so it is worth noting that it causes its own type of stress with journalism faculty. In Fall 2000, Voakes, Beam, and Ogan (2003) surveyed 58 administrators and 403 ACEJMC members about their attitudes toward new technology and the tech-related stress they feel. Essentially, faculties were tech-stressed out. Almost everyone said it was important for journalism programs to keep up with and reflect industry technology changes, with more than three-fourths who said students should be taught with the latest technology. Around 90 percent said they felt confident about learning new technologies (most already learned one to two new programs each year), but only half were willing to teach a course that utilized new software. Everyone used programs for word processing, Web browsing, and e-mail, and 60 percent used desktop publishing, but only 41 percent used Web page design, and 18 percent used video editing software. One-third of both faculty and administrators perceived more stress from technology compared to five years ago, although more faculty members say they see it in their colleagues. The researchers found faculty rank to be a better predictor of tech-stress than age, with tenured, associate professors saying they felt it the most. A gender-focused analysis of the same data (Ogan & Chung, 2003) found more confidence among women about new computers and software, and significantly more women teaching skills courses, but also more tech-stress among women who teach those courses. Tech-stressed convergence faculties will not likely receive help soon. As many as 64 percent of administrators say they lack money to buy new equipment (Tanner & Duhé, 2005) and, much like in journalism newsrooms, about a third lack funding to train faculty (Wenger, 2005a). It seems the academy and industry may not be so different after all.

## Conclusion

One thing is clear—convergence has become part of the modern lexicon of journalism, and both the academy and industry will feel its influence, in some way, for decades to come. Representatives from both these groups essentially agree about what constitutes convergence and its prevalence. In that respect, perception may become reality. Even if convergence is not practiced in as many newsrooms as we thought, some realization of multiplatform journalism will happen if editors (or owners and publishers) and educators see convergence as needed or wanted. Journalism educators have no shortage of issues when considering convergence: the industry, curriculum, sequences, accreditation, skills, faculty, and budgets. So much depends on factors beyond the academy's control, like news industry adoption, the economy, and the always-fickle audience—whom, supposedly, these changes are meant to better serve and inform. Yet it must be members of the journalism academy who balance hope and vision with demand and reality to plot the course into an uncertain media future—and face the long-term consequences of their decisions.

This discussion shows that convergence is complicated, and teaching about it will be even more so. Journalism education enjoys a history of demonstrating enthusiasm and adaptation, and it will need all it can muster for convergence education to be an overall success. Some administrators have shown due caution about steaming full-ahead into an area filled with so many curricular icebergs and fraught with so many still-unresolved issues, with so few compelling reasons to make the voyage. If leaders in journalism education truly value the sweeping perspectives of the liberal arts, then they should recognize that media convergence, at its most basic level, applies these time-honored concepts in ways ancient scholars could scarcely imagine. They should also consider convergence not as a fad, but as a philosophy about how media interact with modern society and the effects on journalism education's mission. Attempting to realize that philosophy will be empowering, as well as a bit terrifying.

Convergence will certainly demand more from faculty, who must experience a paradigm shift of their own. While most express interest in teaching convergence, many spent years working for newspapers and broadcast outlets in what now could be considered a different media reality, one ruled by medium and not multiplatform messages. Media have changed, so teaching about media must also change, which requires time, energy, and hard work. Faculty will, no doubt, experience some discomfort with their classroom and research routines. They may not admit it, but many have internalized the academy's slow, steady, and stable pace, which seems somewhat out of sync with modern reality. As convergence becomes newsroom reality, faculty need opportunities to spend quality time back in industry, opportunities that are financially supported and make realistic demands on their lives and schedules. At the very least, program chairs should consider single course releases during regular semesters that allow faculty to essentially take themselves back to school.

When faculty come banging on industry's front door to learn about convergence and keep their skills fresh, they should be welcomed with open arms, a desk to work, and a key to the snack room. As the journalism industry's fortunes wane, professionals should realize they have no stronger advocates than journalism educators. Unfortunately, many practitioners do not understand, nor appreciate, the delicate and difficult role journalism schools actively play in not just their continued professional success, but utter survival in today's media information feeding frenzy. Convergence may be about modern media, but the cold yet demanding shoulder industry often shows journalism education is more than a century old. The success of convergence in both industry and education relies heavily on improving what, ironically, both groups say is their specialty—communication.

Many professors and practitioners are focused on the sweat and toil needed to make convergence happen. They should also consider they have front seats on an exciting ride that will be remembered as a pivotal time for media and information, and how they are not just living history, but shaping the future. By the way, this future may include boat-cars after all, but not the retro-1960s model now prized only by collectors. *Time* magazine's "Coolest Inventions of 2003" included the Gibbs Aquada, a James Bond–like roadster that does 100 mph on land, sports a jet engine on water, and challenges your bank account at $250,000. With enough time, ingenuity, effort and money, convergence may not just float—it could cause journalism's equivalent of a sonic boom.

## References


Accrediting Council for Education in Journalism and Mass Communication. (2001, November). Council revises Standard 3: Curriculum. *ACEJMC Ascent, 8*(3), 1.

Accrediting Council for Education in Journalism and Mass Communication. (2004, September). ACEJMC accrediting standards. Retrieved July 5, 2007, from http://www2.ku.edu/~acejmc/PROGRAM/STANDARDS.SHTML

Accrediting Council for Education in Journalism and Mass Communication. (2007, April 25). ACEJMC Accredited Programs, 2006–2007. Retrieved July 5, 2007, from http://www2.ku.edu/~acejmc/STUDENT/PROGLIST.SHTML

Association for Education in Journalism and Mass Communication. (2001, March). Journalism and mass communication education: 2001 and beyond. Retrieved July 5, 2007, from http://www.aejmc.org/_scholarship/_publications/_resources/_reports/jmc_beyond.php

Becker, L. B., Huh, J., Vlad, T., & Mace, N. L. (2003, August 1). Monitoring change in journalism and mass communication faculties 1989–2001. Retrieved July 5, 2007, from http://www.grady.uga.edu/ANNUALSURVEYS/Enrollment01/Supplemental/summary02.htm.

Becker, L. B., Vlad, T., Pelton, R., & Papper, R. A. (2006, August 7). 2005 Survey of editors and news directors. Retrieved July 5, 2007, from http://www.grady.uga.edu/ANNUALSURVEYS/AnnualSurvey2005Reports/Editorreport2005_merged_v5.pdf

Becker, L. B., Vlad, T., Tucker, M., & Pelton, R. (2006, Autumn). 2005 Enrollment report: Enrollment growth continues, but at reduced rate. *Journalism and Mass Communication Educator, 61*(3), 297–327.

Birge, E. (2004). Teaching convergence—But what is it? *Quill, 92*(4), 10–13.

Birge, E. (2006, August). The great divide. *Quill, 94*(6), 20–24.

Brian, D. (2001). *Pulitzer: A life*. New York: John Wiley.

Criado, C. A., & Kraeplin, C. (2003, August). *The state of convergence journalism: United States media and university study*. Paper presented at the meeting of the Association for Education in Journalism and Mass Communication, Kansas City, MO.

Dailey, L., Demo, L., & Spillman, M. (2005). The convergence continuum: A model for studying collaboration between media newsrooms. *Atlantic Journal of Communication, 13*(3), 150–168.

Dickson, T. (2000). *Mass media education in transition: Preparing for the 21st century*. Mahwah, NJ: Lawrence Erlbaum.

Editorial Staff. (1999, August 14). J-schools facing September exams. *Editor and Publisher,* p. 18.

Fenno, J. W. (1789, March 4). *Gazette of the United States*.

Fitzgerald, M. (2003, September 8). School daze. *Editor and Publisher,* n.p.

Haiman, R. J. (2002, December 18). Can convergence float? Retrieved July 5, 2007, from http://poynteronline.org/content/content_view.asp?id=14540.

Hoskins, R. L. (1985, July). *A new accreditation problem: Defining the liberal arts and sciences*. Paper presented at the meeting of the Association for Education in Journalism and Mass Communication, Portland, OR.

Huang, E., Davison, K., Shreve, S., Davis, T., Bettendorf, E., & Nair, A. (2006). Bridging newsrooms and classrooms: Preparing the next generation of journalists for converged media. *Journalism Communication Monographs, 8*(3), 221–262.

International Amphicar Owners Club. (2006). History of the Amphicar. Retrieved July 5, 2007, from http://www.amphicar.com/history.htm.

John S. and James L. Knight Foundation. (2002). Newsroom training: Where's the investment? Retrieved July 5, 2007, from http://www.knightfdn.org/publications/journalismtraining/Newsroom_Training_Book.pdf

Kittross, J. M. (1989). *Six decades of education for broadcasting . . . and counting*. Boston: Emerson College.

Kraeplin, C., & Criado, C. A. (2005). Building a case for convergence journalism curriculum. *Journalism and Mass Communication Educator, 60*(1), 47–56.

Lowery, W., Daniels, G. L., & Becker, L. B. (2005). Predictors of convergence curricula in journalism and mass communication programs. *Journalism and Mass Communication Educator, 60*(1), 32–46.

Martin, E. F., Jr., Wenger, D. H., South, J. C., & Otto, P. I. (2004, August). *Walking in step to the future: Views of journalism education by practitioners and educators*. Paper presented at the meeting of the Association for Education in Journalism and Mass Communication, Toronto, Canada.

McClelland, J. (2004, Spring). Diverse field needs a mix of backgrounds: Thousands of J-majors will never find real work. *Masthead, 56,* 15–17.

O'Dell, D. F. (1935). *The history of journalism education in the United States*. New York: Teachers College, Columbia University.

Ogan, C., & Chung, D. (2003, Winter). Stressed out! A national study of women and men journalism and mass communication faculty, their uses of technology, and levels of professional and personal stress. *Journalism and Mass Communication Educator, 57*(4), 352–368.

Online News Association. (2001). *Digital journalism credibility study*. Retrieved March 31, 2003, from http://www.cyberjournalist.net/news/000148.php.

Outing, S. (1999, January 2). Preparing J-school students for new media convergence. *Editor and Publisher, 132*(1), 49.

Paul, N. (2001, March). Online industry cites needs. In *Journalism and mass communication education: 2001 and beyond* (pp. 25–29). Columbia, SC: Association for Education in Journalism and Mass Communication.

Pavlik, J., Morgan, G., & Henderson, B. (2001, March). Information technology: Implications for the future of journalism and mass communication education. In *Journalism and mass communication education: 2001 and beyond* (pp. 16–20). Columbia, SC: Association for Education in Journalism and Mass Communication.

Phipps, J. L. (2000, October 30). New media powers J-school programs;
Colleges receive more funding and facility upgrades. *Electronic Media, 19,* 12.

Pryor, L. (2006, February 12). Teaching the future of journalism [Electronic version]. *Online Journalism Review*. Retrieved July 5, 2007, from http://www.ojr.org/ojr/stories/060212pryor/

Pulitzer, J. (1904). The college of journalism. *North American Review, 178*(50), 641–680.

Santayana, G. (1905). *Reason in common sense* (vol. 1). New York: Scribner's.
Sloan, W. D. (1990). In search of itself: A history of journalism education. In W. D. Sloan (Ed.), *Makers of the media mind: Journalism educators and their ideas*. Hillsdale, NJ: Lawrence Erlbaum, p. 1–22.

South, J., & Nicholson, J. (2002). Cross training: In an age of news convergence, schools move toward multimedia journalism. *Quill, 90*(6), 10–15.

Southern Association of Colleges and Schools; Commission on Colleges. (2001, December). *Principles of accreditation: Foundations for quality enhancement.* Retrieved July 5, 2007, from http://www.sacscoc.org/pdf/PrinciplesOfAccreditation.PDF

Tanner, A., & Duhé, S. (2005). Trends in mass media education in the age of media convergence: Preparing students for careers in a converging news environment [Electronic version]. *Studies in Media and Information Literacy Education,* 5(3), Article 66. Retrieved July 5, 2007, from http://www.utpjournals.com/simile/issue19/tanner_fullarticle.html

Thelen, G. (2002). Convergence is coming. *Quill, 90*(6), 16.

Coolest Inventions 2003: Fast and amphibious. (2003). *Time* [Electronic version]. Retrieved July 5, 2007, from http://www.time.com/time/2003/inventions/invaquada.html

U. S. Department of Labor, Bureau of Labor Statistics. (2006, August 4). News analysts, reporters, and correspondents. *Occupational Outlook Handbook, 2006–07 Edition.* Retrieved July 3, 2007, from http://www.bls.gov/oco/ocos088.htm.

Voakes, P. S., Beam, R. A., & Ogan, C. (2003, Winter). The impact of technological change on journalism education: A survey of faculty and administrators. *Journalism and Mass Communication Educator, 57*(4), 318–334.

Warner, C., & Liu, Y. (1990). Broadcast curriculum profile: A freeze-frame look at what BEA members offer students. *Feedback, 31*(3), 6–7.

Weaver, D. H. (2003). Journalism education in the United States. In R. Fröhlich & C. Holtz-Bacha (Eds.), *Journalism education in Europe and North America: An international comparison* (pp. 49–64). Cresskill, NJ: Hampton Press.

Wenger, D. H. (2005a, Fall). Convergence: Who's doing what [Electronic version]. *ASJMC Insights,* 10–15. Retrieved July 25, 2007, from http://www.asjmc.org/publications/fallinsightsinside.pdf

Wenger, D. H. (2005b). Cross-training students key in multimedia world. *Quill, 93*(2), 28–29.

Wenger, D. H., & VanSlyke Turk, J. (2005, Fall). Convergence: Where are we? [Electronic version]. *ASJMC Insights,* 3–9. Retrieved July 25, 2007, from http://www.asjmc.org/publications/fallinsightsinside.pdf

CHAPTER 17

# The Future of Media Convergence

Charles Bierbauer

## How Far Have We Come?

It was not that long ago—a few years before this writing—that we would spend hours in lengthy academic discussion seeking to define media convergence. "What" may have been more important to the academics than it was to the media professionals, who mostly wanted to know "how" and, of course, "how much?" The professionals' answer to the question of "who" was doing convergence—the competition, perhaps—strongly influenced the "why"—because the other guy was doing it.

"The 'anything, anytime, anywhere' paradigm is really going to shift the world of media," observed Liberty Media chairman and convergence advocate John C. Malone. But he also cautioned that "there will be a tough, grinding transition for an awful lot of businesses" (Hansell, 2006).

Lewis and Clark may have had a better sense of where they were headed two centuries earlier on their Voyage of Discovery, even though they had no road map, indeed no road. On the evolutionary trail of convergence, having notionally accepted convergence in one form or another, we now wonder what its future holds. Is there a future for something that has been so elusively difficult to pin down? Was it a fad? Did it ever coalesce? Does it matter? And should we decide that it does matter, how do we teach it and put it into profitable practice?

The chapters preceding this lay out the case for media convergence in the context of the digital revolution, the explosion of consumer choices and the search for a management model—all right, can we make a profit on this thing—that makes it worth the effort. In some more encyclopedic history of the media, convergence may prove to warrant a chapter, not a book. But convergence is a core chapter of our contemporary media behavior being written, scratched out, edited and revised each day as we seek to gather these thoughts about the direction of 21st-century media.

We should not be distraught if convergence proves evanescent. Moses was the last "journalist" who produced a story engraved in stone.

There is no single future for media convergence, just as there has been no common present form of convergence. The "backpack journalist" traipsing some backstreet, backwater or battlefield with a digital video camera, laptop, and satellite phone has a very different sense of convergence—"I am it!"—than an editor who may be flipping copy and streaming video from the newspaper or newscast onto a Web page. The blogger with a camera in her cell phone might whimsically think herself a converged journalist, though in fact she may be more accurately described as an electronic diarist with a multimedia bag of tricks.

## Inside the Industry

Some of us have had journalism careers converged by coincidence, some as a convenience, and only the most recent practitioners by design. My own career serves as example, having meandered, though not without purpose, from print to radio to wire service to free-lance jack-of-a-couple-of-trades to parallel but nonconverged radio and print assignments to television to cable. My 20 years at CNN tracked CNN's evolution, a cable television news network at its inception in 1980, adding radio, then wire, then Web. "Don't forget to file radio, and Internet wants a chat at four," became a standard admonition to CNN correspondents. As of 2007, 27 years after its inception, CNN promoted its "22 branded networks and services... available to more than 2 billion people in more than 200 countries and territories." Those include U.S.-based services, international services, Web sites, content distribution, international partnerships, and joint ventures (Turner Pressroom, 2007).

In 2006, the nation's largest newspaper publisher, Gannett, began transforming all its newsrooms to Information Centers. To change the newsroom mind-set, Gannett changed the names and reorganized functions, creating a data desk, a digital delivery desk, and a multimedia desk. But the biggest change, according to Gannett officials, was the determination to break news online. "The new journalist will write for the Internet and update for newspapers," Gannett Chairman Craig Dubow (2007) said, describing the approach as "platform agnostic" at the Gannett Journalism Educators Symposium.

In Tampa, Florida, Media General's television, newspaper and online troika have often been cited as one of, if not the most converged news operations. One multilevel newsroom circles the *Tribune*, WFLA and TBO—Tampa Bay Online—around a common central news desk. It's not perfect convergence. The Tampa operation continues to evolve. One recent step in the experiment created a single sports editor for all three components, but had not designated a converged editor for every portion of the news spectrum (G. Thelen, personal communication, April 2004).

Tampa area advertisers could maximize eyeballs and lower rates when placing ads across platforms, but ad buyers had remained selective and not necessarily been seduced by the convergence package. In another dimension, the migration of automobile advertising to the Internet has compelled newspapers across the country to adjust to Web-based advertising to hold onto the vital auto ad revenues.

Media General's greatest advantage in Tampa is single ownership of all three delivery modes. In many markets, that simply cannot happen because of cross-media ownership bans enforced by the Federal Communications Commission. The FCC cannot directly constrain newspaper ownership, but it can and does determine the scale and scope of broadcast ownership as well as cross-ownership of print and broadcast entities within a single market. A 2002 goal of the Bush administration to lift the limits would have opened the door for greater media consolidation and, if only as a consequential consideration, many more convergence experiments. Single-market ownership of both broadcast stations and newspapers has been barred in the United States since 1975. Where cross-ownership situations exist, they were either grandfathered at the time of the prohibition or granted an exemption. The Telecommunications Act of 1996 sought to create a "pro-competitive, deregulatory national policy framework" by narrowing constraints, raising ownership caps, and instituting a biennial review of the ownership rules.

In 2003, the Commission attempted to lift the ban on cross-ownership in the largest markets (those with more than eight television stations), liberalized it in mid-size markets (four to eight television stations), and retained the ban only in the smallest markets. The Commission held that because newspapers and broadcast stations do not compete in the same economic market, elimination of the ban could not harm competition Federal Communications Commission, 2002)].

But the U.S. Third Circuit Court of Appeals in 2004 blocked several aspects of the Telecommunications Act, finding "the commission has not sufficiently justified its particular chosen numerical limits for local television ownership, local radio ownership, and cross-ownership of media within local markets" (*Prometheus Radio Project v. FCC*, 2004, p. 15). "All have the same essential flaw: an unjustified assumption that media outlets of the same type make an equal contribution to diversity and competition in local markets," the court's ruling said (*Prometheus Radio Project v. FCC*, 2004, p. 121). In June 2005, the U.S. Supreme Court denied a petition to review the lower court's *Prometheus* ruling.

## Technology

Just as much journalism is reactive to events, media convergence has been substantially reactive to advances in noncommunications fields. Technology has been our driving force, most of it devised for other purposes and adapted for media. Satellites and silicone chips came to us from the space program. The Internet was a research tool designed by DARPA, the Pentagon's Defense Advanced Research Projects Agency, launched in response to the Soviet Union's 1958 orbiting of Sputnik. We would not be talking about convergence without these and thousands of other technological advances. None were invented with CNN, Google, MySpace, or YouTube in mind. Entrepreneurs and a few wild-eyed visionaries latched on to the technological advances of the final quarter of the 20th century, literally snatched some of them from the ether and created a panoply of media that themselves were spun and woven into new iterations and applications. The networks and webs that have resulted may be loosely knit or tightly woven. Ironclad, on the other hand, is far too intransigent and permanent to describe any of the results.

Think about it. We're wired for wireless. We're hooked on being unhooked. Yet we are tethered to technology. Walk through an airport and you think people are talking to themselves when, in fact, they are talking to an electronic twig lodged over their ear. Wonder where the kids are? Call them on their cell phones. They'll each have downloaded a different ringtone. We seek individuality in the most mass of mass media. How can there be any single future in convergence?

The future may have already been bifurcated into real and virtual representations via *Second Life,* "a 3D online digital world imagined, created and owned by its residents." Real corporations, as wary of being left out as they were when the Internet dawned, are purchasing "real" estate in this new world and stocking it with avatars to conduct their business. The media are there, too.

The likelihood is that journalists will evolve a set of roles for any future multimedia environment based on their unique and often multiple skill sets. Businesses will find another set of roles, prospering or failing depending on their adaptability to fickle markets and demanding stockholders. Ownership will move in its own direction, merging, converging, diverging. Consumers will have more choices, theoretically more input—voluntary or involuntary —through "cookie" chains. Perhaps, in this wave of mass manipulation, the consumers will have the most significant impact if they opt out, en masse.

## Trends

The media are not going away. Not even newspapers, though the hand wringing over declining circulation has endured for decades. Presumably smart newspaper people would not pay $6 billion, as McClatchy did for the Knight-Ridder chain in 2006, for a dying business. (In the ensuing year, McClatchy did have second thoughts about some of the properties it had acquired.) The intimate interactivity of the Internet must be more than a fad. Or why would Rupert Murdoch have spent $580 million (also in 2006) to buy MySpace, the online hangout so popular primarily, though not exclusively, with teen Internet users?

MySpace may be only the pig in the Murdoch's News Corp. snake, a piece of business peristalsis. Media organizations are serial practitioners of merger, acquisition, and divestiture. "In this business, you're either a buyer or a seller," Liberty Communications president Jim Keelor told me over lunch a few years ago. (J. Keelor, personal communication, 2004.). Prophetically, it turned out. Liberty, a midsize, South Carolina–based holder of 15 television stations (not to be confused with Malone's Denver-based Liberty Media) was erased from the media map in 2005 when it was acquired by Raycom Media.

In the 1990s it seemed the media mavens all wanted to be big, linking content to distribution. Networks without cable subsidiaries lacked the opportunity to defray newsgathering costs across multiple outlets. So Disney bought ABC. NBC, a part of General Electric since its inception, linked with Microsoft in its MSNBC cable project. CBS was acquired by Westinghouse, a broadcast pioneer more often recognized as an appliance manufacturer, in a merger that retained the CBS name. CBS (Westinghouse) was then bought by Viacom and later spun off as a parallel entity to Viacom's other holdings. In corporate suites, convergence is only a way station on the way to profit maximization and stockholder satisfaction. The much-told tale of Ted Turner is instructive. Upstart, maverick, cable-before-cable-was-cool Turner parlayed CNN, TBS, TCM, Headline News, CNN International, and the Cartoon Network into an empire based on the premise that having many small slices of the pie was just about as good as owning the whole pie. Still Turner, loaded with ambition and chutzpah, yet with at least a tinge of a Southern inferiority complex, wanted the respect that owning a major network might bring. "Why did they have to call us 'Chicken Noodle News?' I never made fun of them," Turner was known to grumble about the denigration of his CNN in its early days (T. Turner, personal communication, about 1995). That stopped, certainly when CNN became the global network of record for its unmatched coverage of the 1991 war in the Persian Gulf.

Having failed in 1985 to acquire CBS, Turner bought into the marriage of cable content and cable distribution systems, even though it meant sacrificing personal control. In 1996, Turner's operation was acquired by Time Warner, and in 2001, online phenomenon AOL merged with Time Warner. In short order, the business and cultural mismatch of AOL Time Warner became the how-not-to exemplar for media mergers. Turner personally lost billions. Investors lost millions. Thousands lost jobs. And the ballyhooed synergies, CNN and *Time* for one, that were supposed to accrue from the sequence of merger and acquisition were abandoned as Time Warner bosses shucked the AOL name and reverted to an "everyman for himself" business model.

Newspapers were no less volatile. In 2006 McClatchy swallowed Knight Ridder's 32 newspapers and spit back a dozen onto the market to be gobbled up by others who saw profitable pickings.

Chicago's Tribune Company picked up the *Los Angeles Times* and other Times Mirror properties in 2000. Yet in 2007, the Tribune put its whole empire on the market, including the Chicago Cubs.

*The New York Times* acquired a valuable asset in the *Boston Globe*, echoing the way the New York Yankees once bought Babe Ruth from the Boston Red Sox. (There are not many new stories in journalism, just variations on about six themes.)

When network news star Katie Couric swapped morning hours on NBC's *Today* show for the evening anchor's chair on CBS in 2006, her new network discovered old media, touting the novelty that the network's newscast could be simulcast and multicast. "The CBS Evening News is going to be showcased on the radio," said CBS News and Sports President Sean McManus, "which is new and different" (Seibel, 2006.) To the chagrin of the network's pantheon of radio heroes Murrow, Sevareid, Trout, Shirer, Smith, Hottelet, LeSueur, Collingwood (need I go on?), it was as though CBS had forgotten it had a radio network. And Katie would be on the Web, too! (There are not many new PR fillips, just variations on about six themes.) And for all that, the *CBS Evening News* with Katie Couric foundered and fell even farther behind ABC and NBC. For all I know, you could be saying "Katie who?"

Let's face it; the medium is no longer the message. No matter what the 1960s guru Marshall McLuhan might have thought of television's sway, the medium is only the venue du jour. Podcasts, text messaging, camera phones, blogs are all 21st-century manifestations for dissemination via new technology. E-mail, or so today's tech teens say, is "so last century." What's next? Wired at birth? Just a small implanted receptor. No earbuds to lose. Cue the channel from your Dick Tracy–like watch. Or just say "Katie Couric," and presto! There's Katie. Or her avatar. Too Orwellian? We're already post-Orwellian.

## Bottom Lines

Make no mistake, all media are businesses. But not all businesses are moving in the same direction. AOL became a power among Internet providers by building its online paid subscriber base to 26 million, then scrapping the paid-for-service model in 2006. The new business model provided free Internet service supported by advertising revenue to a smaller client base, but allowed AOL to dump 5,000 customer service employees and project more than $1 billion in savings by cutting marketing and overhead costs while raising increasing ad revenue. Web business models, though, are as multidirectional as the Web is multifaceted. You can say much the same for cellular phones, which have as many pay permutations as you have friends on your speed dial, probably more. They'll also happily sell you tones and tunes and games. But don't try to lay a pattern over the business models.

Even as Internet service providers—relatively indistinguishable from one another—evolved from paid services to advertiser-supported services, online content providers looked to charge for their unique content. *The New York Times* will give you what's on its news pages, but you'll have to pay to read its syndicated columnists online.

Television, on the other hand, started as a free, advertiser-supported medium. Now, most Americans pay handsomely to receive television via cable or satellite. And they pay still more for premium channels, digital, and high-definition video. Most viewers still endure the commercials, unless they're willing to pay an additional fee for DVR or TiVo-type recording capabilities that facilitate commercial skipping. Of course, if you skip the commercials, you risk missing the more creative side of television.

Another chapter in technology's influence on media will be written after the 2009 digital television transition. Congress mandated the shift from analog to digital television programming. Digital television offers higher resolution and picture quality and the capacity for multicasting several streams of material simultaneously. The transition will also free up the current analog broadcast spectrum for other services and, in part, auction to telephone, cable, and satellite operators. The anticipated auction and distribution of this valuable digital real estate has itself been a source of as yet unresolved controversy and competition (Markoff, 2007).

Newspapers come in all types, relying in varying proportions on advertising, subscription, or newsstand sales. On urban street corners or in suburban shopping malls, ranks of paper boxes for paid newspapers stand side by side with free shoppers. In Bluffton, South Carolina, in 2005, Morris Communications began delivering its community-centered newspaper *Bluffton Today* free to each of 16,000 local addressees. Tied to a vigorous online companion, the *Bluffton* convergence works because of its compact distribution area and its "hyperlocal" commitment to community news, much of it citizen journalism, in a fashion no other medium can match.

Localization may well be the nub of convergence trends. New media are relatively inexpensive to create and maintain. Consider what a small town newspaper can do with a Web site and a town full of contributors. With the aid of a University of South Carolina journalism professor and a $12,000 J-Lab grant, the bi-weekly *Hartsville Messenger* in 2005 took a "community storyteller" approach, inviting readers to be contributors to a Web mélange of reports, Moblogs, video, and audio.

"This project is vitally important to smaller papers because if newspapers don't 'get there first' in the communities we serve, you can be sure someone else will," publisher Graham Osteen observed in a year-one report on the *Hartsville Today* project. (Fisher & Osteen, 2006, p. 62). By mid-2007, page views were up over 25,000 a month.

It remains, in good measure, a journalistic endeavor. The challenge, as faculty partner Doug Fisher notes, is "learning how to sell it, how to work it into newsroom routines, and how to modify the production mentality present in many news operations to accommodate the flexibility needed in a digital workflow" (D. Fisher, personal communication, July 2007).

Digital photos, text-messaged copy, blogs, and streaming video from the Friday night high school football games quickly add up to a mosaic of close-to-home news and activities. It's a bulletin board, but in creative hands far more content-rich and meaningful. The nightly network news can only paint with a broad brush compared to the pointillist detail of a local endeavor. The major dailies can saturate you with coverage of politics and policy. But, "all politics is local," as U.S. House Speaker Tip O'Neill emphatically put it. And local politicians respond to local media, or certainly should.

In how many places across this country and in other countries is the Hartsville experience being replicated? That's what the John S. and James L. Knight Foundation wanted to stimulate in 2006 when it initiated its Knight News Challenge with the intention of funding $25 million worth of innovative projects over a span of five years. The Knight invitation for grant proposals took a simple, direct approach to escape the predictable and pedestrian: "We're looking for great new digital projects that will improve news and for people in specific geographic communities. Anyone, anywhere, of any age may apply." Knight received 1,650 proposals in the first year of the challenge and funded projects originated by individuals, universities and corporations (John S. and James L. Knight Foundation, 2007).

Multimedia environments, if only because they can be fractured into such differentiated niches, assure an audience for just about anything from Al-Jazeera to old jazz. Where there's an audience—and it doesn't have to be huge—and a business plan, there's pretty sure to be a medium catering to its interest.

Consider Al-Jazeera, for a moment, because convergence is not just multimedia. It should also be seen as bringing multiple perspectives to media audiences. Founded in 1996, Al-Jazeera owes its existence to the global recognition of the CNN effect, that is, the ability to stimulate interest in a story through the ubiquitous and unceasing nature of the all-news networks. The Qatar-based Al-Jazeera applied that principle to stories from an Arab perspective, capitalizing on its ability to gain access where Western media may not have been welcome or might have proceeded only at great risk. Al-Jazeera alternately draws praise for broadening the media spectrum and criticism for its Arab voice. In good new media style, Al-Jazeera raised that voice first in Arabic broadcasts with, of course, a Web component. In 2006 it launched an English-language service, based in part in Washington, where it hoped to find an audience, perhaps small, but certainly influential. By reaching more broadly across the United States and the non-Arab world it hoped to persuade audiences that there is more than one way to look at events and emotions in the tempest of the Middle East. Those ideas may be millennia old, but technology has put new audiences in reach.

Convergence and new media are reshaping notions of the traditional audience demographic. As media splinter, splinter audiences become more distinguishable and more attractive to target. The notion that only 18–34-year- olds matter to advertisers is rapidly vanishing. Some day they may be the vestigial domain of ESPN Classic and the Outdoor Life Network, though I'm probably misrepresenting their demographics.

The conventional wisdom about who has and spends money has been changing. Advertisers have long known that kids don't have money, but they influence what moms buy. The 18–34-year-olds, particularly while single or childless, may still have disposable income and carefree habits. But their boomer parents are riding into retirement on the biggest bow wave of wealth, earned and inherited, with the health and appetites to still enjoy it. The sheer volume of ads for Nexium, Lipitor, Viagra, and all the cosmetic appurtenances of aging speak loudly to the demographics of media consumption—print, broadcast and Web. Or are those ads simply concentrated on the media that I and others well over 34 years old consume? While we focus much of our attention on convergence from the journalist's perspective, there is strong evidence advertisers grasp what's happening. Ads that pop up on our computer screens are more than repurposed versions of what we see on television or in print. They move, they swirl, they play games with us. They offer us prizes and premiums. They also get in the way, obscure what we're looking for, annoy unmercifully and seem impervious to the computer's delete function. But then advertisers have always known that soft, cute and inoffensive are not strong sales approaches. There's no other way to explain the jangling, pounding volume of car ads on television. When you need a car, you are likely first to remember the dealer's name that's imprinted on your tympanum.

## Teaching Convergence

My colleagues have tackled the teaching of convergence earlier in this work and in journalism programs across the country. Some have embraced it wholeheartedly, some gingerly, and some not at all. Some have leapt into convergence and then crept back. Some teach convergence as an entity in and of itself, others as something pervasive to all disciplines.

I would steer you to the Web—where else?—to know what is going on. Check what's been happening at Missouri, Ball State, Brigham Young, Columbia, Temple, USC—take your pick, the one in Southern California or the one in South Carolina. We all approach convergence a bit differently, but recognize it has a role in our current media spectrum and will influence our students' future as journalists, or more broadly as communicators. Though academic programs have long periods of gestation, nothing precludes us from making constant adjustments within courses and curricula.

When I am in the classroom, I don't teach media technology, though I am cognizant of it. I teach perspective. How, for example, do media and government or politics interface? How might that interface be affected by the medium in which you are working? Do officials talk one way to newspaper reporters and another to television correspondents? Does it matter if the Web is the reporter's first stop on the way back to the newsroom? Is multitasking now basic training?

Nonhierarchical newsrooms create new challenges for reporters and editors. Filing fast to the Web may be an imperative of the new media paradigm. But it requires installing the fact checkers and the editorial gatekeepers at multiple sites in the newsroom. Journalism remains a collective process that involves gathering, synthesizing, checking—and double-checking, before publishing or broadcasting. Online interviews have been gaining currency. Interviewees like having a record of the dialogue lest they be misquoted. Interviewers are wary of unseen interviewees, lest they misrepresent themselves.

Those are not new things to teach, but things to teach in a new context. Our students eagerly want to now how. We need to keep reminding them to ask why.

We need to recognize and teach convergence across the board because the multimedia environment will be just as important to public relations and advertising students as it will be to those studying journalism. We must put more emphasis on creating graduates who understand the management aspects of the multimedia newsroom.

## Cautionary Lessons

Have we been here before, contemplating the ephemeral nature of some element of our environment? The horse and buggy are gone, at least as mass transit. Romanticism, impressionism, Cubism? Eight-track? Beta? Stick shift? Baking from scratch?

Television changed the newspapers, but didn't kill them. Well, perhaps it did do in the evening newspaper editions. The movies survived television, though DVD and piracy are making inroads on theater ticket sales. Cable news altered the network news and the morning newspaper. We don't want to read today about what we already saw yesterday, unless the newspapers can add perspective. They can. Convergence? It's not what it was, or what people thought it was, only a few years ago. We don't even much like the name. Think, instead, multimedia, cross-media, transmedia. Call it what you want.

Adaptive media is a term worth considering. We have adapted the media to our purposes and adapted to the opportunities and strengths of the new media as they have become available to us. And since you are reading this in a book—book!—consider what changes may have occurred between my writing and your reading.

Some of us are early adopters, the first on our block or in our business to put the new media in play. Some of us are cautious, some reluctant and some Luddites. The truth is, we don't all need all that's out there. Certainly not as consumers. You can be quite well informed with a local newspaper and a radio—well, perhaps not "talk radio." If the Internet only leads you to e-Bay or an interactive "World of Warcraft" game, you're not going to stay up on the news and events.

But if we, as journalists and teachers of journalism, are doing a good job of teaching persuasive communication we will find the modality of any medium. It will shape us a bit, and we will shape it to suit the public needs. Meanwhile, we need to keep our eye on the immutable basics even as we learn different mechanics. We will be able to do more with the expanded capabilities of new media and multiple applications only if we have something meaningful to communicate.

Editors and news directors tell us with fair consistence tell us that they want to hire young journalists with strong basic skills. Those include strong writing, clear copy editing, and the ability to tell a good story. If the aspiring journalist emerging from our programs can also take pictures—stills and video all emerging from one palm-sized digital camera—or do Web design, that's a plus. Many television news directors say they'd like to have one or two backpack journalists, but do not expect all their reporters to do the multiple jobs of shooting, writing, and editing. In reality, a journalist multitasking is only paying a fraction of the attention needed to digging up stories, asking the right questions, and pulling disparate elements into a cohesive whole. That takes a team. And because we are journalists, temperamental and cantankerous by nature, counterintuitive by intuition, skeptical by virtue of experience, but preferably not cynical by dint of disillusionment, we are unlikely to agree on where we are or where we're going. Since we are writing separately, I'd venture that you won't find complete agreement in the works of those colleagues whose chapters precede mine. I'd be disappointed if there were not fairly disparate views on how riveted we should be by convergence. Certainly, the journalism and communications programs around the country have sweated their way to varying approaches to convergence, ranging from those that have created full convergence curricula to those with a sense that convergence should be woven through curricula, yet not be as pervasive and pernicious as kudzu along a South Carolina highway.

As much as I believe convergence is a reality in all communication disciplines, I'm camped in the latter corps. Our students will be more marketable journalists if their resumes show they have multimedia skills. But they will be better journalists if their professors have drummed the basics into them—writing, editing, storytelling, and the ethics of it all. We also cannot shirk a responsibility to acquaint them with business realities. One of the worst things a professor can hear from an otherwise competent recent graduate struggling in the job market is "How come no one told me about that?" One thing we do tell them is that any place they find a job—and that includes nonprofits—there is a bottom line. Employees with diverse skills and content that can be purposed over diverse media or roles are likely to enhance the bottom line.

To be overly prescriptive here would risk limiting the shelf life of this text to about six months and incurring the wrath of my fellow authors. It's also just plain hard to say where this technology train will take us. The media are and should be vigorously adaptive to changing technologies and circumstances. Journalism should remain just as vigorously committed to its core values of accuracy, fairness and integrity. Those of us who are educators should reassert our dedication to teaching not only the "how" but especially the "why" of what we expect journalists to do, regardless of venue or medium.

As for "media convergence," Google it in about five years. If Google is still an operative term.

## References


Dubow, C. (2007, April 11). Remarks at Gannett Journalism Educators Symposium, McLean, VA.

Federal Communications Commission. (2002). 2002 Biennial Review Order 13748–49, Paras. 331–32.

Fisher, D., & Osteen, G. (2006). *Hartsville Today*, the first year of a small-town citizen-journalism site. Retrieved July 31, 2007, from http://www.j-lab.org/HVTDyear1.pdf

Hansell, S. (2006, January 25). As gadgets get it together; media makers fall behind. *New York Times*. Retrieved March 3, 2008, from http://www.nytimes.com/2006/01/25/technology/techspecial2/25converge.html?_r=1&oref=slogin

John S. and James L. Knight Foundation (2007, April 2). The first-year winners of the Knight News Challenge will be announced May 23 at the Interactive Media Conference & Tradeshow 2007. Retrieved July 30, 2007, from http://www.knightfdn.org/default.asp?story=news_at_knight/releases/2007/2007_04_02_newschallenge.html

Markoff, J. (2007, May 22). Google proposes innovation in radio spectrum auction *The New York Times,* p. C-10.

*Prometheus Radio Project v. Federal Communications Commission*. (2004). United States Court of Appeals for the Third Circuit. *United States of America (No. 03–3388)*. Retrieved July 30, 2007, from http://www.ca3.uscourts.gov/indexsearch/archives.asp

Seibel, D.S. (2006, July 17). When Couric broadcasts the news, some radio and Web sites will too. *New York Times*. Retrieved July 30, 2007, from http://www.nytimes.com/2006/07/17/business/media/17cbs.html?ex=1185940800&en=fc0ad34fab1

*Telecommunications Act of 1996*. (1996). 104 Pub. L. 104, 110 Stat. 56, 11.

Turner Pressroom (2007). CNN. Retrieved July 1, 2007 from http://www.turnerinfo.com/companyinfo.aspx?P=CNN

# CONTRIBUTORS

**Timothy E. Bajkiewicz** began working in Detroit radio in 1987. He earned his BA in anthropology (1993) and MA in mass communications (1995) from the University of Florida, and PhD from the University of North Carolina at Chapel Hill (2002). In 2001 he became an assistant professor in the University of South Florida's School of Mass Communications. There he led team-teaching in a cutting-edge undergraduate convergence course with Media General's News Center in Tampa, where students produced content for the *Tampa Tribune*, NBC-affiliated WFLA-TV, and TBO.com. In 2004–2005 USF named him an Outstanding Undergraduate Teacher and in 2005 he took second place in AEJMC's Promising Professor competition. In 2008 he became an associate professor in the School of Mass Communications at Virginia Commonwealth University, where he teaches broadcast journalism and graduate multimedia journalism. His research interests include news convergence, new media, and media literacy.

**Charles Bierbauer** is Dean of the College of Mass Communications and Information Studies at the University of South Carolina. As a journalist, Charles Bierbauer was a foreign correspondent for ABC and the *Chicago Daily News*, a Washington-based correspondent for CNN, a wire service reporter for the Associated Press, and a radio and television reporter for Westinghouse. He's reported for TV, radio, wire, and web–often all in the same day–as an early practitioner of multimedia journalism. As an academic, Dean Bierbauer has embraced and encouraged teaching across platforms in all communications disciplines. He believes convergence or multimedia work lies on a continuum of adaptation for students, faculty, and professionals as new media emerge, provided there is no divergence from core values.

**Carrie Anna Criado** earned her JD from the University of Houston and worked with Camille Kraeplin to complete their first longitudinal study examining the state of convergence journalism in the media industries while they were colleagues at Southern Methodist University in Dallas in 2002–2003. They have published articles in *Journalism and Mass Communication Educator* and *Newspaper Research Journal* based on their convergence research. Criado, now based in Houston, has returned to the legal field, but continues her work in media scholarship.

**George L. Daniels** is an assistant professor of journalism at the University of Alabama. After spending eight years in the local television newsroom working as a news producer at stations in Richmond, Virginia; Cincinnati, Ohio; and Atlanta, Georgia, Daniels moved from the newsroom to the classroom. He's conducted research on diversity issues in the media workplace and change in the television newsroom as well as media convergence. Before going to work in television news, Daniels worked briefly as a freelance writer for *The Richmond Free Press* in his hometown of Richmond, Virginia. A cum laude graduate of Howard University, Daniels received both his master's degree and PhD in mass communication from the Grady College of Journalism and Mass Communication at the University of Georgia.

**Tony DeMars** (PhD, University of Southern Mississippi) is Associate Professor in the Department of Mass Media, Communication, and Theater at Texas A&M University–Commerce. His teaching areas include broadcast journalism, new media, and media and society. His publications include the book *Modeling Behavior From Images of Reality in Television Narratives*, the *Journal of Radio Studies* article "Buying Time to Start Spanish-Language Radio in San Antonio: Manuel Davila and the Development of Tejano Programming," and the article "Local TV Market Multicasting: A New Paradigm for Digital

Rich Media," published in *The Journal of New Media & Culture*. He has held officer positions in the Broadcast Education Association, The Association for Education in Journalism and Mass Communication and the Southern States Communication Association and is a past president of the Texas Association of Broadcast Educators.

**Michel Dupagne** received his PhD in mass communications at Indiana University in 1994. He is an associate professor in the School of Communication at the University of Miami in Coral Gables, Florida. His research interests include communication technologies, media economics, and international communication. He has published numerous articles in communication journals and coauthored *High-Definition Television: A Global Perspective*. He serves on the editorial boards of the *Journal of Broadcasting & Electronic Media*, *Journalism & Mass Communication Quarterly,* and the *Journal of Media Economics*.

**Vincent F. Filak**, PhD, is Assistant Professor in the Department of Journalism, University of Wisconsin–Oshkosh. Prior to that, he was an assistant professor at Ball State University, where he served as the faculty adviser to the *Ball State Daily News*, the university's award-winning newspaper. He has also taught news writing and reporting at the University of Wisconsin and the University of Missouri and worked as a night-side city desk reporter at the *Wisconsin State Journal* in Madison, Wisconsin. He received bachelor's and master's degrees in journalism from the University of Wisconsin and a PhD from the University of Missouri. He is the coeditor of *Convergent Journalism: An Introduction* (Focal Press, 2005) and has published more than a dozen scholarly articles, many of them on convergence and intergroup relations. He has presented more than 25 peer-reviewed conference papers, including six award winners. In 2006, he received the Distinguished Researcher Award from Ball State's College of Communication, Information, and Media. Filak serves on the editorial advisory board of *Journalism and Mass Communication Educator* and as a reviewer for the *Atlantic Journal of Communication* and the *Newspaper Research Journal*. He is the vice president of College Media Advisers and the executive director of the Indiana Collegiate Press Association.

**Holly A. Fisher**, MMC, is the Research Editor for South Carolina-based SC Biz News LLC, publisher of the *Charleston Regional Business Journal* and *SCBIZ* magazine. She oversees the company's Web sites and e-mail products. She was previously the company's supplements editor, writing and editing special projects and publications. Fisher is a former regional director and board member for the Society of Professional Journalists. She also has served as an adjunct professor at the College of Charleston. Previously, she worked in the newspaper business in Indiana, Texas, and South Carolina. She has a bachelor's degree in journalism from Ohio University and a master's of mass communication from the University of South Carolina.

**Bruce Garrison** received his PhD in journalism at Southern Illinois University in 1979. He is a journalism professor in the School of Communication at the University of Miami in Coral Gables, Florida. He has conducted research about new communication and computer technologies and their uses by journalists in gathering news and information. He has written numerous books, journal articles, and papers about online news use, computer-assisted news reporting, feature writing, media ethics, international journalism, and sports journalism. He serves on the editorial boards of *Journalism & Mass Communication Quarterly* and the *Newspaper Research Journal*.

**August E. ("Augie") Grant** is Associate Professor and Newsplex Academic Liaison at the School of Journalism and Mass Communications at the University of South Carolina. Grant is a technology futurist who specializes in research on new media technologies and consumer behavior, with an interest in integrating both quantitative and qualitative research. He has been published in every major academic journal in the communication field. His publications deal with adoption and use of emerging communications technologies, broadband services, audience behavior, and theories of new media. In addition to his academic research, Grant advises technology companies on strategies for introducing new technologies. Grant is the creator and coeditor of *Communication Technology Update* (in its 11th edition), founding editor of *The Convergence Newsletter*, and coauthor of *Principles of Convergent Journalism*.

**Michael Holmes** is Associate Director, Insight & Research, for the Center for Media Design at Ball State University. Holmes' research interests include emerging media, visual communication, and media measurement. He is a coprincipal investigator in the Middletown Media Studies. His research has appeared in *The Police Chief, Management Science, Human Communication Research, Health Communication*, and *Behavior Research Methods*.

**Janet Kolodzy** is an associate professor in journalism at Emerson College in Boston, where she specializes in broadcast and convergence-oriented journalism. She has developed and taught most of the cross-media undergraduate and graduate courses at Emerson and is the author of the text, *Convergence Journalism: Writing and Reporting across the News Media*. Before entering academia in 1998, Kolodzy worked as a print and broadcast journalist. She was an education reporter and an editor for daily newspapers in Little Rock, Arkansas and Cleveland, Ohio. She then worked for 11 years at CNN and CNN International as writer, copyeditor, producer, and senior producer.

**Susan Keith** worked for 16 years as a newspaper reporter, copyeditor/designer, or assigning editor at daily newspapers in Alabama and Florida. She earned her PhD from the School of Journalism and Mass Communication at the University of North Carolina in 2003 and teaches editing, media ethics and law, and PhD courses about theory and research methods at Rutgers University. Her research focuses on journalistic practice, especially the ethical, legal, and practical issues resulting from media transformation. Her research has been published in *Journalism & Mass Communication Quarterly, Journalism Studies,* the *Journal of Mass Media Ethics,* the *Journal of Broadcasting and Electronic Media,* and *Newspaper Research Journal*.

**Kenneth C. Killebrew** has been writing professionally for more than 30 years. He began his career as a newspaper reporter, but after three years moved into broadcasting and became a reporter/anchor. He later worked as an assignment editor. During his decade in television reporting, Killebrew won four awards, including two awards for best investigative reporting, and a second place finish for best reporter in the state of Illinois. Killebrew is the author of *Managing Media Convergence: Pathways to Journalistic Cooperation* from Blackwell Press, one of the first books on convergence management ever published. Killebrew holds a PhD in Journalism from the University of Tennessee and a master's degree in organizational and applied communication research. His bachelor's degree in journalism is from Bradley University, where he also earned a minor in economics.

**Van Kornegay**, MA, is Associate Professor at the School of Journalism and Mass Communications at the University of South Carolina, where he is Chair of the Visual Communications sequence. Professionally, he has worked in higher education public relations and as a freelance writer, photographer, and graphic artist. He has also worked as a graphics journalist for the Associated Press Graphics Service in New York and the Gannett Graphics Service in Washington, DC. He served as a public affairs officer for an international relief agency in Albania and Kosovo, and his reportage from the Balkan states after the fall of communism and during the conflict in Kosovo appeared in regional and national newspapers and magazines.

**Camille Kraeplin** holds a PhD from the University of Texas and worked with Carrie Anna Criado to complete their first longitudinal study examining the state of convergence journalism in the media industries while they were colleagues at Southern Methodist University in Dallas in 2002–2003. Kraeplin oversaw the second survey in 2004–2005. Kraeplin and Criado have published articles in *Journalism and Mass Communication Educator* and *Newspaper Research Journal* based on their convergence research. Kraeplin, now an assistant professor of journalism at SMU, is currently working on a book about the way young women use and are portrayed by the mass media.

**Jennifer H. Meadows** is a professor in the Media Arts Option of the Department of Communiction Design at California State University, Chico. She earned her MA in radio, television, and motion pictures at the University of North Carolina at Chapel Hill and a PhD in radio-television-film from the University of Texas at Austin. Her teaching and research integrate practical and theoretical concerns related to television production, the relationships between media and users, and changing media behavior in the new media marketplace. In addition to her teaching and research, she is coeditor of *Communication Technology Update* and is a consultant to many different media organizations.

**Steven McClung** is an associate professor in the Integrated Marketing Communication division of the College of Communication at Florida State University. He has degrees from Marshall University and the University of Tennessee. His research interests are Internet marketing, sports marketing, radio, and audio media. McClung teaches marketing communication management, marketing, media planning, consumer behavior, and sports marketing.

**Bryan Murley** has advised college media for six years after a long and varied career in print journalism. He now teaches new and emerging media classes at Eastern Illinois University and advises the award-winning Daily Eastern News web site. He is also director for innovation at the Center for Innovation in

College Media, where he writes the Innovation in College Media Web log. He is a PhD student at the University of South Carolina, studying Web logs and the emerging church movement.

**Robert ("Bob") Papper** is Professor and Chair of the Department of Journalism, Media Studies, and Public Relations at Hofstra University. A graduate of Columbia College and the Columbia Graduate School of Journalism, he's worked as a producer, writer, and manager at television stations in Minneapolis (WCCO-TV), Washington, DC (WRC-TV), San Francisco (KPIX-TV), and Columbus, Ohio (WSYX-TV). He's the author of *Broadcast News Writing Stylebook* and has won more than a hundred state, regional, and national awards, including four regional Edward R. Murrow Awards and a DuPont-Columbia for "Outstanding Contributions to Television." He's a past president of the Maine Association of Broadcasters and a longtime member of the national education committee of the Radio Television News Directors Association. In 2006, he was honored as the Ball State University Researcher of the Year. Bob is in his 15th year overseeing the annual research for RTNDA and is the coeditor of *Electronic News: A Journal of Applied Research & Ideas*.

**Mark Popovich** is Professor Emertius at Ball State University, where he taught graduate research methodology courses and media law. He is a former chair of the Department of Journalism at BSU and has served as director of the university's London Centre program. He is a past president of the International Society for the Scientific Study of Subjectivity, an organization for which he is currently treasurer.

**Varsha Sherring** is a doctoral candidate at Regent University in Virginia Beach, Virginia. Varsha received her first master's degree in TV production from Poona University in India. Her second master's degree was in film editing and directing from Regent University in 1999. Shortly thereafter, she joined the Christian Broadcasting Network (CBN) office in India as Producer of Special Programming. In 2005, Varsha moved to the United States to pursue her doctorate in communications.

**B. William (Bill) Silcock**'s qualitative research focuses on newsroom culture to deepen our understanding of how journalists, especially TV newscast producers, determine "what is news." His media convergence studies scholarship and teaching focuses on the linkages between newsroom cultures—gatekeepers and their routines—across multiple platforms. He also conducts research on media ethics and war coverage. Recent article topics include media convergence, Iraq War images, producers as mythmakers, media, journalism ethics, and changing media systems in the Balkans. Focused internationally, Silcock has traveled to 20 nations, his travels resulting in articles for *Journalism Quarterly*, the *Journal of Broadcast & Electronic Media, Journalism, Electronic News, Visual Communication Quarterly,* and the *Journal of Mass Media Ethics*. His first coauthored textbook *Managing Television News: A Handbook for Ethical and Effective Producing* has been adopted by universities across America. With a passion for global news, Silcock has taught journalism students at Brigham Young University, Missouri School of Journalism, and the Walter Cronkite School at Arizona State University. He offers global media training overseas on diverse topics such convergent newsrooms, investigative reporting, and on-camera performance.

**E. Jordan Storm** is a doctoral candidate at the S. I. Newhouse School of Public Communications at Syracuse University. She earned her master's degree from the University of South Carolina, applying theory to inform her thesis research on convergent journalism practices in the newspaper industry. Her current research interests focus on the ways media organizations serve their various publics, including participatory media, communication for development, and feminist media.

**Jeffrey S. Wilkinson** is Associate Dean and Professor of International Journalism at United International College in Zhuhai, China. A specialist in broadcast news and new media technology, Wilkinson combines practical skill with academic theory to train future journalists. His expertise in news reporting, writing, and production is balanced with quantitative social-psychological perspectives and training. An award-winning reporter in the 1980s, Wilkinson taught broadcast journalism at the University of Tennessee, where he helped initiate and bring streaming technology in 1997. In 1998, Wilkinson moved to Hong Kong and taught six years in the communication program at Hong Kong Baptist University. He specialized in new media technologies, digital audio and video production, international communication, and mass media and society. Wilkinson has been published in some of the major journals in his profession, including *Journalism & Mass Communication Quarterly, Journal of Broadcasting & Electronic Media,* and *Journal of Mass Media Ethics*. Wilkinson has won a number of conference top-paper awards, is a regular reviewer for various organizations, and served as Associate Editor for *Journal of Broadcasting & Electronic Media*.

# INDEX